Nonlinear Dynamics of Structures Under Extreme Transient Loads

The effect of combined extreme transient loadings on a structure is not well understood – whether the source is man-made, such as an explosion and fire, or natural, as an earthquake or extreme wind loading. A critical assessment of current knowledge is timely (with Fukushima-like disasters or terrorist threats).

The central issue in all these problems is structural integrity, along with their transient nature, their unexpectedness, and often the uncertainty behind their cause. No single traditional scientific discipline provides full answer, but a number of tools need to be brought together: nonlinear dynamics, probability theory, some understanding of the physical nature of the problem, as well as modeling and computational techniques for representing inelastic behavior mechanisms.

The book covers model building for different engineering structures and provides detailed presentations of extreme loading conditions. A number of illustrations are given: quantifying a plane crash or explosion induced impact loading, quantifying the effects of strong earthquake motion, quantifying the impact and long-duration effects of strong stormy winds - along with a relevant framework for using modern computational tools. The book considers the levels of reserve in existing structures, and ways of reducing the negative impact of high-risk situations by employing sounder design procedures.

Nonlinear Dynamics of Structures Under Extreme Transient Loads

By

Adnan Ibrahimbegovic and Naida Ademović

CRC Press
Taylor & Francis Group
Boca Raton London New York

CRC Press is an imprint of the
Taylor & Francis Group, an **informa** business

CRC Press
Taylor & Francis Group
6000 Broken Sound Parkway NW, Suite 300
Boca Raton, FL 33487-2742

First issued in paperback 2020

ISBN 13: 978-0-367-72878-6 (pbk)
ISBN 13: 978-1-138-03541-6 (hbk)

Library of Congress Cataloging-in-Publication Data

Names: Ibrahimbegovic, Adnan, author. | Ademovic, Naida, author.
Title: Nonlinear dynamics of structures under extreme transient loads / Adnan Ibrahimbegovic and Naida Ademovic.
Description: First edition. | Boca Raton, FL : CRC Press/Taylor & Francis Group, [2019] | Includes bibliographical references and index. |
Identifiers: LCCN 2019001405 (print) | LCCN 2019002310 (ebook) | ISBN 9781351052498 (Adobe PDF) | ISBN 9781351052474 (Mobipocket) | ISBN 9781351052481 (ePub) | ISBN 9781138035416 (hardback : acid-free paper) | ISBN 9781351052504 (ebook)
Subjects: LCSH: Live loads. | Load factor design. | Structural dynamics. | Nonlinear mechanics.
Classification: LCC TA654.3 (ebook) | LCC TA654.3 .I27 2019 (print) | DDC 624.1/72--dc23
LC record available at https://lccn.loc.gov/2019001405

Visit the Taylor & Francis Web site at
http://www.taylorandfrancis.com

and the CRC Press Web site at
http://www.crcpress.com

Contents

PART II

Authors

Adnan Ibrahimbegovic obtained his PhD in Structural Engineering from the University of California at Berkeley and Habilitation in Mechanics from University Pierre Marie Curie in Paris. He is currently a Full Professor in Mechanical Engineering at the Université de Technologie Compiègne in France. He holds the Chair for Computational Mechanics for interdisciplinary research within the Sorbonne Universités group, and positions of Senior Member of the Institut Universitaire de France and Head of Joint Mechanics-Mathematics Research Platform at UTC. He was formerly Head of teaching and research in Civil Engineering at the Ecole Normale Supérieure (ENS-Cachan), one of elite engineering schools of French system of higher education. He is a Fellow of the International Association for Computational Mechanics, recipient of Humboldt Research Award in Technischen Mechanik from Germany, Claude Levy Strauss Award from Brazil, Asgard Award from Norway, Research Agency Award from Slovenia, and was invited professor and short course instructor at KAIST (South Korea), IPN (Mexico), USP (Brazil), Universities of Innsbruck (Austria), TU Tampere (Finland), Luxembourg, TU Braunschweig (Germany), Pavia (Italy), Ljubljana (Slovenia). He has written over 500 publications, including 8 books, 29 book chapters, 24 special issues/proceedings and close to 200 scientific papers in refereed journals.

Naida Ademović is an Associate Professor in Civil (Structural) Engineering at the Faculty of Civil Engineering at the University of Sarajevo in Bosnia and Herzegovina, which she joined in 2001. She is also an Associate Professor at the Center of Interdisciplinary Studies, University of Sarajevo, at the new Masters program "Natural Disasters Risk Management in Western Balkan Countries."

Naida received her 5-year Diploma (Bachelor of Science) in Structural Engineering from the University of Sarajevo. She received a Master of Science degree in "Computational Engineering" from Rühr University in Bochum, Germany, in 2004, and a second Masters degree (Advanced Masters in Structural Analysis of Monuments and Historical Constructions - SAHC) from the University of Minho, Portugal and University of Padova, Italy, in 2011. The SAHC program she was involved in won an EU Prize for Cultural Heritage/Europa Nostra Award in 2017. Naida received her PhD in 2012 in the field of earthquake engineering and masonry structures at the University of Sarajevo.

She has co-authored two books and published over 70 peer-reviewed scientific papers in refereed journals and international conferences.

Chapter 1

Initial boundary value problem

1.1 NEWTON'S SECOND LAW—STRONG FORM OF ELASTODYNAMICS

The first general equation of motion was developed by Newton and it is known as Newton's Second Law of Motion. This law enforces for the motion momentum either conservation, in the absence of external force, or change, due to an applied external force. Newton's Second Law applies in the same manner to point-like particles and to any point in a rigid body, apart from their constrained motion. It is also applicable to each point in a mass continua, like deformable solids or fluids, for either small or large motion, but the latter requires the motion to be accounted for by material derivative. The essential restriction concerns the mass conservation hypothesis; hence, if the mass is not constant, it is necessary to make some modifications to Newton's Second Law, which are consistent with the conservation of momentum.

The model problem of one-dimensional (1D) elasticity in large displacements is selected with the aim to demonstrate the finite element method applied to nonlinear dynamics. It is important to note that the main solution steps for the chosen model problem will remain the same for a number of other problems in nonlinear dynamics.

The reference frame in which the initial configuration of the 1D solid corresponds to a closed interval $\bar{\Omega} = [0, l]$ is selected. Each particle of this configuration is identified by its position vector:

$$x \in \bar{\Omega} \equiv [0, l] \tag{1.1}$$

The dynamic loads are applied in terms of time-dependent distributed loading $b(x,t)$, producing the motion of the body that can be described by the trajectory of each particle from its initial position x to the current position at time t, denoted as $\varphi(x,t)$. Alternatively, the motion can also be described by the displacement field specifying the displacement value along x for each particle:

$$x^{\varphi} = x + d(x,t) =: \varphi(x,t) \equiv \varphi_t(x) \tag{1.2}$$

The deformed configuration is defined as the assembly of all particles in their current position (see Figure 1.1):

$$x^{\varphi} \in \bar{\Omega}_t^{\varphi} = [0, l_t^{\varphi}] \tag{1.3}$$

Velocity and acceleration of motion can be written in either a spatial or material description, by using x^{φ} or x as the independent variable, respectively. The point transformation

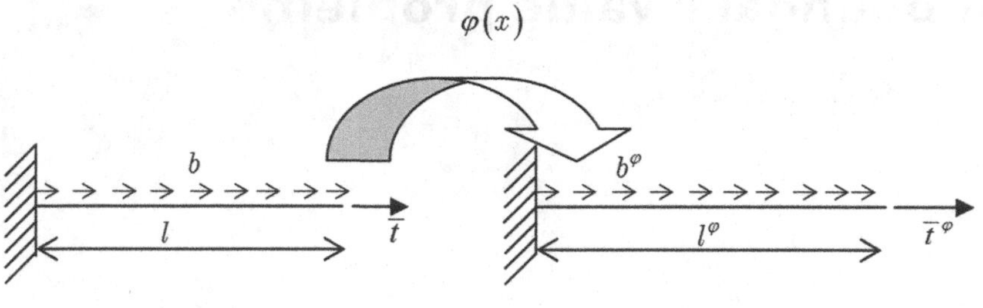

Figure 1.1 Mapping from the initial to current configuration.

in (1.2) allows one to easily switch from one description to other. This also being applicable for inertia effects which are important for dynamic problems.

The velocity in the *material description* can simply be computed as the partial derivative of motion in (1.2) with respect to time while keeping the particle material position fixed; this results with:

$$\upsilon(x,t) = \left.\frac{\partial\varphi(x,t)}{\partial t}\right|_{\bar{x}} \tag{1.4}$$

It is important to understand that the velocity vector in (1.4) acts in the current configuration, but that it is parameterized in the material description with respect to the particle position in the initial configuration through the point transformation $x^{\varphi} = \varphi_t(\bar{x})$. By assuming that the motion is sufficiently regular to define the inverse of this transformation, φ_t^{-1}, the velocity vector can also be expressed by the spatial description:

$$\hat{\upsilon}\left(\varphi_t^{-1}(x^{\varphi}),t\right) =: \upsilon^{\varphi}(x^{\varphi},t) \equiv \left.\frac{\partial\varphi\left(\varphi_t^{-1}(x^{\varphi}),t\right)}{\partial t}\right|_{\bar{x}^{\varphi}} \tag{1.5}$$

Further, in the same manner, one can continue on with the relationship between spatial and material descriptions of the particle acceleration. Namely, in a material description, the latter can be expressed simply as the second partial derivative of the motion with respect to time:

$$a(x,t) = \left.\frac{\partial\upsilon(x,t)}{\partial t}\right|_{x} \equiv \left.\frac{\partial^2\phi(x,t)}{\partial t^2}\right|_{x} \tag{1.6}$$

On the other hand, the spatial description of the acceleration vector will result in a so-called convective term:

$$a^{\varphi}\left(x^{\varphi},t\right) = \frac{\partial\upsilon^{\varphi}\left(x^{\varphi},t\right)}{\partial t} + \frac{\partial\upsilon^{\varphi}\left(x^{\varphi},t\right)}{\partial x^{\varphi}}\upsilon^{\varphi}\left(x^{\varphi},t\right) \tag{1.7}$$

The *mass conservation principle* postulates that in any deformed configuration of a deformable solid body, Ω^{φ}, the total mass of the body will remain constant. Therefore, the mass density $\rho^{\varphi}(x^{\varphi},t)$ will change in each particular deformed configuration Ω^{φ}, but the total mass will not:

$$m\left(\Omega^{\varphi}\right) = \int_{\Omega^{\varphi}} \rho^{\varphi}\left(x^{\varphi},t\right)\mathrm{d}x^{\varphi} \tag{1.8}$$

Denoting the mass density as $\rho(x)$, in the initial configuration, the principle of mass conservation allows one to write the following:

$$\int_{\Omega^\varphi} \rho^\varphi\left(x^\varphi,t\right)\mathrm{d}x^\varphi = \int_{\Omega} \rho(x)\mathrm{d}x \tag{1.9}$$

By considering that the mass conservation principle also applies to an arbitrary sub-domain, the local form of this mass conservation principle can be acquired as:

$$\rho^\varphi\left(\underbrace{x^\varphi,t}_{\rho_t(x)}\right) = \frac{\rho(x)}{\lambda(x,t)};\lambda = \frac{\mathrm{d}x^\varphi}{\mathrm{d}x} \tag{1.10}$$

The last result derived for the 1D case can easily be generalized to the three-dimensional (3D) case taking into account a proper change of infinitesimal volume through the determinant of the deformation gradient:

$$\rho^\varphi dV^\varphi = \rho dV \Rightarrow \rho^\varphi = \frac{\rho}{J}; J = \det[\mathbf{F}] \tag{1.11}$$

Taking into account these results, it is possible to deliver the spatial and material descriptions of the local form of equations of motion. The former, which applies to the current position of the particle "x^φ" in a given deformed configuration, reads:

$$\rho^\varphi\left(x^\varphi,t\right)a^\varphi\left(x^\varphi,t\right) = \frac{\partial\sigma^\varphi\left(x^\varphi,t\right)}{\partial x^\varphi} + b^\varphi\left(x^\varphi\right);\ \text{in } dx^\varphi \tag{1.12}$$

The driving force of motion in the spatial description is defined by the expression equivalent to the static equilibrium equations written in term of the Cauchy or true stress.

As is well known, static equilibrium equations can also be written in the material description by using the first Piola–Kirchhoff stress $P(x,t)$ and the distributed loading $b(x,t)$ per unit volume of the initial configuration. This kind of transformation can be applied to the problem in dynamics by adding the inertia term in the material description and the principle of mass conservation defined in (1.6) and (1.10), respectively. The final result is the material description of the equations of motion, which can be written:

$$\rho(x)a(x,t) = \frac{\partial P(x,t)}{\partial x} + b(x,t);\ \text{in } dx \tag{1.13}$$

It is important to note that the material description results in the linear inertia term, which follows from using the fixed reference frame where $a(x,t) = (\partial^2\phi(x,t)) / (\partial t^2)$. The linear form of the inertia terms in the material description of the equations of motion also applies to the 3D case, which is written by using the direct tensor notation:

$$\rho(\mathbf{x})\mathbf{a}(\mathbf{x},t) = \operatorname{div}\mathbf{P}(\mathbf{x},t) + \mathbf{b}(\mathbf{x},t); \forall\mathbf{x} \in \Omega \,\&\, \forall t \in [0,T] \tag{1.14}$$

1.2 D'ALEMBERT PRINCIPLE—WEAK FORM OF EQUATIONS OF MOTION

The D'Alembert principle is the dynamic analog to the principle of virtual work for applied forces in a static system. It is more general than Hamilton's principle, avoiding the restriction to holonomic systems (Morris 1997). In this respect, the D'Alembert principle is chosen as the starting point in the development of the weak form of the equations of motion.

The snap-shot of motion taken at time "t" can be described formally with the equilibrium equations. However, unlike the statics problem, these equilibrium equations should also include an extra external load in terms of the inertia force; the latter, denoted as $f_{\text{inertia}} = -\rho\left(x^{\varphi},t\right)a\left(x^{\varphi},t\right)$, is proportional to the mass and directed opposite to acceleration. Thus, it can be written:

$$0 = -\rho^{\varphi}\left(x^{\varphi},t\right)a^{\varphi}\left(x^{\varphi},t\right)+\frac{\partial\sigma^{\varphi}\left(x^{\varphi},t\right)}{\partial x^{\varphi}}+b^{\varphi}\left(x^{\varphi},t\right) \tag{1.15}$$

This point of view of D'Alembert provides an important conceptual advantage in allowing one to reduce a new problem of describing the motion in dynamics to a familiar problem of equilibrium in statics. As stated above, the later can then be treated by the principle of virtual work, stating that the work of internal forces on chosen virtual displacement should be equal to the virtual work of external forces, including the inertia force among them. In this approach, the time is kept fixed at the chosen value $\bar{t}$ corresponding to a particular deformed configuration; hence, the virtual displacement field is independent of time and can be denoted: $w^{\bar{\varphi}} = \hat{w}^{\varphi}\left(\varphi(x,\bar{t})\right)$, see Figure 1.2.

Equations of motion can be multiplied by the chosen virtual displacement (or weighting function) and integrated over the whole domain. We then use the integration by parts to remove derivatives on the stress field and to provide the final weak form of the 1D problem in elastodynamics:

$$\begin{aligned} 0 = G_{\text{dyn}}\left(\varphi_t;w^{\bar{\varphi}}\right) &:= \int_{\Omega^{\varphi}} w^{\bar{\varphi}}\left[\rho^{\varphi}\left(x^{\varphi},t\right)a^{\varphi}\left(x^{\varphi},t\right)-\frac{\partial\sigma^{\varphi}\left(x^{\varphi},t\right)}{\partial x^{\varphi}}-b^{\varphi}\left(x^{\varphi},t\right)\right]dx^{\varphi} \\ &= \int_{\Omega^{\varphi}}\left[w^{\bar{\varphi}}\rho^{\varphi}\left(x^{\varphi},t\right)a^{\varphi}\left(x^{\varphi},t\right)+\frac{dw^{\bar{\varphi}}}{dx^{\varphi}}\sigma^{\varphi}\left(x^{\varphi},t\right)-w^{\bar{\varphi}}b^{\varphi}\left(x^{\varphi},t\right)\right]dx^{\varphi}-w^{\bar{\varphi}}(l_t)\bar{t}^{\varphi}(t) \end{aligned} \tag{1.16}$$

Taking into account the result obtained in (1.10), together with the material parameterization of the virtual displacement field $w(x) = w^{\varphi}(\varphi(x,\bar{t}))$, the weak form of the equations of motion can be recast in the material description:

$$0 = G_{\text{dyn}}\left(\varphi_t(x);w(x)\right) := \int_{\Omega}\Big[w(x)\rho(x)a(x,t)\mathrm{d}x \; \underbrace{+\frac{\mathrm{d}w(x)}{\mathrm{d}x}P(x,t)-w(x)b(x,t)\Big]\mathrm{d}x-w(l)\bar{t}(t)}_{G_{\text{stat}}} \tag{1.17}$$

where $G_{\text{stat}}(\cdot)$ is the result of the 1D elastostatics. The same kind of result for the 3D case can be achieved, and by using the direct tensor notation it reads:

$$0 = G_{dyn}\left(\varphi_t(\mathbf{x});\mathbf{w}(\mathbf{x})\right) := \int_{\Omega}\mathbf{w}(\mathbf{x})\cdot\rho(\mathbf{x})\mathbf{a}(\mathbf{x},t)dV + G_{\text{stat}}\left(\varphi_t(\mathbf{x});\mathbf{w}(\mathbf{x})\right) \tag{1.18}$$

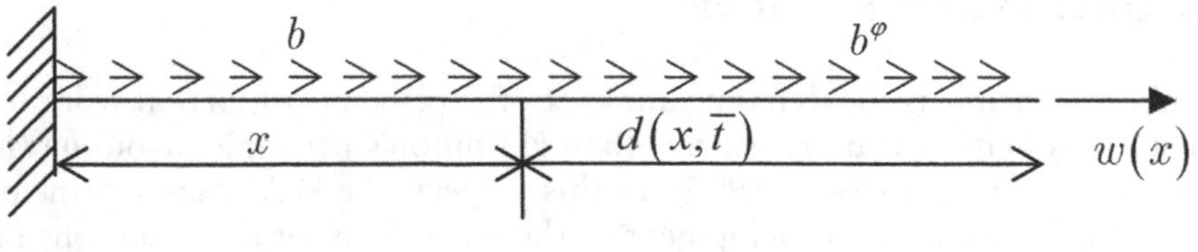

Figure 1.2 Application of the D'Alembert principle.

1.3 HAMILTON'S PRINCIPLE—ENERGY CONSERVATION

The weak form of equations of motion can be expressed in terms of variational formulation, which applies to hyperelastic materials. Here, the strain energy density $\psi(\varphi)$ and external loads together define the potential energy functional $\Pi(\varphi_t(x))$, which can be written:

$$\Pi(\varphi) = \int_{\Omega} \psi(\varphi)\,dx - \Pi_{ext}(\varphi) \tag{1.19}$$

The first variation of such a functional is equivalent to the weak form of the equilibrium equations in statics, and, moreover, a positive value of the second variation indicates that the equilibrium state remains stable:

$$\begin{aligned} 0 &= G_{stat}\left(\varphi_t(x); w(x)\right) := \frac{d}{ds}\left[\Pi\left(\varphi_{t,s}(x)\right)\right]\Big|_{s=0} \\ 0 &< \frac{d^2}{ds^2}\left[\Pi\left(\varphi_{t,s}(x)\right)\right]\Big|_{s=0} \end{aligned} \tag{1.20}$$

It is possible to generalize this result to dynamics, by appealing to the Hamilton variational principle employing the total energy functional, which consists of both potential and kinetic energy:

$$H\left(\varphi(x,t), \upsilon(x,t)\right) := T\left(\upsilon(x,t)\right) + \Pi\left(\varphi(x,t)\right) \tag{1.21}$$

In the material description, the kinetic energy can be written as a quadratic form in velocities:

$$T\left(\upsilon(x,t)\right) = \int_{\Omega} \frac{1}{2}\rho(x)\left[\upsilon(x,t)\right]^2 dx \tag{1.22}$$

One can appeal to the principle of least action, to postulate that in the dynamic motion of a hyperelastic body over a time interval $[t_1, t_2]$, the zero value of the first variation of total energy (or yet referred to Hamiltonian) will produce the equations of motion, whereas a positive value of the second variation will confirm the motion stability:

$$\begin{aligned} 0 &= G_{dyn} := \frac{d}{ds}\left[\int_{t_1}^{t_2} H\left(\varphi_{t,s}(x)\right)dt\right]_{s=0} \quad ; \\ 0 &< \frac{d^2}{ds^2}\left[\int_{t_1}^{t_2} H\left(\varphi_{t,s}(x)\right)dt\right]_{s=0} \end{aligned} \tag{1.23}$$

It is easy to confirm the last result by using the directional derivative computation. In fact, the only new term with respect to statics concerns the directional derivative of the kinetic energy in (1.22), which can be written:

$$\begin{aligned} &\frac{d}{ds}\left[\int_{t_1}^{t_2} T\left(\frac{\partial\left(\varphi_{t,s}(x)\right)}{\partial t}\right)\right]_{s=0} dt = \int_{t_1}^{t_2} \frac{\partial T(\upsilon)}{\partial \upsilon}\frac{\partial^2 \varphi_{t,s}}{\partial s \partial t}\,dt = \\ &= \left[\frac{\partial T(\upsilon)}{\partial \upsilon}\frac{\partial \varphi_{t,s}}{\partial s}\Big|_{s=0}\right]_{t_1}^{t_2} - \int_{t_1}^{t_2} \frac{\partial}{\partial t}\left[\frac{\partial T(\upsilon)}{\partial \upsilon}\right] w\,dt \\ &= \left[\frac{\partial T\left(\upsilon(t_2)\right)}{\partial \upsilon}\underbrace{w(t_2)}_{0} - \frac{\partial T\left(\upsilon(t_1)\right)}{\partial \upsilon}\underbrace{w(t_1)}_{0}\right] - \int_{t_1}^{t_2}\int_{\Omega} w\rho\dot{\upsilon}(x,t)\,dx\,dt \end{aligned} \tag{1.24}$$

The last result is obtained by imposing a supplementary admissibility condition on zero value of the variations at both limits of the time interval, with $w(t_1)=0$ and $w(t_2)=0$, which allows for the recovering of the inertia term of equations of motion from the first variation of the kinetic energy. The last result along with the variation of the potential energy producing will confirm that the first variation of the Hamiltonian corresponds to the equations of motion (1.17):

$$0=\frac{d}{ds}\left[\int_{t_1}^{t_2} H\left(\varphi_{t,s}(x)\right)dt\right]_{s=0} \equiv G_{\text{dyn}} \tag{1.25}$$

In the absence of external load with $\Pi_{\text{ext}}\left(\varphi(x,t)\right)=0$, the Hamilton principle implies that the total energy of the motion remains conserved:

$$0=\frac{d}{ds}\left[\int_{t_1}^{t_2} H\left(\varphi_{t,s}(x)\right)dt\right]_{s=0} \Rightarrow H(t)=\text{cst.} \tag{1.26}$$

This is an important observation that has to be taken into account in the development of time-integration schemes.

1.4 MODAL SUPERPOSITION METHOD

Generally, superposition is applicable when there is a linear relationship between external forces and corresponding structural displacements. For the dynamic analysis of linearly elastic systems, modal superposition is a powerful idea of obtaining solutions.

The modal superposition procedure is also applicable to dynamic linearly elastic systems having continuously distributed properties, which have an infinite number of degrees of freedom requiring that their equations of motion are written in the form of partial differential equations. This is all in the sense that for obtaining practical solutions only a limited number of the lower modes of vibration can be considered.

It is applicable to both free-vibration and forced-vibration problems. This method has often been used in the analysis of linear response, and it has seldom been applied in nonlinear problems (Morris 1997).

The modal superposition procedure can be used to obtain an independent SDF equation for each mode of vibration of the undamped structure. Thus, the use of the normal coordinates serves to transform the equations of motion from a set of N simultaneous differential equations, which are coupled by the off-diagonal terms in the mass and stiffness matrices, to a set of N independent equations in normal coordinates. The dynamic response, therefore, can be obtained by solving separately for the response of each normal (modal) coordinate and then superposing these to obtain the response in the original geometric coordinates (Clough and Penzien 2003).

The equations of motion for a free undamped system of vibration can be written in the form:

$$\mathbf{M}\ddot{\mathbf{v}}+\mathbf{K}\mathbf{v}=0 \tag{1.27}$$

In order to determine in which case the conditions set in (1.27) will be satisfied, we have to solve the problem of free-vibrations. It can be assumed that the free-vibration motion is a simple harmonic function, expressed as:

$$\mathrm{v}(t)=\phi\sin\left(\omega t+\theta\right) \tag{1.28}$$

where ϕ is the shape of the system (amplitude) and θ is a phase angle.

Taking the second derivative of equation (1.28) and inserting it and the equation (1.28) into (1.27) one obtains:

$$-\omega^2 \mathbf{M}\phi \sin(\omega t + \theta) + \mathbf{K}\phi \sin(\omega t + \theta) = 0 \tag{1.29}$$

As the sine term is arbitrary it can be omitted, so the equation now reads:

$$\left[\mathbf{K} - \omega^2 \mathbf{M}\right]\phi = 0 \tag{1.30}$$

which is a well-known characteristic equation of the system, where ϕ_i and ω_i are natural modes of free-vibrations and natural frequencies.

The free-vibration mode shapes ϕ_i have certain special properties which are very useful in structural dynamics analyses. The orthogonality relationships state that the vibrating shapes are orthogonal with respect to the stiffness matrix as well as with respect to the mass matrix.

The equation of motion for the system (1.29) can be written as:

$$\begin{aligned} \mathbf{K}\phi &= \omega^2 \mathbf{M}\phi \\ \mathbf{f}_s &= -\mathbf{f}_I \end{aligned} \tag{1.31}$$

where:

$\mathbf{f}_s$ is the elastic-resisting force factor and

$-\mathbf{f}_I$ is the applied-inertia-load vector.

Using Betti's law on the two vibration modes, one obtains, as indicated in Figure 1.3:

$$-\mathbf{f}_{\mathrm{I}m}^T \phi_n = -\mathbf{f}_{\mathrm{I}n}^T \phi_m \tag{1.32}$$

Combining equations (1.31) and (1.32), one obtains:

$$\omega_m^2 \phi_m^T \mathbf{M} \phi_n = \omega_n^2 \phi_n^T \mathbf{M} \phi_m \tag{1.33}$$

As the mass matrix is symmetric and a scalar is obtained as the result of the matrix operations, it is evident that equation (1.33) can be presented in the form:

$$\left(\omega_m^2 - \omega_n^2\right)\phi_m^T \mathbf{M} \phi_n = 0 \tag{1.34}$$

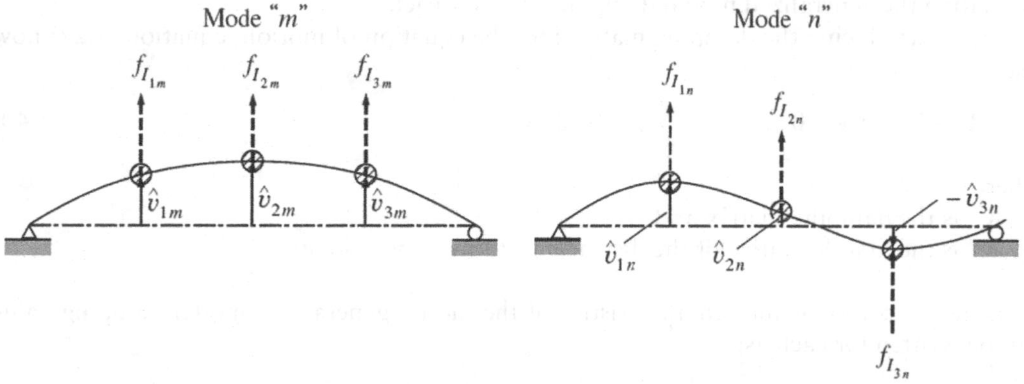

Figure 1.3 Vibration mode shapes and resulting inertial forces (Clough and Penzien 2003).

As the two frequencies are not the same, one obtains the first orthogonality conditions, which reads:

$$\phi_m^T \mathbf{M} \phi_n = 0 \quad \omega_m \neq \omega_n \tag{1.35}$$

The second orthogonality condition states that

$$\phi_m^T \mathbf{K} \phi_n = 0 \quad \omega_m \neq \omega_n \tag{1.36}$$

For a free-vibration problem in linear dynamics, with d_0 and v_0 as the initial displacement and velocity, the solution is obtained with mode superposition

$$d_{\text{free}}(t) = \sum_{i=1}^{n} \phi_i \left(\phi_i^T \mathbf{M} d_0 \cos \omega_i t + \left(\frac{1}{\omega_i} \right) \phi_i^T M v_0 \sin \omega_i t \right) \tag{1.37}$$

The mode superposition method can also be employed to construct the solution to a forced-vibration problem in linear dynamics, where the forcing term is supplied by Duhamel's integral with:

$$d_{\text{forced}}(t) = d_{\text{free}}(t) + \sum_{i=1}^{m} \phi_i \left(\int_0^t \left(\phi_i^T M f(s) \right) / \omega_i \sin \omega_i (t-s) ds \right) \tag{1.38}$$

1.4.1 Rayleigh Damping

In order to apply the modal analysis of undamped systems to damped systems, it is common to assume the proportional damping, a special type of viscous damping. The proportional damping model can be represented as a linear combination of the mass and stiffness matrices, leading to well known Rayleigh damping, or classical damping model.

$$\mathbf{C} = \alpha \mathbf{M} + \beta \mathbf{K} \tag{1.39}$$

where α and β are real scalars, known as proportionality constants and they have units 1/sec and sec units respectively. Modes of classically damped systems preserve the simplicity of the real normal modes as in the undamped case. The properties of mode orthogonality as stated in (i) and (j) are used here. These are called mass proportional and stiffness proportional damping, and the damping behavior associated with them may be recognized by evaluating the generalized modal damping value for each.

When introducing the damping matrix into the equation of motion, equation (1.27) now reads:

$$\mathbf{M}\ddot{\mathbf{v}} + \mathbf{C}\dot{\mathbf{v}} + \mathbf{K}\mathbf{v} = 0 \tag{1.40}$$

where:

$\mathbf{C}$ is the damping matrix and

$\dot{\mathbf{v}}$ is the first derivative of the displacement, meaning velocity.

Using the orthogonality characteristics of the modes generalized modal damping value can be written for each as:

$$C_n = \phi_n^T \mathbf{C} \phi_n = \alpha \phi_n^T \mathbf{M} \phi_n \tag{1.41a}$$

or:

$$C_n = \phi_n^T C \phi_n = \beta \phi_n^T \mathbf{K} \phi_n \tag{1.41b}$$

Combining (1.41a) and (1.41b) with (1.42):

$$2\omega_n \mathbf{M}_n \xi_n = \alpha \mathbf{M}_n \ \text{ or } \mathbf{K}_n = \omega_n^2 \mathbf{M}_n \tag{1.42}$$

one obtains:

$$\xi_n = \frac{\alpha}{2\omega_n} \text{ or } \xi_n = \frac{\beta \omega_n}{2} \tag{1.43}$$

From (1.43), it is obvious that for mass proportional damping, damping is inversely proportional to the frequency on one hand, and on the other for stiffness proportional damping it is directly in proportion with the frequency. Combining these two gives the above mentioned Rayleigh's damping (see Figure 1.4). The two damping factors α and β can be evaluated using the solution of a pair of simultaneous equations if the damping ratios ξ_m and ξ_n connected with the two specific modes of frequencies ω_m and ω_n are known.

As damping is a rather uncertain parameter and its variation with frequency is usually unknown, in most practical applications it is feasible to assume the same damping ratio for the tow control frequencies. The proportional factors α and β then are given as:

$$\begin{Bmatrix} \alpha \\ \beta \end{Bmatrix} = \frac{2\xi}{\omega_m + \omega_n} \begin{Bmatrix} \omega_m \omega_n \\ 1 \end{Bmatrix} \tag{1.44}$$

1.4.2 Nonproportional Damping and Fast Computation Methods

A nonproportional damping matrix is formed as an assembly of proportional matrices for each part of the structure, with characteristic materials and specified damping ratios for each segment of the structure.

In this situation, the assembly of the stiffness, mass and damping property matrices will be formulated as

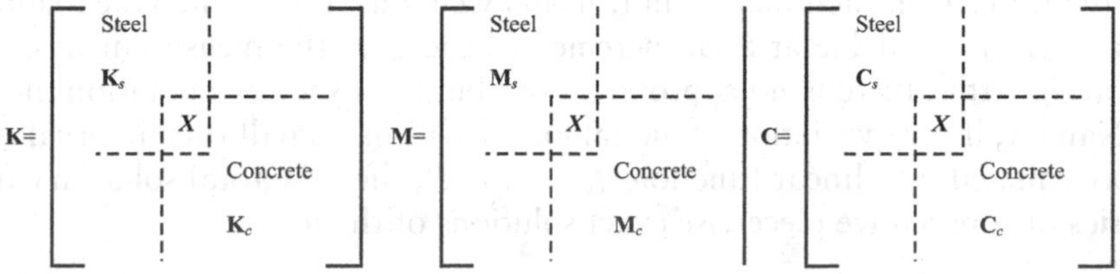

where X DoF are common to both parts of the structure. The same procedure as explained above is used for solving this problem.

The vast majority of engineering structures in practical applications are characterized by nonproportional damping. Many practical examples with nonproportional damping matrices can be given, starting from the frictional support systems, over mechanical structures with energy-absorbing or vibration-isolation devices and coupled systems, to soil-structure and fluid–structure interaction problems. Dynamic response of the systems with nonproportional damping can be obtained by direct integration or by the calculation of the complex eigenvectors of the complete system, which are then used to transform the system into an uncoupled set of complex modal equations (Veletsos and Ventura 1986; Chen 1987).

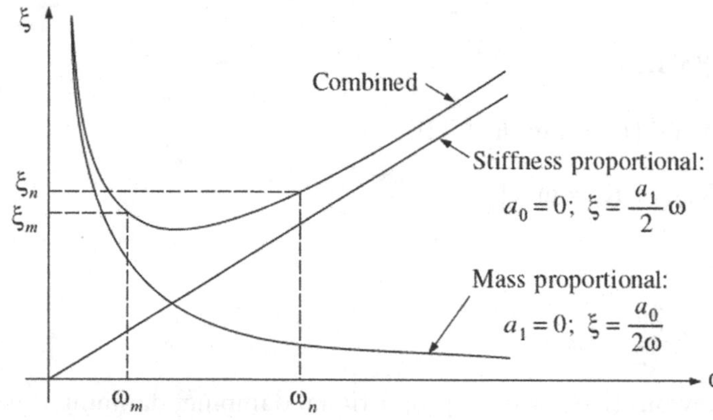

Figure 1.4 **Relationship between damping ratio and frequency (for Rayleigh damping) (Clough and Penzien 2003).**

For a nonproportionally damped system, the equations of motion in the modal coordinates are coupled through the off-diagonal terms of the modal damping matrix and consequently the system possesses complex modes instead of real normal modes. This has been investigated by many authors (Ibrahimbegović et al. 1991; Phani 2004; Adhikari and Woodhouse 2001; Udwadia 2009). Researchers have proposed different methods to overcome this issue and to take into account nonlinearities (for further details see Bernal 1994; Hall 2006; Charney 2005; Medina and Krawinkler 2004; Zareian 2006). However, the proposed solutions by all the researchers above have real limitations; for example, Bernal's (Bernal 1994) approach does not preclude spurious damping forces appearing when masses are assigned to rotational degrees of freedom. Charney's (Charney 2005) proposal is questionable once the overall tangent stiffness of the system becomes negative due to second-order, P-Δ effects and/or significant material strength and stiffness deterioration.

A general method for handling nonproportional damping was proposed by Ibrahimbegović and Wilson (1989), which is characterized by high accuracy and low computational cost.

The main idea of this solution method is to make an approximation in loading time variation—making it linear in each increment (this fits well with the earthquake loading, where one measurement is available at each increment; hence, all the measurement can be connected linearly). After, there is no approximation, but the exact computation of the modal response. Namely, if time variation of modal load f_i within a small time interval $\Delta t = t_1 - t_0$ can be approximated by a linear function ($f_i = a + b \cdot t$), then the total solution can be written as a series of consecutive piecewise exact solutions of the form:

$$y_i(t_i) = A_0 + A_1\Delta t + A_2 \exp(-\xi_i\omega_i\Delta t)\cos\omega_{Di}\Delta t + A_3 \exp(-\xi_i\omega_i\Delta t)\sin\omega_{Di}\Delta t \tag{1.45}$$

where the coefficients in equation (1.45) above are:

$$A_0 = \frac{a}{\omega_i^2} - \frac{2\xi_i b}{\omega_i^3} \tag{1.46}$$

$$A_1 = \frac{b}{\omega_i^2} \tag{1.47}$$

$$A_2 = y_i(t_0) - A_0 \tag{1.48}$$

$$A_3 = \frac{1}{\omega_{Di}}\left[\dot{y}_i(t_0) + \xi_i\omega_i A_2 - A_1\right] \tag{1.49}$$

The pseudo-force concept can be applied in a very similar way to account for modal coupling effects in the analysis of systems with nonproportional damping. In this case, the crucial step is to separate the damping matrix into a proportional component that is uncoupled by the modal coordinate transformation, plus a nonproportional component that is transferred to the right-hand side of the equations of motion, where its effects are represented as pseudo-forces. In this context, we will assume that the adequate nonproportional viscous damping matrix is denoted as C and that it has been formulated to represent the system's actual damping mechanism. Now modal coordinate transformation will be applied, and the damping matric will be transformed into diagonal elements which are representing the proportional damping contribution and the nonproportional damping effects are presented by the off-diagonal elements. Thus, we introduce the additive split of the modal damping matrix:

$$\mathrm{C}_m = \mathrm{C}_d + \hat{\mathrm{C}} = \mathrm{diag}(2\xi_i\omega_i) + \hat{\mathrm{C}} \tag{1.50}$$

The damping coupling between the modes is expressed by these off-diagonal coefficients and are considered as pseudo-forces applied on the right side of the equation of motion.

$$\ddot{y}_i + 2\xi_i\omega_i\dot{y}_i + \omega_i^2 y_i = f_i - \hat{C}_{ij}\dot{y}_j, i = 1,2,...m. \tag{1.51}$$

The contribution of the off-diagonal modal damping coefficients to these pseudo-forces can be defined in an iterative manner:

$$\ddot{y}_i^{(k)} + 2\xi_i\omega_i\dot{y}_i^{(k)} + \omega_i^2 y_i^{(k)} = f_i - \sum_{j=1}^{mm}\hat{C}_{ij}\dot{y}_j^{(k-1)}, i = 1,2,\ldots m. \tag{1.52}$$

This method is characterized by high accuracy and low computational cost thanks to its unconditionally stable iterative procedure with strong convergence properties. The important aspect is that the procedure uses an exact solution for piecewise linear loading excitation; hence, its accuracy depends on the proper representation of real loading variation in time. In addition, the algorithm can easily be extended to systems subjected to periodic loading (Ibrahimbegović and Wilson 1989).

1.5 TIME-INTEGRATION SCHEMES

In this case, we make an approximation not only in loading time variation, but also response function approximation (e.g. sin or cos functions replaced by polynomials), and we arrive at the time-integration schemes.

The final product of semi-discretization procedure carried out by the finite element method in nonlinear dynamics is a system of nonlinear ordinary differential equations in time. Any such nonlinear dynamics problem is quite different from the nonlinear equilibrium problem in statics, which is governed by a set of nonlinear algebraic equations parameterized by pseudo-time. Despite this difference, in constructing the solution to a nonlinear dynamics problem by using the time-integration schemes, all the main ingredients of the incremental/iterative solution procedure for quasi-static problems are present. The traditional methods for solving linear dynamics problems, such as modal superposition method (e.g. Clough and Penzien 2003; Géradin and Rixen 1992) are of small interest for nonlinear dynamics problems since the free-vibration modes would change with each incremental/iterative modification of the tangent stiffness matrix. A few words will be given in the following subsection regarding modal superposition.

In order to solve the nonlinear dynamics problems by using the time-integration schemes, a few steps have to be followed. In this respect, the time interval of interest $[0, T]$ is subdivided into the chosen number of time steps, which will specify the time instants where the selected time-integration scheme should deliver the solution:

$$0 < t_1 < t_2 < \ldots < t_n < t_{n+1} < \ldots < T \tag{1.53}$$

This corresponds to the incremental analysis for quasi-statics problems; however, contrary to statics, the complete solution requires not only the nodal values of displacements but also of velocities and accelerations. For example, the solution at time t_n is defined with:

$$\mathbf{d}_n = \mathbf{d}(t_n); \mathbf{v}_n = \dot{\mathbf{d}}(t_n); \mathbf{a}_n = \ddot{\mathbf{d}}(t_n) \tag{1.54}$$

1.5.1 Central Problem of Dynamics (Time Evolution)

In order to simplify the computer code architecture, in general, one-step time-integration schemes are used that construct the solution over a single step at the time. For a typical time step starting at time t_n, the time-integration scheme should deliver the nodal values of displacements, velocities and acceleration at time t_{n+1}, which verify the equations of motion; this can formally be stated in terms of:

Central problem in nonlinear *dynamics for* $[t_n, t_{n+1}]$

Given: $\mathbf{d}_n, \mathbf{v}_n, \mathbf{a}_n$
Find: $\mathbf{d}_{n+1}, \mathbf{v}_{n+1}, \mathbf{a}_{n+1}$
such that: $M\mathbf{a}_{n+1} + \hat{f}^{\text{int}}(\mathbf{d}_{n+1}) = f_{n+1}^{\text{ext}}$

One possibility for solving the central problem in dynamics is by exploiting the one-step time-integration schemes. In fact, it is always possible to recast a set of second-order differential equations that characterize the nonlinear dynamics in terms of the set of the first-order equations containing twice as many equations. In this way, we switch to the so-called state space form of the system, where both displacements and velocities are considered as independent variables:

$$\dot{\mathbf{z}}(t) = h(\mathbf{z}(t)); \mathbf{z} = \begin{bmatrix} d \\ v \end{bmatrix}; h(z) = \begin{bmatrix} v \\ M^{-1}\left(f^{\text{ext}} - \hat{f}^{\text{int}}(d)\right) \end{bmatrix} \tag{1.55}$$

Thus one can directly apply the time-integration schemes for integrating the first-order systems (heat transfer equation or internal variable evolution equations). However, this would not lead to the most efficient implementation. The computational efficiency can be improved significantly with the time-integration schemes applicable directly to the original form of the second-order differential equations. Several time-integration schemes of this kind, both explicit and implicit, are presented subsequently.

1.5.2 Central Difference (Explicit) Scheme

The central difference scheme is based upon the *second-order approximation* of the differential equations of motion, with a truncated series development employing only the first derivative of displacements and velocities in (1.47). This kind of approximation can be constructed in a systematic manner by using the central difference operator (e.g. Dahlquist and Björck 1974, p. 353); the corresponding approximation of the first differential equation in (1.29) is obtained, which can be written:

$$\dot{\mathbf{d}}(t) = \mathbf{v}(t) \Rightarrow \frac{d_{n+1} - d_{n-1}}{2h} = v_n \tag{1.56}$$

The last result confirms the explicit form of the central difference approximation since in computing the displacement d_{n+1} only the displacement and the velocity in two previous steps is required. However, for simplicity of computer code architecture, it is preferred that the numerical implementation of the central difference scheme, where the only values needed for computing the displacement at time t_{n+1}, are those from the previous time t_n. In order to ensure such a format of the central difference scheme, the corresponding approximations of the velocity at the mid-point of each time step is used, which allow one to write:

$$\left.\begin{aligned} \frac{1}{h}\left(d_{n+1} - d_{n-1}\right) = v_{n+1/2} \\ \frac{1}{h}\left(d_n - d_{n-1}\right) = v_{n-1/2} \end{aligned}\right\} \Rightarrow \frac{1}{2h}\left(d_{n+1} - d_{n-1}\right) = \frac{1}{2}\left(v_{n+1/2} + v_{n-1/2}\right) \tag{1.57}$$

Combining equation (1.56) and (1.57), the corresponding approximation of velocity at time t_n can be written as:

$$v_n = \frac{1}{2}\left(v_{n+1/2} + v_{n-1/2}\right) \tag{1.58}$$

The same kind of central difference approximation based upon the mid-point values can also be constructed for the acceleration vector:

$$\mathbf{a}(t) = \dot{\mathbf{v}}(t) \Rightarrow a_n = \frac{v_{n+1/2} - v_{n-1/2}}{h} \tag{1.59}$$

The corresponding approximation for the velocity vector in the middle of the interval of interest is obtained by adding up equations (1.58) and (1.59), and it can be written:

$$v_{n+1/2} = v_n + \frac{h}{2}\mathbf{a}_n(t) \tag{1.60}$$

By inserting (1.60) into (1.57), the final form of the central difference approximation for the displacement vector at time t_{n+1} is obtained, and it reads:

$$d_{n+1} = d_n + hv_n + h^2 a_n \tag{1.61}$$

It is evident that this kind of approximation for displacement is fully explicit in that it only employs the known values of displacements, velocities and accelerations at time t_n.

The final form of the central difference approximation of velocity at time t_{n+1} can be obtained as a linear combination of the results in (1.59) and (1.61) above, as well as the equivalent result for the subsequent time step,[1] which results in:

$$v_{n+1} = v_n + \frac{h}{2}\left(a_n + a_{n+1}\right) \tag{1.62}$$

1 $v_{n+1} = \frac{1}{2}\left(v_{n+3/2} + v_{n+1/2}\right)$ and $v_{n+3/2} = v_{n+1/2} + \frac{h}{2}a_{n+1}$

Table 1.1 Central Difference Scheme in Nonlinear Dynamics

Initialize: $d_0, \upsilon_0 \rightarrow a_0 = M^{-1}\left(f_0^{ext} - \hat{f}^{int}(d_0)\right)$

Compute for each step: $n = 0,1,2...\left(\text{given} : d_n, \upsilon_n, a_n \text{ and } h\right)$

$d_{n+1} = d_n + hv_n + h^2 a_n$

$Ma_{n+1} + \hat{f}^{int}\left(d_{n+1}\right) = f_{n+1}^{ext} \Rightarrow a_{n+1} = M^{-1}\left(f_{n+1}^{ext} - \hat{f}^{int}\left(d_{n+1}\right)\right)$

$\upsilon_{n+1} = \upsilon_n + \frac{h}{2}\left(a_n + a_{n+1}\right)$

Next: $n \leftarrow n+1$

In the last expression, the acceleration vectors a_n and a_{n+1} are computed directly from the equations of motion at times t_n and t_{n+1}, respectively. The most efficient implementation of the central difference scheme is presented in Table 1.1.

The most costly phase in the proposed implementation of the central difference scheme clearly pertains to the solution of a set of algebraic equations for computing the acceleration vector a_{n+1}. However, this cost can be reduced considerably by using a diagonal form of the mass matrix, which allows for the solution to be obtained with the number of operations equal only to the number of equations "n." With this kind of computational efficiency, we can easily accept a very small time step that is often required to meet the conditional stability of the explicit scheme. For example, for the 1D hyperelastic bar (with free energy density $\psi(\lambda)$) and 2-node finite element approximations (with a typical element length l^e), the conditional stability of the central difference scheme requires (e.g. Oden and Fost 1973) the time step "h" no larger than:

$$h \leq \frac{l^e}{\sqrt{6}c_{\max}}; c_{\max} = \max_{\forall \lambda > 0}\left[\frac{\tilde{C}(\lambda)}{\rho}\right]^{1/2}; \tilde{C}(\lambda) = \frac{d^2\psi(\lambda)}{d\lambda^2} \tag{1.63}$$

Interestingly enough, by using a diagonal form of the mass matrix, the largest acceptable time step which guarantees the stability of the central difference scheme increases to:

$$h \leq \frac{l^e}{\sqrt{2}c_{\max}} \tag{1.64}$$

However, in a number of cases of practical interest, the restriction placed by the conditional stability of the explicit central difference scheme can be too severe, forcing us to take a time step which is too small with respect to the required result accuracy. For that reason, the central difference scheme is mostly used in applications with dynamics phenomena of short duration, such as impact problems, explosions or wave propagation with very short duration; in any such problem, the computed response would have a significant contribution of high-frequency modes and would require very small time steps, which can easily be handled with the central difference scheme.

Remarks:

1. For nonlinear inelastic behavior in dynamics, the central difference scheme provides a very interesting alternative to fully implicit schemes. Namely, we can solve the equations of motion by the explicit central difference scheme, combined with an implicit scheme solution of a small set of the evolutions equations for internal variables at each numerical integration point. This kind of explicit–implicit approach to nonlinear dynamics problems with inelastic behavior can be implemented within the

framework of the operator split procedure. We, thus, obtain the highest computational efficiency in solving a large set of equations of motion that provides the best value of displacements $d_{n+1}^{(i)}$ as well as the corresponding value of the total deformation field $\varepsilon_{n+1}^{(i)}$, along with results reliability for the local computations of internal variables and admissible value of stress that is carried out by an implicit scheme (e.g. the backward Euler scheme). The latter should be implemented in a manner which guarantees the convergence of the local computation that enforces the stress admissibility constraint in agreement with chosen plasticity or damage criteria (e.g. the stress value cannot be larger than given plasticity or damage threshold). It is important to note that only stress values are needed for this kind of strategy, with no need to compute the tangent moduli.

2. We have implemented this kind of strategy for a coupled damage-plasticity constitutive model suitable for representing the behavior of concrete under impact (see Hervé, Gatuingt and Ibrahimbegović 2005). This model employs a judicious combination from the damage model of Mazars (see Mazars 1986) for representing the concrete cracking with the threshold related to the principal elastic strains, and the Gurson-like (see Gurson 1977 or Needleman and Tvergaard 1984) viscoplasticity model, capable of representing the concrete hardening behavior under compressive stress. A graphic illustration of the elastic domain defined by the proposed criterion is represented in Figure 1.5. We note that there is a considerable complexity of the constitutive model of this kind with a fairly long list of internal variables (such as the damage variable d_{n+1}, viscoplastic strains ε_{n+1}^{vp}, the porosity f^*, as well as the hardening variables; e.g. Hervé, Gatuingt and Ibrahimbegović 2005), which would represent quite a significant challenge for guaranteeing the convergence of the fully implicit schemes. However, the explicit computations of equations of motion, accompanied by only implicit computations of internal variables does not have any such difficulty. A couple of illustrative results which concern the impact computation on a concrete slab, both for the cases with and without perforation, are presented in Figures 1.6 and 1.7.

1.5.3 Trapezoidal Rule or Average Acceleration (Implicit) Scheme

For a large number of problems in low-velocity dynamics, such as the problems typical of earthquake engineering with significant damping, the dynamic response is to a large extent governed by fairly few, low-frequency modes. One finds that the contribution of

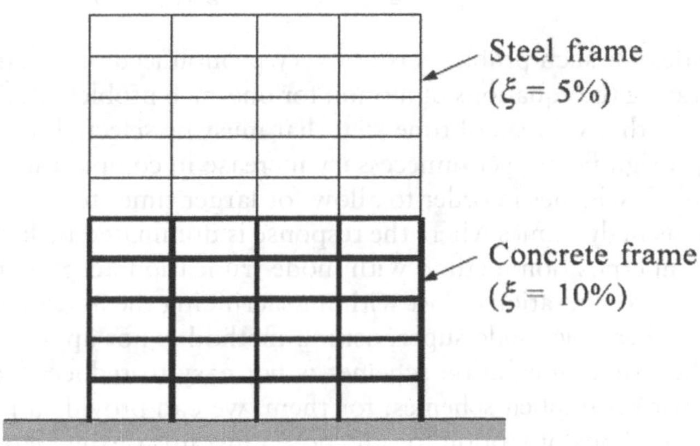

Figure 1.5 Steel and concrete frame structure (Clough and Penzien 1993).

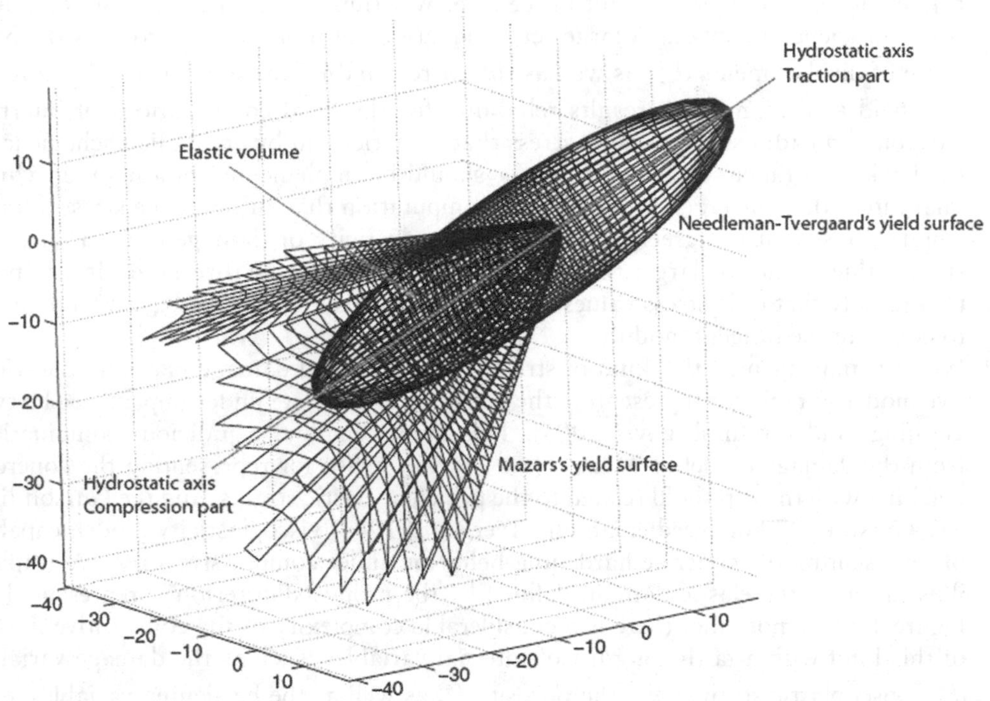

Figure 1.6 Elastic domain for a coupled damage-plasticity model of concrete under impact (see Hervé, Fabrice Gatuingt and Ibrahimbegović 2005).

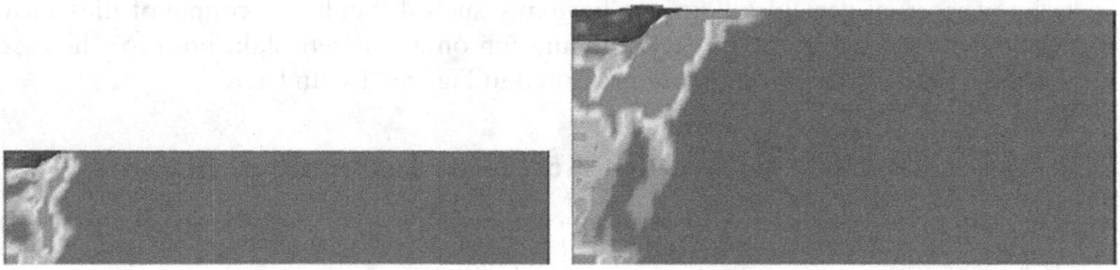

Figure 1.7 Impact on a reinforced concrete slab: contours of damage variable for the cases with and without perforation (see Hervé, Fabrice Gatuingt and Ibrahimbegović 2005).

high-frequency modes for such problems is not very pronounced and that they are quickly damped out. Integrating the equations of motion for one such problem in dynamics by using an explicit scheme, with a very small time step that must be selected to guarantee the stability, would imply a significant, yet unnecessary increase in computational cost. We, thus, have to turn to implicit schemes in order to allow for larger time steps.

For linear problems in dynamics where the response is dominated by low frequencies, one can use the modal superposition method with mode truncation keeping a few modes only, in order to reduce the computational cost without sacrificing the accuracy. For a nonlinear dynamics problem, where the mode superposition method is not applicable, the computational cost of implicit time-integration schemes is not easy to reduce. However, this does not mean we will discard implicit schemes; for them, we can provide a more reliable solution satisfying the equations of motion for all chosen instants of the incremental sequence.

Any implicit time-integration scheme for nonlinear dynamics will lead to a set of coupled nonlinear algebraic equations, which requires an iterative solution scheme with a significant increase in computational cost at each step. The only manner of compensating for this increase of computational cost is by using (much) larger time steps with respect to those typical of explicit schemes.

The trapezoidal rule, sometimes called the average acceleration method, is the first implicit scheme we will discuss subsequently. By applying the trapezoidal rule (e.g. Dahlquist and Björck 1974, p. 347) to the equations of motion, we obtain the second-order approximation to evolution equations for displacement and velocity, which can be written:

$$\begin{aligned} \dot{d}(t) = v(t) &\Rightarrow d_{n+1} - d_n = \frac{h}{2}(v_n + v_{n+1}) \\ \dot{v}(t) = a(t) &\Rightarrow v_{n+1} - v_n = \frac{h}{2}(a_n + a_{n+1}) \end{aligned} \tag{1.65}$$

Rewriting the result in $(1.59)_1$ we can obtain the corresponding approximation for the velocity vector in terms of displacement increment:

$$v_{n+1} = -v_n + \frac{2}{h}(d_{n+1} - d_n) \tag{1.66}$$

With this result in hand, we can then obtain from $(1.59)_2$ the same kind of approximation of the acceleration vector at time t_{n+1} which reads:

$$a_{n+1} = -a_n - \frac{4}{h}v_n + \frac{4}{h^2}(d_{n+1} - d_n) \tag{1.67}$$

Note that both of these approximations are implicit in the sense that they depend upon the displacement value at time t_{n+1}. Combining these velocity and acceleration approximations with the equations of motion written at time t_{n+1}, we can obtain the most suitable implementation of the trapezoidal rule as presented in Table 1.2.

In terms of computational cost, the most demanding phase in the proposed implementation of the trapezoidal rule pertains to iterative solution of the set of nonlinear algebraic equations with displacements at time t_{n+1} as unknowns. If Newton's method is employed for

Table 1.2 Trapezoidal Rule or Average Acceleration Scheme for Nonlinear Dynamics

Initialize: $d_0, v_0 \rightarrow M^{-1}\left(f_0^{ext} - f^{int}(d_0)\right)$

At each step: $n = 0, 1, 2\ldots$ (given: d_n, v_n, a_n and h)

Find $d_{n+1}, v_{n+1}, a_{n+1}$

Iterate $(i) = 1, 2\ldots$

$$a_{n+1}^{(i)} = a\left(d_{n+1}^{(i)}\right) := \frac{4}{h^2}\left(d_{n+1}^{(i)} - d_n\right) - \frac{4}{h}v_n - a_n$$

$$M\hat{a}\left(d_{n+1}^{(i)}\right) + \hat{f}^{int}\left(d_{n+1}^{(i)}\right) = f_{n+1}^{ext} \rightarrow d_{n+1}^{(i+1)}$$

$$v_{n+1}^{(i)} = \frac{2}{h}\left(d_{n+1}^{(i)} - d_n\right) - a_n$$

IF

IF $\left\| f_{n+1}^{ext} - Ma_{n+1}^{(i)} - f^{int}\left(d_{n+1}^{(i)}\right)\right\|$ THEN $(i) \rightarrow (i+1) \rightarrow \bar{d}_{n+1}$

ELSE $n \rightarrow n+1$

this computation, we have to solve, at each iteration, the consistently linearized form of the system, which can be written:

$$\left[\frac{4}{h^2}M + K_{n+1}^{(i)}\right]u_{n+1}^{(i)} = f_{n+1}^{\text{ext}} - f^{\text{int}}\left(d_{n+1}^{(i)}\right) - Ma\left(d_{n+1}^{(i)}\right) \tag{1.68}$$

The matrix $K_{n+1} = (\partial f^{\text{int}}(d_{n+1})) / (\partial d_{n+1})$above is the tangent stiffness already defined for nonlinear problems of static equilibrium. The tangent operator in dynamics is yet referred to as the effective tangent stiffness, which includes both the true tangent stiffness for statics and the corresponding contribution from the mass matrix. The presence of the mass matrix (with full rank) in the effective tangent stiffness has the regularizing effect, which ensures the correct rank. We can thus always obtain the unique solution for incremental displacement and carry out the displacement update according to:

$$d_{n+1}^{(i+1)} = d_{n+1}^{(i)} + u_{n+1}^{(i)} \tag{1.69}$$

This iterative procedure continues until the chosen convergence tolerance is reached, providing the final displacement value $\bar{d}_{n+1}$ for that particular step. For linear dynamics problems, where the internal force can be written as $\hat{f}^{\text{int}}\left(d_{n+1}^{(i)}\right) = Kd_{n+1}$, the trapezoidal rule guarantees the unconditional stability of computations regardless of the time step size. Moreover, the trapezoidal rule ensures energy conservation in linear dynamics problems, for the free-vibration phase of motion in the absence of external loading. Namely, with the time derivative of total energy identical to the equations of motion, it is easy to conclude that:

$$\frac{d}{dt}\left[E(t)\right] = \underbrace{\dot{d}(t) \times \left(M\ddot{d}(t) + Kd(t)\right)}_{G_{\text{nnn}}(\varphi,v)=0} = E(t) = \text{cst.} \tag{1.70}$$

The discrete approximation of the free-vibration problem in linear dynamics provided by the trapezoidal rule will also ensure the conservation of the total energy with:

$$\begin{aligned} 0 &= \frac{d_{n+1} - d_n}{h} \cdot M \frac{a_{n+1} + a_n}{2} + \frac{d_{n+1} - d_n}{h} \cdot K \frac{d_n + d_{n+1}}{2} = \\ &= \frac{1}{2}\left(v_{n+1} + v_n\right) \cdot M\left(v_{n+1} - v_n\right) + \frac{1}{2}\left(d_{n+1} - d_n\right) \cdot K\left(d_{n+1} - d_n\right) = \\ &= \left(\frac{1}{2}v_{n+1} \cdot Mv_{n+1} + \frac{1}{2}d_{n+1}Kd_{n+1}\right) - \left(\frac{1}{2}v_n \cdot Mv_n + \frac{1}{2}d_n \cdot Kd_n\right) = \\ &= E_{n+1} - E_n \end{aligned} \tag{1.71}$$

The result of this kind is an additional confirmation of the unconditional stability of the trapezoidal rule method for linear dynamics problems. The unconditional stability of the same scheme in nonlinear dynamics is guaranteed only (see Belytschko and Schoeberle 1975) in the case where the discrete approximation of the internal energy verifies the following condition:[1]

$$\psi_{n+1} := \psi_n + u_{n+1} \cdot \left(\hat{f}^{\text{int}}\left(d_{n+1}\right) + \hat{f}^{\text{int}}\left(d_n\right)\right)/2 \geq 0 \tag{1.72}$$

[1] We note in passing that the inertia term in the Lagrangian formulation of nonlinear dynamics still remains linear; hence, the stability condition for nonlinear dynamics only concerns the internal energy.

Such a condition is violated (e.g. see Hughes 1977) for nonlinear dynamics problems with softening behavior.[1] Moreover, the trapezoidal rule can no longer ensure the energy conservation for nonlinear dynamics problems, which is one of the main motivations for the study of other implicit time-integration schemes, such as mid-point rule.

1.6 MOTIVATION FOR DEVELOPMENTS PRESENTED IN PART I AND PART II

1.6.1 Damping Model Characterization

For structural systems where a reasonable degree of homogeneity in the energy loss mechanism exists, modeling damping as proportional or classical is suitable and convenient. However, for structures made up of more than a single type of material, where the different materials provide drastically differing energy loss mechanisms in various parts of the structure, the distribution of damping forces will not be similar to the distribution of the inertial and elastic forces; in other words, the resulting damping will be nonproportional (Clough and Penzien 1993).

An easier remedy to this problem would be to use the material damage as the source of damping. Here, one needs to define nonlinear inelastic models, which pertains to any particular material. In Part I of this book, we present detailed studies for the most frequently used materials in engineering applications, such as steel, concrete and masonry.

During vibration motion (earthquake, etc.) of each and every structure, there would be some energy loss. The latter pertains to damping phenomena in vibrating structures, which in many expositions on the subject have been defined with great uncertainty with no clear recommendations on how to define the damping forces. Damping capacity is defined as the ratio of the energy dissipated in one cycle of oscillation to the maximum amount of energy accumulated in the structure. *Material damping* and *interface damping* are the most frequently present damping sources in the structure among many others. The *material damping* contribution comes from different damage mechanisms and their interaction within the material from which the structure is built. Thus, the damping is dependent upon the type of material, as well as on manufacturing methods (Kareem and Gurley 1996). Taking all this into account and the fact that the material characteristics vary due to material heterogeneities, providing the reliable estimation of material damage on structural damping is perhaps the most reliable path to constructing predictive models. This is illustrated in Part I, for most frequently used structure materials.

The interface damping mechanism is Coulomb friction between members and connections of a structural system and nonstructural components like partitions, facades and other frictional mechanisms still being unknown and not precisely identified. It is interesting to note that many experiments have shown that most of the energy dissipation mechanisms through the structure are dependent on displacement amplitude rather than the frequency of the structure. The mathematical formulation of the Coulomb friction is written as:

$$f_d = \mu \frac{\dot{u}}{|\dot{u}|} = \mu \operatorname{sgn}(\dot{u})$$

where:

μ is the friction coefficient of contacting surfaces and
$\dot{u}$ is the first derivative of displacement.

[1] The softening behavior implies the stress decrease with increasing strain.

It is clear that a nonlinear analysis is required here which most probably will consume much computational time and as such is not commonly used in the engineering practice. This is simplified into a linear viscous damping model represented in the well-known formulation:

$$f_d = c\dot{u}$$

In contrast to the real structure's behavior (see above), viscous damping is frequency dependent and not displacement dependent. It is clear that nonlinearity is not covered by this model.

1.6.2 Extreme Loading Representation

Once we develop a more detailed model of damping phenomena, we can significantly increase the model's predictive capabilities. Further increase can also be achieved through the more reliable representation of extreme loading conditions. In Part II, we provide the corresponding developments for some of the extreme loads applied to engineering structures. We include among them the most frequently encountered types of loading such as earthquakes and fires, as well as the impact loads applied to structures, either from another structure (e.g. airplane) or fluid (e.g. dam break).

Any of these representations for extreme loading applications require a more general framework, which pertains both to improving the structure formulations and to additional field computations (e.g. temperature, fluid flow or ground or other structure motion). The large diversity of such applications leads to multi-physics and interaction problems that need to be solved in order to sufficiently obtain a predictive model for engineering structures under extreme loading conditions. Several illustrative examples presented in Part II, dealing with earthquakes, airplane impact, fire and dam break, are most likely to be useful for engineering practice.

REFERENCES

Adhikari, S. and J. Woodhouse. 2001. "Identification of Damping: Part 1. Viscous Damping". *Journal of Sound and Vibration* 243 (1): 43–61. doi:10.1006/jsvi.2000.3391.

Belytschko, T. and D. F. Schoeberle. 1975. "On the Unconditional Stability of an Implicit Algorithm for Nonlinear Structural Dynamics". *Journal of Applied Mechanics* 42 (4): 865. doi:10.1115/1.3423721.

Bernal, Dionisio. 1994. "Viscous Damping in Inelastic Structural Response". *Journal of Structural Engineering* 120 (4): 1240–54. doi:10.1061/(asce)0733-9445(1994)120:4(1240).

Charney, F. A. 2005. "Consequences of Using Rayleigh Damping in Inelastic Response History Analysis". *Congreso Chileno de Sismologia e Ingenieria Antisismica*. Conception. Paper No. A10–17.

Chen, H. C. 1987. *Solution Methods for Damped Linear Systems*. UCB/SEMM Rep. 87/08, Berkeley: University of California.

Clough, Ray W. and Joseph Penzien. 1993. *Dynamics of Structures*. First edition.New York: McGraw-Hill Education (ISE Editions).

Clough, Ray W. and Joseph Penzien. 2003. *Dynamics of Structures*. Second edition.New York: McGraw-Hill Education (ISE Editions).

Dahlquist, G. and A. Björck. 1974. *Numerical Methods*. Englewood Cliffs, NJ: Prentice-Hall.

Géradin, M. and D. Rixen. 1992. *Théorie des vibrations: Applications à la dynamique des structures*. Paris: Masson.

Gurson, A. L. 1977. "Continuum Theory of Ductile Rupture by Void Nucleation and Growth: Part I —Yield Criteria and Flow Rules for Porous Ductile Media". *Journal of Engineering Materials and Technology* 99. doi:10.1115/1.3443401.

Hall, John F. 2006. "Problems Encountered from the Use (or Misuse) of Rayleigh Damping". *Earthquake Engineering & Structural Dynamics* 35 (5): 525–45. doi:10.1002/eqe.541.

Hervé, Guillaume, F. Gatuingt and Adnan Ibrahimbegović. 2005. "On Numerical Implementation of a Coupled Rate-Dependent Damage-plasticity Constitutive Model for Concrete in Application to High-rate Dynamics". *Engineering Computations* 22 (5/6).

Hughes, Thomas J. R. 1977. "A Note on the Stability of Newmark's Algorithm in Nonlinear Structural Dynamics". *International Journal for Numerical Methods in Engineering* 11 (2): 383–86. doi:10.1002/nme.1620110212.

Ibrahimbegović, A., H. C. Chen, E. L. Wilson and R. L. Taylor. 1991. "Ritz Method for Dynamic Analysis of Large Discrete Linear System with Non-Proportional Damping". *Earthquake Engineering and Structural Dynamics* 19: 887–99.

Ibrahimbegović, A. and E. L. Wilson. 1989. "Simple Numerical Algorithms for the Mode Superposition Analysis of Linear Structural Systems with Non-Proportional Damping". *Computers & Structures* 33 (2): 523–31. doi:10.1016/0045-7949(89)90026-6.

Ibrahimbegović, Adnan and Edward L. Wilson. 1989. "Simple Numerical Algorithm for the Mode Superposition Analysis of Linear Structural Systems with Non-Proportional Damping". *Computer and Science* 33: 523–33.

Kareem, A. and K. Gurley. 1996. "Damping in Structures: Its Evaluation and Treatment of Uncertainty". *Journal of Wind Engineering and Industrial Aerodynamics* 59 (2–3): 131–57. doi:10.1016/0167-6105(96)00004-9.

Mazars, Jacky. 1986. "A Description of Micro- and Macroscale Damage of Concrete Structures". *Engineering Fracture Mechanics* 25 (5–6): 729–37. doi:10.1016/0013-7944(86)90036-6.

Medina, R. A. and H. Krawinkler. 2004. *Seismic Demands for Non-Deteriorating Frame Structures and their Dependence on Ground Motions Berkeley*. PEER Report, Berkeley, CA: Pacific Earthquake Engineering Research Cente.

Morris, Nicholas F. 1997. "The Use of Modal Superposition in Nonlinear Dynamics". *Computers & Structures* 7 (1): 65–72. doi:10.1016/0045-7949(77)90061-x.

Needleman, A. and V. Tvergaard. 1984. "An Analysis of Ductile Rupture in Notched Bars". *Journal of the Mechanics and Physics of Solids* 32 (6): 461–90. doi:10.1016/0022-5096(84)90031-0.

Oden, J. T. and R. B. Fost. 1973. "Convergence, Accuracy and Stability of Finite Element Approximations of a Class of Non-Linear Hyperbolic Equations". *International Journal for Numerical Methods in Engineering* 6 (3): 357–65. doi:10.1002/nme.162006030.

Phani, A. S. 2004. *Damping Identification In Linear Vibrations*. PhD Thesis, Cambridge: Cambridge University Engineering Department.

Udwadia, Firdaus E. 2009. "A Note on Nonproportional Damping". *Journal of Engineering Mechanics* 135 (11): 1248–56. doi:10.1061/(asce)0733-9399(2009)135:11(1248).

Veletsos, Anestis S. and Carlos E. Ventura. 1986. "Modal Analysis of Non-Classically Damped Linear Systems". *Earthquake Engineering & Structural Dynamics* 14 (2): 217–43. doi:10.1002/eqe.4290140205.

Zareian, F. 2006. *Simplified Performance-Based Earthquake Engineering*. PhD Dissertation, Stanford, CA: Department of Civil & Environmental Engineering, Stanford University.

Part I

Chapter 2

Steel structures

In the following sections the governing equations of a boundary value problem are developed for the case where the computed strain field can be highly non-homogeneous, with one or more sub-domains where the peak resistance is passed leading to a strain softening regime with very large strains, while the rest of the structure still remains in either elastic or elastoplastic strain hardening regime with fairly small strains. The model developed herein for representing this kind of problem is capable of describing both strain hardening and strain softening regimes, as well as identifying the stress state that corresponds to the passage from hardening to softening. In particular, the latter is considered to coincide with the satisfaction of the localization condition (e.g. Hill 1962; Rice 1976) modified for the presence of hardening.

2.1 ESSENTIAL INGREDIENTS OF PLASTICITY MODEL

As it is well known, plasticity theory began with Tresca in 1864, when he undertook an experimental program into the extrusion of metals and published his famous yield criterion. Further advances with yield criteria and plastic flow rules were made by different researchers such as Saint-Venant, Levy, von Mises, Hencky and Prandtl. It was in the 1980s and 1990s that a more rigorous foundation based on thermodynamic principles was developed opening the door to many numerical and computational aspects of the plasticity problem. It is still the most well understood constitutive model with internal variables capable of describing several observed phenomena of inelastic behavior. In the following section, the plasticity model with hardening and introduced softening will be elaborated in detail. This is used as a basis for implementation in structures and further in push-over analysis. At the end of the chapter, a refined model is presented that represents a novelty to the current practice. The chapter closes with several examples.

2.1.1 1D Plasticity with Hardening and Softening

The plasticity model with hardening is just a minor modification of the perfect plasticity model, in the sense that an increase of the plasticity threshold is enabled beyond the limit of elasticity σ_y as a function of accumulated plastic deformations. On the other hand, softening can be seen as a decrease of the stress. In this sense, it can be stated that the sign of the yield stress changes; an increase would lead to hardening and a decrease (or negative) hardening would be equivalent to softening. For simplicity, a 1D model for a simple shear test is considered. This model is represented by a bar of length l (of a unit section), fixed at one end and submitted to an imposed shear force at the other end. (Figure 2.1).

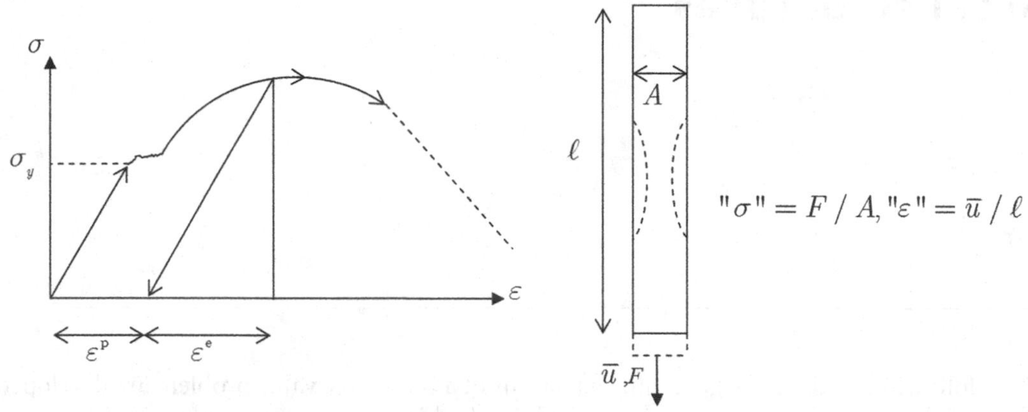

Figure 2.1 Simple tension test.

As displacements are considered to be infinitesimal, the small displacement gradient theory is valid, so the application of standard kinematics and equilibrium equations are valid:

$$\bar{\varepsilon} = \frac{\partial \bar{u}}{\partial x} \tag{2.1}$$

$$\frac{\partial \sigma}{\partial x} + b = 0 \tag{2.2}$$

where $\bar{u}, \bar{\varepsilon}$ and σ, are, respectively, transverse displacement, infinitesimal deformation and the Cauchy stress field, b is the external distributed loading. All the fields with a superposed bar indicated in (2.1) are considered as smooth, which tacitly implies hardening plasticity. Setting the external loading b represented in (2.2) to zero, it is obvious that the stress σ will also be smooth (constant) along the bar.

Taking the essential ingredients, one can construct the constitutive model of plasticity. The total strain can be decomposed into elastic $\bar{\varepsilon}^e$ and plastic components $\bar{\varepsilon}^p$:

$$\bar{\varepsilon} = \bar{\varepsilon}^e + \bar{\varepsilon}^p \tag{2.3}$$

The thermodynamic for plasticity with hardening is represented by the potential strain energy function depending upon the elastic strain and hardening variable $\bar{\xi}$:

$$\bar{\psi}\left(\bar{\varepsilon}^e, \bar{\xi}\right) = \frac{1}{2} \bar{\varepsilon}^e E \bar{\varepsilon}^e + \bar{\Xi}\left(\bar{\xi}\right) \tag{2.4}$$

And finally, for the developments to follow, the yield criterion specifying the admissible values of stress and stress-like hardening variable $\bar{q}$ is represented in terms of variables $\sigma, \bar{q}$:

$$\bar{\phi}(\sigma, \bar{q}) = |\sigma| - \left(\sigma_y - \bar{q}\right) \leq 0 \tag{2.5}$$

where σ_y denotes the yield stress.

All the remaining ingredients of the plasticity model can be obtained from standard thermodynamic considerations and the principle of the maximum plastic dissipation (Lubliner 1990). Namely, the second law of thermodynamics (Truesdell and Noll 1965; Lubliner 1990;

Maugin 1992) states that the dissipation always remains non-negative, which can be written by making use of the results from (2.3) and (2.4) as:

$$\begin{aligned} 0 \leq \mathcal{D} &= \sigma\dot{\bar{\varepsilon}} - \frac{d}{dt}\bar{\psi}\left(\bar{\varepsilon}^e, \bar{\xi}\right) \\ &= \left(\sigma - \frac{\partial \bar{\psi}}{\partial \bar{\varepsilon}^e}\right)\dot{\bar{\varepsilon}}^e + \frac{\partial \bar{\psi}}{\partial \bar{\varepsilon}^e}\dot{\bar{\varepsilon}}^p - \frac{\partial \bar{\Xi}}{\partial \bar{\xi}}\dot{\bar{\xi}} \end{aligned} \tag{2.6}$$

It is clear from (2.6) that for any stress state in the elastic domain, the values of internal variables will not change, $\left(\dot{\bar{\varepsilon}}^p = 0, \dot{\bar{\xi}} = 0\right)$ and dissipation remains equal to zero. Then the stress-like variables can be defined through the following constitutive equations:

$$\begin{aligned} &\bar{\phi} < 0, \dot{\bar{\varepsilon}}^p = 0, \dot{\bar{\xi}} = 0 \\ &\mathcal{D} = 0 \Rightarrow \sigma = \frac{\partial \bar{\psi}}{\partial \bar{\varepsilon}^e}; \bar{q} = -\frac{\partial \bar{\Xi}}{\partial \bar{\xi}} \end{aligned} \tag{2.7}$$

By assuming the same constitutive relation in (2.7) remains valid for plastically admissible stress on the yield surface, the corresponding (positive) value of plastic dissipation from the second principle of thermodynamics can be obtained according to:

$$\bar{\phi} = 0, 0 < \mathcal{D}^p = \sigma\frac{\partial \bar{\varepsilon}^p}{\partial t}\sigma + \bar{q}\frac{\partial \bar{\xi}}{\partial t} \tag{2.8}$$

The principle of maximum plastic dissipation can be used in order to obtain the evolution equations of internal variables for such a process. One looks into all the plastically admissible variables of stress satisfying (2.5) (for which $\bar{\phi}\left(\sigma^*, \bar{q}^*\right) = 0$), for the true solution $(\sigma, \bar{q})$ that will maximize plastic dissipation, or otherwise:

$$(\sigma, \bar{q}) = \arg\left\{\min_{\left[\bar{\phi}\left(\sigma^*, \bar{q}^*\right)\right]}\left[-\mathcal{D}^p\left(\sigma^*, \bar{q}^*\right)\right]\right\} \tag{2.9}$$

The Lagrange multiplier method can be used to handle constraint and define the corresponding unconstrained minimization problem:

$$\begin{aligned} &\mathcal{L}^p\left(\sigma, \bar{q}, \dot{\bar{\gamma}}\right) = \max_{\forall \dot{\bar{\gamma}}^* > 0} \min_{\forall\left(\sigma^*, \bar{q}^*\right)} \mathcal{L}^p\left(\sigma^*, \bar{q}^*, \dot{\bar{\gamma}}\right); \\ &\mathcal{L}^p\left(\sigma, \bar{q}, \dot{\bar{\gamma}}\right) = -\mathcal{D}^p(\sigma, \bar{q}) + \frac{\partial \bar{\gamma}}{\partial t}\bar{\phi}(\sigma, \bar{q}) \end{aligned} \tag{2.10}$$

The Kuhn–Tucker optimality conditions (e.g. Luenberger 1984; Strang 1986) of this problem will provide the evolution equation of the plastic strain and hardening variable.

$$\begin{aligned} &\left.\frac{\partial \mathcal{L}^p}{\partial \sigma^*}\right|_\sigma = 0 \Rightarrow \frac{\partial \bar{\varepsilon}^p}{\partial t} = \dot{\bar{\gamma}}\frac{\partial \bar{\phi}}{\partial \sigma} = \dot{\bar{\sigma}} sign(\sigma) \\ &\left.\frac{\partial \mathcal{L}^p}{\partial \bar{q}}\right|_{\bar{q}} = 0 \Rightarrow \frac{\partial \bar{\xi}}{\partial t} = \frac{\partial \bar{\gamma}}{\partial t}\frac{\partial \bar{\phi}}{\partial \bar{q}} = \dot{\bar{\gamma}} \end{aligned} \tag{2.11}$$

The constraint equation is obtained as well, which can be combined with the corresponding value in (2.8) in an elastic case in order to provide the final form of the loading/unloading conditions:

$$\dot{\gamma} \geq 0, \bar{\phi} \leq 0, \dot{\gamma}\bar{\phi} = 0 \tag{2.12}$$

If one recalls the model of perfect plasticity, it is clearly observed that the presented model gives the same evolution equation for plastic deformations and the same stress constitutive equation for plasticity deformation and the same stress constitutive relation.

The appropriate value of the plastic multiplier can be obtained from the consistency conditions to guarantee the admissibility of stress for subsequent states, meaning the only case of plastic loading that gives a non-zero value of $\dot{\gamma}$:

$$\dot{\sigma} = E\left(\dot{\varepsilon}^e - \underbrace{\dot{\gamma}\frac{\partial\bar{\phi}}{\partial\sigma}}_{\dot{\varepsilon}^p}\right)$$

$$\dot{\bar{q}} = -\underbrace{\frac{d^2\Xi}{d\xi^2}}_{\bar{K}}\underbrace{\dot{\xi}}_{\dot{\gamma}\frac{\partial\bar{\phi}}{\partial\bar{q}}}$$

$$\begin{aligned} 0 = \frac{d}{dt}\bar{\phi} &= \frac{\partial\bar{\phi}}{\partial\sigma}\frac{\partial\sigma}{\partial t} + \frac{\partial\bar{\phi}}{\partial\bar{q}}\frac{\partial\bar{q}}{\partial t} \\ &= \frac{\partial\bar{\phi}}{\partial\sigma}E\frac{\partial\bar{\varepsilon}}{\partial t} - \frac{\partial\gamma}{\partial t}\left(\frac{\partial\bar{\phi}}{\partial\sigma}E\frac{\partial\bar{\phi}}{\partial\sigma} + \frac{\partial\bar{\phi}}{\partial\bar{q}}\bar{K}\frac{\partial\bar{\phi}}{\partial\bar{q}}\right) \end{aligned} \tag{2.13}$$

$$\dot{\gamma} = \frac{\partial\bar{\gamma}}{\partial t} = \frac{\dfrac{\partial\bar{\phi}}{\partial\sigma}E\dfrac{\partial\bar{\varepsilon}}{\partial t}}{\dfrac{\partial\bar{\phi}}{\partial\sigma}E\dfrac{\partial\bar{\phi}}{\partial\sigma} + \dfrac{\partial\bar{\phi}}{\partial\bar{q}}\bar{K}\dfrac{\partial\bar{\phi}}{\partial\bar{q}}}$$

Further, the stress rate form of the stress–strain constitutive equations can be expressed according to:

$$\dot{\sigma} = \begin{cases} E\dot{\varepsilon} \\ \left[E - \dfrac{E\dfrac{\partial\bar{\phi}}{\partial\sigma}E}{\dfrac{\partial\bar{\phi}}{\partial\sigma}E\dfrac{\partial\bar{\phi}}{\partial\sigma} + \dfrac{\partial\bar{\phi}}{\partial\bar{q}}\bar{K}\dfrac{\partial\bar{\phi}}{\partial\bar{q}}}\dfrac{\partial\bar{\phi}}{\partial\sigma}\right]\dot{\varepsilon}, \; \dot{\gamma} > 0 \end{cases} \tag{2.14}$$

Figure 2.2 clearly indicates linear isotropic vs. saturated hardening. In particular, for the yield criterion set in (2.5), leading to $\partial\bar{\phi}/\partial\sigma = \text{sign}(\sigma)$ and $\partial\bar{\phi}/\partial\bar{q} = 1$, a simplified form of the stress rate equation for plastic loading can be defined as:

$$\dot{\sigma} = \frac{E\bar{K}}{E+\bar{K}}\dot{\varepsilon} \quad \dot{\gamma} > 0 \tag{2.15}$$

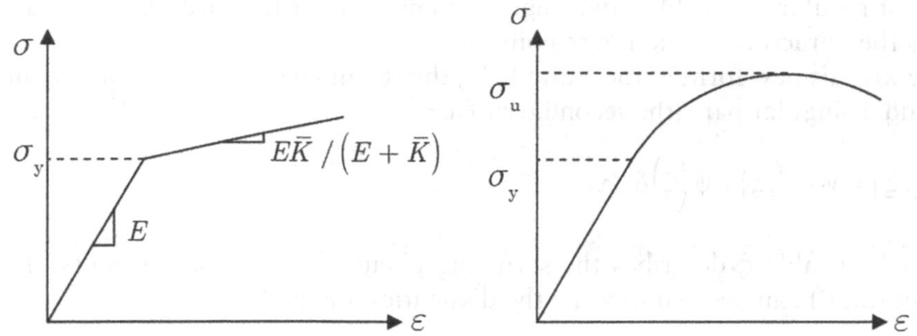

Figure 2.2 (a) Linear isotropic hardening (b) Saturated hardening.

This clearly reveals that a further increase of stress is allowed beyond elastic limit σ_y with increasing values of strains only as long as one is kept in a strain hardening regime with $\bar{K} = \left(d^2\Xi / d\xi^2\right) > 0$. The peak resistance in the stress–strain diagram is identified at:

$$C^{ep} = \frac{E\bar{K}}{E+\bar{K}} = 0 \Rightarrow \bar{K} = \frac{d^2\Xi}{d\xi^2} = 0 \tag{2.16}$$

Assuming that $\bar{K}(\xi_u) = 0$, defining the ultimate value of stress as $\sigma_u = \sigma_y - \bar{q}(\xi_u)$ and if then the softening phase is entered, with $\bar{K} < 0$, according to (2.15) the increase in strain will lead to a decrease in stress. Namely, in the latter case, the 1D problem under consideration with zero external load and constant stress state will lead to the gradual spreading of the plasticity throughout the domain, starting from the first plastified section, since the stress will keep increasing. In the same constant stress state and strain softening case, the first section that passes the peak (typically a slightly weakened section that is supposed to transform a bifurcation into a limit load problem) will reduce the stress level leading to unloading in all other sections where peak is not passed; all subsequent inelastic deformation will be accumulated in the particular section where the shear band is created. For this specific problem, taking into account that the shear band thickness tends to zero, the displacement field can be written as:

$$u(x,t) = \bar{u}(x,t) + \bar{\bar{\eta}}(t)M_{\bar{x}}(x) \tag{2.17}$$

with

$$M_{\bar{x}}(x) = H_{\bar{x}}(x) - N_a(x);$$

$$H_{\bar{x}}(x) = \begin{cases} 1, & x > \bar{x} \\ 0, & x < \bar{x} \end{cases}; \tag{2.18}$$

$$N_a(x) = \begin{cases} 1, & x = l \\ 0, & x = 0 \end{cases};$$

In (2.17), $\bar{u}(x,t)$ is the smooth displacement field, and $\bar{\bar{\eta}}(t)$ is the displacement discontinuity which appears at the shear band location $\bar{x}$.

By differentiating (2.18), the corresponding strain field is obtained:

$$\varepsilon(x,t) = \underbrace{\bar{\varepsilon}(x,t) + \bar{\bar{\eta}}(t)G(x)}_{\tilde{\varepsilon}} + \bar{\bar{\eta}}(t)\delta_{\bar{x}}(x) \tag{2.19}$$

It consists of regular (smooth) $\tilde{\varepsilon}$ and singular components (the third term in (2.19)), where $\delta_{\bar{x}}$ denotes the Dirac-delta function at point $\bar{x}$.

In line with this new form of the strain field, the strain energy consists of a regular, as set in (2.4), and a singular part (the second term) as:

$$\psi\left(\varepsilon,\bar{\xi},\bar{\bar{\xi}}\right) = \bar{\psi}\left(\bar{\varepsilon}^{e},\bar{\xi}\right) + \bar{\bar{\psi}}\left(\bar{\bar{\xi}}\right)\delta_{\bar{x}}(x) \tag{2.20}$$

The internal variable $\bar{\bar{\xi}}$ describes the softening phenomena on discontinuity. The results obtained in (2.20) can be used to write the dissipation inequality as:

$$\begin{aligned} 0 \le \mathcal{D}^{\mathrm{loc}}_{\Omega} &= \int_{\Omega}\left[\sigma\dot{\varepsilon} - \frac{d}{dt}\psi\left(\varepsilon,\bar{\xi},\bar{\bar{\xi}}\right)\right]dx \\ &= \int_{\Omega}\left[\sigma\dot{\tilde{\varepsilon}} - \frac{d}{dt}\bar{\psi}\left(\tilde{\varepsilon}^{e},\bar{\xi}\right)\right]dx + \left[t\dot{\bar{\bar{\eta}}} - \frac{d}{dt}\bar{\bar{\psi}}\left(\bar{\bar{\xi}}\right)\right]_{\bar{x}} \end{aligned} \tag{2.21}$$

The definition of the Dirac-delta function in (2.19) is used in order to obtain the traction at discontinuity as:

$$t = \sigma\big|_{\bar{x}} \tag{2.22}$$

Comparing the expression in (2.21) with the one in (2.8), it can be concluded that the dissipation in the post-localization phase can be reduced to the sum of the regular and singular components which can be written as:

$$\mathcal{D}^{\mathrm{loc}}_{\Omega} = \underbrace{\int_{\Omega}\sigma\dot{\varepsilon}^{p} + q\dot{\bar{\xi}}\,dx}_{\mathcal{D}_{\Omega}} + \left[\bar{\bar{q}}\dot{\bar{\bar{\xi}}}\right]_{\bar{x}} ;\bar{\bar{q}} = -\frac{\partial\bar{\bar{\psi}}}{\partial\bar{\bar{\xi}}} \tag{2.23}$$

with the orthogonality condition set in (2.24), which should be imposed and solved simultaneously with the global equilibrium equations, the following condition applies:

$$\int_{\Omega} G(x)\sigma dx + t = 0 \tag{2.24}$$

The contribution of the discontinuity to the total dissipation is defined in (2.21) and represents the second term in the equation. In order to further develop the exact nature of such a contribution, the yield criterion at discontinuity is postulated, which defines the corresponding admissible traction values according to:

$$\bar{\bar{\phi}}\left(t,\bar{\bar{q}}\right) := |t| - \left(\sigma_{u} - \bar{\bar{q}}\right) = 0 \tag{2.25}$$

The principle of maximum plastic dissipation applied to the case where only discontinuity remains active can be written by introducing the Lagrange multiplier in the form:

$$\gamma = \bar{\bar{\gamma}}\delta_{\bar{x}} \tag{2.26}$$

which further restricts the corresponding Lagrangian to the discontinuity according to:

$$\mathcal{L}^{p}\left(\bar{\bar{q}},\bar{\bar{\gamma}}\right) = \left[-\bar{\bar{q}}\dot{\bar{\bar{\xi}}}\right]_{\bar{x}} + = \int_{\Omega}\underbrace{\bar{\bar{\gamma}}\delta_{\bar{x}}}_{\gamma}\,\bar{\bar{\phi}}dx = \left[-\bar{\bar{q}}\dot{\bar{\bar{\xi}}} + \bar{\bar{\gamma}}\bar{\bar{\phi}}\right]_{\bar{x}} \tag{2.27}$$

This reduces the Kuhn–Tucker optimality conditions to:

$$\dot{\bar{\bar{\xi}}} = \dot{\bar{\bar{\gamma}}}$$
$$\dot{\bar{\bar{\gamma}}} \geq 0, \; \bar{\bar{\phi}} \leq 0, \; \dot{\bar{\bar{\gamma}}}\bar{\bar{\phi}} = 0 \tag{2.28}$$

The time derivative of this yield criterion provides the consistency condition for computing the corresponding value of the plastic multiplier with:

$$0 = \text{sign}(t)\dot{t} + \underbrace{\frac{\partial \bar{\bar{q}}}{\partial \bar{\bar{\xi}}}}_{-\bar{\bar{K}}}\dot{\bar{\bar{\xi}}} \Rightarrow \dot{\bar{\bar{\gamma}}} = \frac{1}{\bar{\bar{K}}}\dot{t}\,\text{sign}(t) \tag{2.29}$$

where the result in (2.28) is exploited. The rate of change of the traction at discontinuity can be obtained from the time derivative of the orthogonality condition in (2.23), leading to:

$$\dot{\bar{\bar{\gamma}}} = \frac{1}{\bar{\bar{K}}}\int_{\Omega} G\dot{\sigma}dx \tag{2.30}$$

The presence of discontinuity does not at all affect the elastic response as the rigid-plastic traction displacement law is employed. The corresponding coefficients of the traction displacement law to be chosen are the traction yield value $t_y = \sigma_u$, which is chosen in accordance with the ultimate values of stress in the hardening regime, and the softening modulus which depends on a particular choice of softening law.

The total external work (G_f) expended in the process of driving the effective flow traction $t_y^{\text{eff}} = t_y - \bar{\bar{q}}$ is set to zero. If a linear softening law is chosen, the corresponding value of the softening modulus $\bar{\bar{K}}$ can be obtained from the following condition:

$$0 = t_y - \left|\bar{\bar{K}}\right|\bar{\bar{\xi}}_u \Rightarrow \bar{\bar{\xi}}_u = \frac{t_y}{\left|\bar{\bar{K}}\right|}$$
$$G_f = \frac{1}{2}t_y\bar{\bar{\xi}}_u = \frac{1}{2}\frac{t_y^2}{\left|\bar{\bar{K}}\right|} \Rightarrow \left|\bar{\bar{K}}\right| = -\frac{1}{2}\frac{t_y^2}{G_f} \tag{2.31}$$

Any other softening regime, instead of the linear softening law chosen in (2.31), can be selected, and in this respect adequate modification should be made. For example, with an exponential softening response it leads to:

$$G_f = \int_0^{\infty} t_y \exp\left(-\alpha\bar{\bar{\xi}}\right)d\bar{\bar{\xi}} = -\frac{t_y}{\alpha}\left[\exp\left(-\alpha\bar{\bar{\xi}}\right)\right]_0^{\infty} \Rightarrow t_y = -\bar{\bar{q}}\left(\bar{\bar{\xi}}\right) = t_y \exp\left(-\frac{t_y}{G_f}\bar{\bar{\xi}}\right) \tag{2.32}$$

This can be further extended to the 2D/3D case, with the tensor notation as the only novelty. For further details see Ibrahimbegović (2009).

2.2 LOCALIZED FAILURE PLASTICITY MODEL

In the spirit of the classical works on localization problems (Hill 1962; Mandel 1996; Rice 1976), it is assumed that the bifurcation phenomena in an elastoplastic response can be interpolated as the difference between two smooth stress fields, defining the corresponding jump in the stress rate as:

$$[\![\dot{\sigma}]\!] = \mathbf{C}^{ep}\left(\mathbf{m} \otimes \mathbf{n}\dot{\bar{\bar{\eta}}}\right) \tag{2.33}$$

Additionally, according to Newton's Third Law which imposes the continuity of traction across the displacement discontinuity line Γ_s and the Cauchy principle which relates the traction vector to stress tensor, it can be written:

$$0 = [\![\dot{\mathbf{t}}]\!] = [\![\dot{\sigma}]\!]\mathbf{n} + [\sigma]\underbrace{\dot{\mathbf{n}}}_{=0} \tag{2.34}$$

The second term in (2.34) is equal to zero as it is assumed that the direction of the discontinuity remains fixed in time. By combining (2.33) and (2.34), the following is obtained:

$$\begin{aligned} 0 &= \mathbf{C}^{ep}\left(\mathbf{m} \otimes \mathbf{n}\dot{\bar{\bar{\eta}}}\right)\mathbf{n} \\ &== \dot{\bar{\bar{\eta}}}\mathbf{A}^{ep}\mathbf{m} \end{aligned} \tag{2.35}$$

where $\mathbf{A}^{ep}$ is the acoustic tensor. The acoustic tensor in this case remains compatible with the von Mises yield criterion, where the plastic deformation in the softening phase remains incompressible and can be written as:

$$\mathbf{A}^{ep} = \mu\mathbf{1} - \frac{2\mu}{\left(1 + \dfrac{\bar{K}}{3\mu}\right)\|\mathbf{s}\|^2}\mathbf{sn} \otimes \mathbf{sn} \tag{2.36}$$

where μ and $\bar{K}$ are, respectively, the shear and hardening moduli.

The characteristic equations are obtained from the principal values of the acoustic tensor, which can be written in the reference system of the principal values of stress:

$$0 = \det\left[\mathbf{A}^{ep} - \lambda\mathbf{1}\right] = \left(\lambda^* - 1\right)^2\left\{\frac{2}{\left(1 + \dfrac{\bar{K}}{3\mu}\right)\|\mathbf{s}\|^2} \times \left(n_1^2 s_1^2 + n_2^2 s_2^2\right) + \lambda^* - 1\right\}; \quad \lambda^* = \frac{\lambda}{\mu} \tag{2.37}$$

By imposing the localization condition in (2.35) with $\lambda_1^* = 0$ from (2.37) it leads to:

$$\bar{K} = 3\mu\left[\left(\frac{\sqrt{2}\tau}{\|\mathbf{s}\|}\sin 2\theta\right)^2 - 1\right]; \quad \tau = \frac{1}{2}(\sigma_1 - \sigma_2) \tag{2.38}$$

Additionally, assuming that the localization condition occurs at peak resistance where $\bar{K}\left(\bar{\xi}_u\right) = 0$ gives an angle θ under which the slip line is formed:

$$\theta = \frac{1}{2}\sin^{-1}\left(\frac{\|\mathbf{s}\|}{\sqrt{2}\tau}\right) \tag{2.39}$$

By computing the value of $\|\mathbf{s}\|$ in terms of the principal stress value, it can further lead to:

$$1 = \left(\frac{\sqrt{2}\tau}{\|s\|\sqrt{2}} \right)^2 + \left(\frac{p + 3B\varepsilon_{33}^p}{\|s\|\sqrt{2}} \right)^2 \frac{(1-2\nu)^2}{3};$$

$$B = \frac{E}{3(1-2\nu)};\ p = \frac{tr[\sigma]}{3} \tag{2.40}$$

When the pressure p is related to $3B\varepsilon_{33}^p$, or for the case of (quasi-)incompressible material, the slip line will occur at 45° with respect to the maximum principal stress as is clear from (2.39) and (2.40).

Having established that the localization condition coincides with the initiation of the displacement discontinuity, it is necessary to turn to the final phase of the analysis as the discontinuity remains active in this location. The strain field, in this case, is written as:

$$\varepsilon = \nabla^s \mathbf{u} = \underbrace{\nabla^s \bar{\mathbf{u}} + \underbrace{\tilde{\mathbf{G}}\bar{\bar{\eta}}}_{\tilde{\varepsilon}}}_{\bar{\varepsilon}} + \bar{\bar{\eta}}(\mathbf{m} \otimes \mathbf{n})^s \delta_{\Gamma_s} \tag{2.41}$$

such that:

$$\int_{\Omega} \tilde{\mathbf{G}} d\Omega + \int_{\Gamma_s} (\mathbf{m} \otimes \mathbf{n})^s d\Gamma_s = 0 \tag{2.42}$$

The well-known result in distribution theory is used (e.g. Stakgold 1998) so that:

$$\int_{\Omega} \phi \nabla H_{\Gamma_s} d\Omega = \left| \int_{\Gamma_s} \phi \mathbf{n} d\Gamma_s \right. \tag{2.43}$$

In accordance with this form of the strain field, the strain energy can be split into two parts. A smooth part and the discontinuity contribution:

$$\psi\left(\varepsilon, \bar{\xi}, \bar{\bar{\xi}}\right) = \bar{\psi}\left(\bar{\varepsilon}^e, \bar{\xi}\right) + \bar{\bar{\psi}}\left(\bar{\bar{\xi}}\right)\delta_{\Gamma_s} \tag{2.44}$$

The dissipation inequality can thus be written as:

$$\begin{aligned} 0 \le \mathcal{D}_{\Omega}^{\text{loc}} &= \int_{\Omega} \left[\sigma \cdot \dot{\varepsilon} - \frac{d}{dt} \psi\left(\varepsilon, \bar{\xi}, \bar{\bar{\xi}}\right) \right] d\Omega \\ &= \int_{\Omega} \left[\sigma \cdot \dot{\bar{\varepsilon}} - \frac{d}{dt} \bar{\psi}\left(\bar{\varepsilon}^e, \bar{\xi}\right) \right] d\Omega + \int_{\Gamma_s} \left[t \cdot \dot{\bar{\bar{\eta}}} \right] ds - \int_{\Gamma_s} \left[\frac{d}{dt} \bar{\bar{\psi}}\left(\bar{\bar{\xi}}\right) \right] ds \end{aligned} \tag{2.45}$$

If, further, it is assumed that the stress orthogonality condition must be satisfied:

$$\int_{\Omega} \sigma \cdot \tilde{\mathbf{G}} \dot{\bar{\bar{\eta}}} d\Omega + \int_{\Gamma_s} \mathbf{t} \cdot \dot{\bar{\bar{\eta}}} ds = 0 \tag{2.46}$$

From (2.45), the stress constitutive equation is recovered and the total dissipation reduced to the sum of two parts, in the volume and slip line:

$$\mathcal{D}_{\Omega}^{\text{loc}} = \underbrace{\int_{\Omega} \sigma \cdot \dot{\bar{\varepsilon}}^p + \bar{q} \dot{\bar{\xi}} d\Omega}_{\mathcal{D}_{\Omega}^p} + \underbrace{\int_{\Gamma_s} \bar{\bar{q}} \cdot \dot{\bar{\bar{\xi}}} ds}_{\mathcal{D}_{\Gamma_s}^p} \tag{2.47}$$

Once the localized solution develops, the stress orthogonality condition in (2.46) should be joined and solved along with the set of global equilibrium equations. The yield condition controlling the inelastic deformation at discontinuity is set directly in terms of the traction vector component $t_m = \mathbf{t} \cdot \mathbf{m}$:

$$\bar{\bar{\phi}}\left(\mathbf{t}, \bar{\bar{q}}\right) = \left|t_m\right| - \left(t_y - \bar{\bar{q}}\right) \le 0 \tag{2.48}$$

Assuming, further, that the plastic multiplier takes the form $\dot{\gamma} = \dot{\bar{\gamma}} + \dot{\bar{\bar{\gamma}}}\delta\Gamma_s$, the principle of maximum plastic dissipation can be used with:

$$\begin{aligned} &\max_{\forall \dot{\bar{\bar{\gamma}}} > 0} \; \min_{\forall (t^*, \bar{\bar{q}}^*)} \bar{\bar{\mathcal{L}}}^p\left(\sigma^*, \bar{\bar{q}}^*, \dot{\bar{\bar{\gamma}}}\right); \; \bar{\bar{\mathcal{L}}}^p(\cdot) \\ &= -\mathcal{D}_\Omega^{\text{loc}}(\cdot) + \int_\Omega \dot{\bar{\gamma}}\bar{\phi}\, d\Omega + \int_{\Gamma_s} \dot{\bar{\bar{\gamma}}}\bar{\bar{\phi}}(\cdot)\, ds \end{aligned} \tag{2.49}$$

The corresponding Kuhn–Tucker optimality conditions then reads:

$$\begin{aligned} 0 &= \int_\Omega \left(-\dot{\bar{\varepsilon}}^p + \dot{\bar{\gamma}} \frac{\partial \bar{\phi}}{\partial \sigma}\right) d\Omega \\ 0 &= \int_\Omega \left(-\dot{\bar{\xi}} + \dot{\bar{\gamma}} \frac{\partial \bar{\phi}}{\partial \bar{q}}\right) d\Omega \end{aligned} \tag{2.50}$$

together with:

$$0 = \int_{\Gamma_s} \left(-\dot{\bar{\bar{\xi}}} + \dot{\bar{\bar{\gamma}}} \frac{\partial \bar{\bar{\phi}}}{\partial \bar{\bar{q}}}\right) ds \Rightarrow \int_{\Gamma_s} \dot{\bar{\bar{\xi}}} ds = \int_{\Gamma_s} \dot{\bar{\bar{\gamma}}} ds \tag{2.51}$$

It is evident that the combination of two kinds of dissipation is essentially handled by the corresponding value of the plastic multiplier. Namely, for the plastic loading case, it is either $\dot{\bar{\gamma}}$ or $\dot{\bar{\bar{\gamma}}}$, or both which remain positive.

The corresponding value of the latter is obtained from the time derivative of (2.48) by exploiting the results in (2.46) and (2.50) to get:

$$\dot{\bar{\bar{\gamma}}} = \frac{1}{\bar{\bar{K}}} \int_\Omega \tilde{\mathrm{G}} \dot{\sigma}\, d\Omega \tag{2.52}$$

If the localization condition in (2.35) is reached within a given time increment, one can then separate the step in one part where $\dot{\bar{\gamma}}$ evolves and the remaining part, where only $\dot{\bar{\bar{\gamma}}}$, will be modified computing the corresponding dissipation as indicated in (2.52). The corresponding "ultimate" value of hardening variable $\bar{\xi}_u$ occurring in such a time step can as well be computed.

2.3 STRUCTURAL PLASTICITY MODEL

2.3.1 Simple Form

Going further into the structural level, it is of great advantage to transform the weak form of the differential equilibrium equation into a set of algebraic equations with the nodal values of displacements as unknowns using the finite element based discrete approximation (Hughes 1987; Bathe 1996; Zienkiewicz and Taylor 2000).

In this part, the basic idea will be presented with two examples. Further, the application of such a model will be illustrated and implemented in a practical frame structure composed of columns and beams. Why such structures? From experimental tests and seismic actions, it has been observed from failure modes, that these structures fail due to localized failures in critical zones, termed as plastic (inelastic) hinges. So, it has been observed that this model is convenient for such representation.

$$\underset{e=1}{\overset{\mathrm{Nel}}{A}}\left[\mathbf{f}^{\mathrm{int},e}(t)-\mathbf{f}^{\mathrm{ext},e}(t)\right]=0$$
$$\mathbf{f}^{\mathrm{int},e}(t)=\int_{\Omega^e}\mathbf{B}^T\sigma(t)d\mathbf{\Omega}^e, t\in[0,T] \tag{2.53}$$

where $(\cdot)^{\mathrm{ext},e}$ and $(\cdot)^{\mathrm{int},e}$ denotes, respectively, the element contributions toward the external and internal force vector, whereas $\mathbf{\Omega}^e$ is the element domain. Parameter t in (2.53) is the pseudo-time which is used to describe the particular loading program, which is typically described by an incremental sequence.

It is only possible, if the evolution of the plastic strains $\bar{\varepsilon}^p(t)$ and hardening variable $\xi(t)$ is known, to obtain, the complete response and the corresponding value of stress $\sigma(t)$ in (2.53) in the strain hardening phase.

The corresponding value of stress $\sigma(t)$ can be obtained one incremental step at a time by integrating the rate equations such that the loading/unloading conditions in (2.12) remain satisfied. If the backward Euler is used for this kind of integration (Bathe 1996; Simo and Hughes 2000), it is possible to compute the evolution of the plastic strains and hardening variable over any time interval $\Delta t = t_{n+1} - t_n$ with:

$$\bar{\varepsilon}^p_{n+1}=\bar{\varepsilon}^p_n+\bar{\gamma}_{n+1}\frac{\partial\bar{\phi}}{\partial\boldsymbol{\sigma}_{n+1}}$$
$$\bar{\xi}_{n+1}=\bar{\xi}_n=+\bar{\gamma}_{n+1} \tag{2.54}$$

which further implies that the corresponding stress evolution reads:

$$\boldsymbol{\sigma}_{n+1}=\boldsymbol{\sigma}_n+\mathbf{C}\left(\Delta\varepsilon_{n+1}-\bar{\gamma}_{n+1}\frac{\partial\bar{\phi}}{\partial\boldsymbol{\sigma}_{n+1}}\right) \tag{2.55}$$

The plastic multiplier $\bar{\gamma}_{n+1}$ in (2.54) and (2.55) is computed from plastic admissibility condition:

$$\bar{\gamma}_{n+1}>0;\bar{\phi}(\bar{\gamma}_{n+1})=0 \tag{2.56}$$

The computations described in (2.54) to (2.56) are carried out independently from one quadrative point to another for the chosen numerical integration rule (e.g. 2×2 Gauss quadratic) (e.g. Hughes 1987; Bathe 1996) which is employed to compute the internal force integral in (2.53). The main result of such a computation which is needed at the global level for solving the set of equilibrium equations in (2.53) is the plastically admissible value of stress in (2.55), as well as the elastoplastic tangent modulus of the discrete problem:

$$\mathbf{C}^{\mathrm{ep}}_{n+1}=\frac{\partial\boldsymbol{\sigma}_{n+1}}{\partial\bar{\varepsilon}_{n+1}}=\mathbf{C}-\frac{2\mu}{\left(1+\frac{K}{3\mu}\right)\|\mathbf{s}\|^2}\mathbf{s}\otimes\mathbf{s}-\frac{(2\mu)^2\bar{\gamma}_{n+1}}{\|\mathbf{s}_n+2\mu\Delta e_{n+1}\|}$$
$$\left[I-\frac{1}{\|\mathbf{s}_{n+1}\|^2}\mathbf{s}_{n+1}\otimes\mathbf{s}_{n+1}-\frac{1}{3}\mathbf{1}\otimes\mathbf{1}\right] \tag{2.57}$$

The starting point for equivalent computations is the state where the localized conditions in (2.35) are verified for one of the quadrative points where we check the stress values. Using the same computations as in the hardening regime, the following is obtained:

$$\bar{\bar{\phi}}\left(\bar{\bar{\gamma}}_{n+1}\right) = 0; \; t_{m,n+1} = t_y - \bar{\bar{q}}\left(\bar{\bar{\xi}}_{n+1}\right) \tag{2.58}$$

$$\bar{\bar{K}}_{n+1} = \frac{d\bar{\bar{q}}}{d\bar{\bar{\xi}}_{n+1}} \tag{2.59}$$

The choice of the discrete approximation of the displacement field or the choice of element type is detrimental in finite element implementation. Two problems ought to be addressed in this respect: first, the locking phenomena in the plastic hardening behavior of the quasi-incompressible type, which occurs for a given choice of the von Mises yield criterion rendering the plastic deformation incompressible (Nagtegaal, Parks and Rice 1974). Second, the displacement discontinuity representation in the finite element setting. In this respect, a mixed-enhanced strain field is introduced, which, on one side, introduces the independent volume-change deformation measure $\theta(x)$ and, on the other side, the displacement discontinuity mode which produces a purely deviatoric strain field.

The set of equilibrium equations can be written as:

$$\begin{aligned} \mathbf{r} &= \underset{e=1}{\overset{\text{Nel}}{A}}\left[\int_{\Omega^e} \bar{\mathbf{B}}\cdot\sigma d\Omega - \mathbf{f}^{\text{ext}}\right] = 0 \\ \mathbf{h}^e &= \int_{\Omega^e} \tilde{\mathbf{G}}\cdot\sigma d\Omega + \int_{\Gamma_s} \bar{\bar{\eta}}\mathbf{m}\cdot\mathbf{t}d\Gamma = 0, e\in[1,\text{Nel}] \end{aligned} \tag{2.60}$$

where the supplementary equation expresses a weak form of equilibrium over a particular element Ω^e with respect to the variation of the enhanced strains.

In this way, the problem is reduced to its original size without the incompatible modes by appealing to the static condensation procedure (e.g. Ibrahimbegović, Fadi and Lotfi 1998).

2.3.1.1 Numerical Examples of Small Experiments

The finite element models all employ the 4-node quadrilateral element with $\bar{B}$ strain interpolations and embedded discontinuity. Two examples are provided: one is the pure shear test and the other the traction test.

The former is presented in Figure 2.3 and the material properties are given in Table 2.1.

The problem is solved using the arclength method. In order to change the bifurcation problem into a localization one, a band of elements (in gray in Figure 2.3) has been weakened by slightly reducing the limit stress. The sensitivity of the proposed model to mesh refinement and mesh distortion is checked by considering the different meshes as presented in Figure 2.3.

The results are given in Figure 2.4 in terms of the horizontal shear load vs. the horizontal displacement of point A. The numerical results are fully independent of mesh refinement and distortion.

In the latter, a double notched beam submitted to a traction test under a plain strain condition (see Figure 2.5) is considered. The tested sample is 20 cm long and 10 cm high, and the material parameters are kept the same as in the former example with exceptions of $\sigma_y = 0.55\,\text{GPa}$, $\sigma_\infty = 0.75\,\text{GPa}$, $\bar{\bar{K}} = -0.10\,\text{GPa/mm}$. The test is carried out under displacement

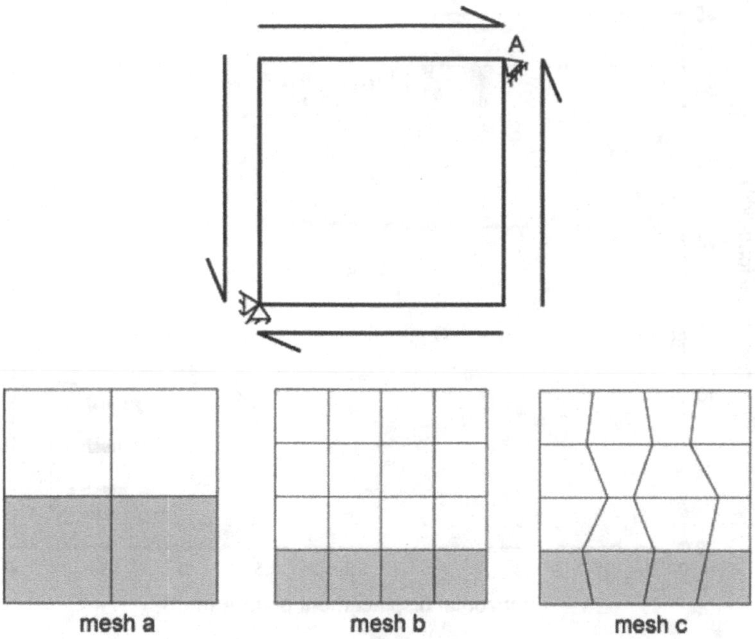

Figure 2.3 Pure shear test.

Table 2.1 Material Properties for Pure Shear Test

Continuum Plasticity Model	
Young's Modulus	210 GPa
Poisson Ratio	0.4999
σ_y	0.35 GPa
σ_∞	0.60 GPa
β	200
Softening Rule Parameter	
$\bar{\bar{K}}$	−0.05 GPa/mm

control. Figure 2.6 shows the corresponding force–displacement diagram computed for this simple shear test.

It is between the two notches of the beam where the shear band develops. Accumulated plastic deformation and discontinuity lines activated in localized elements at the end of the test are illustrated in Figures 2.7 and 2.8 respectively. It is evident from these two illustrations that localized elements correspond to those where plastic deformation has significantly developed.

Figures 2.7 and 2.8 show the contributions of two different dissipative mechanisms. One corresponds to bulk dissipation mechanism that is characterized by the development of plastic deformation in fracture process zone and the other to a surface dissipation mechanism that is characterized by the creation of a plastic slips resulting with displacement discontinuity.

2.3.2 Multi-Scale Model Parameters Identification

The application of the combined hardening and softening constitutive model of plasticity can be applied to structures. One of the possible applications is in metallic frames with thin-walled cross-sections, composed of columns and beams. When studying the full collapse of

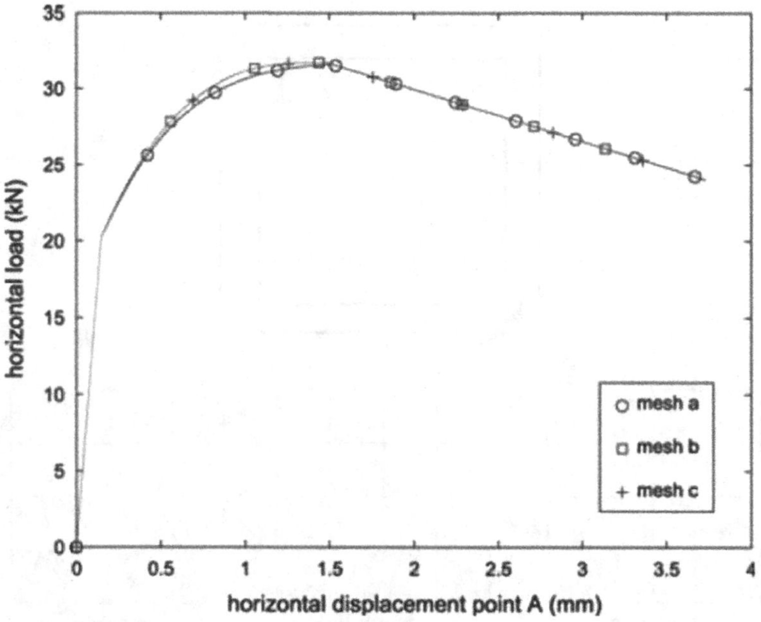

Figure 2.4 Load/displacement results for a pure shear test.

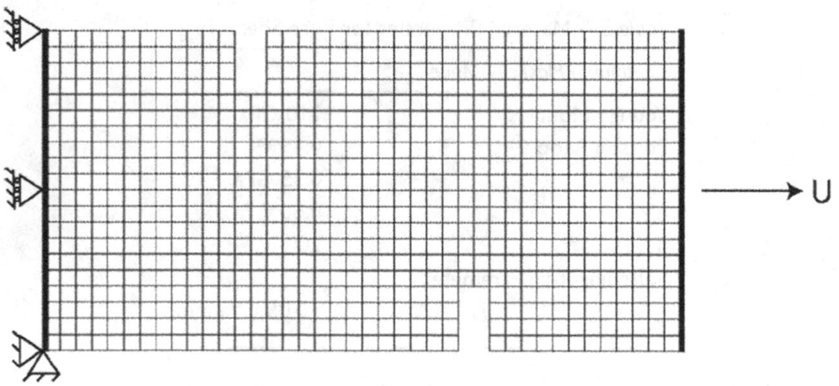

Figure 2.5 Traction test.

a frame, a softening response is observed after reaching its limit capacity; the load reduces with additional frame deformation. This structural softening response can be modeled using an elastoplastic constitutive model with a softening relation between the generalized strain measures and the corresponding stress resultants. As mentioned in the previous subsection, the key point here is the introduction of localized energy dissipation. This is achieved by introducing a strong discontinuity in the kinematic fields. In this case, it would be in axial displacement and beam axis rotation. Second, the local dissipative mechanism will be defined at the discontinuity in terms of a softening cohesive law (e.g. a softening law between the bending moment and the rotation jump). For beams, the introduced discontinuity can be naturally regarded as a softening plastic hinge.

In respect of the previous example, a planar straight stress-resultant beam finite element is developed with the following features: (i) Euler–Bernoulli kinematics, (ii) an elastoplastic

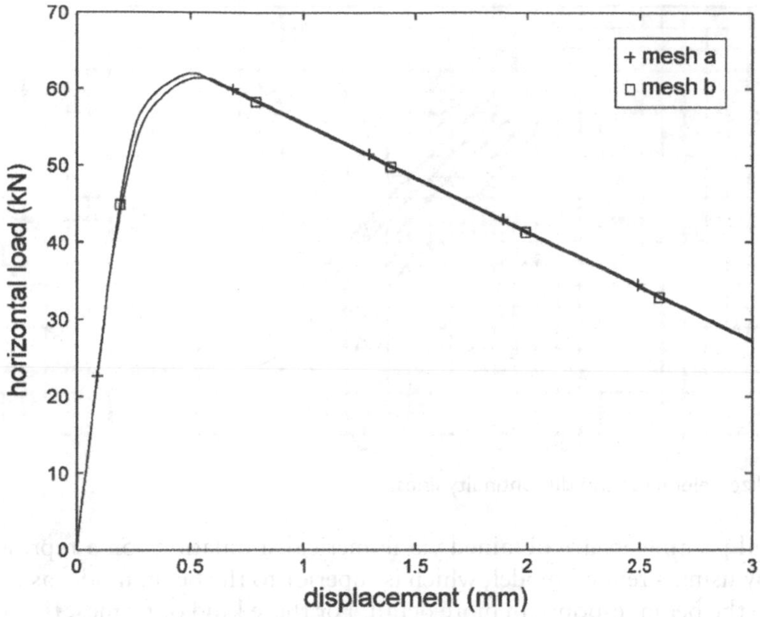

Figure 2.6 **Load/displacement results for a simple shear test.**

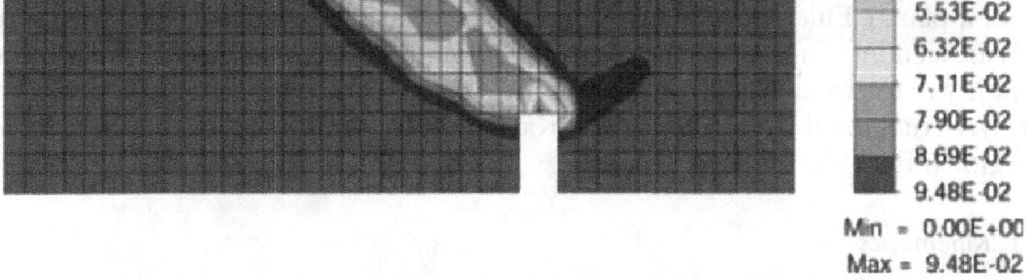

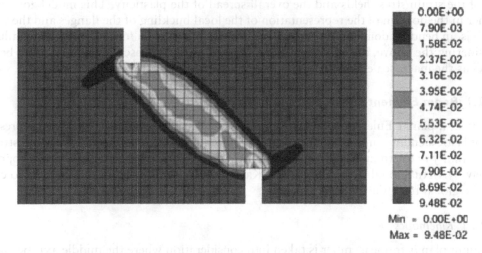

Figure 2.7 **Accumulated plastic deformation.**

stress-resultant constitutive model with isotropic hardening, (iii) a localized softening plastic hinge related to the strong discontinuity in generalized displacements, and (iv) an approximation of the geometrically nonlinear effects by using the von Karman strains for the virtual axial deformations.

The finite element that will be developed in the following sections can be successfully used for the limit load analysis, the push-over analysis and the complete failure analysis of planar frames made of steel. More complex material models are used. In order to describe the beam material behavior, stress-resultant elastoplasticity with hardening is used on one hand and on the other stress-resultant rigid-plastic softening is employed to describe material behavior at the discontinuity.

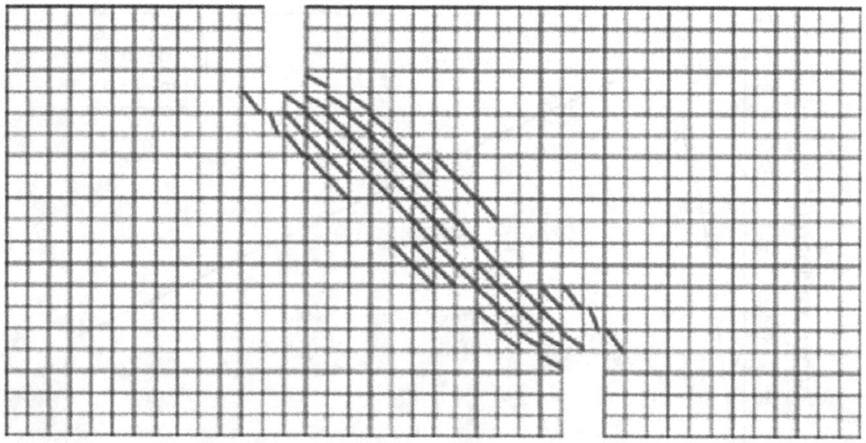

Figure 2.8 Localized elements and discontinuity lines.

The material parameters are obtained via numerical simulations on a representative part of the beam by using a refined model, which is superior to the beam model as it has the ability to describe the beam response in more detail. For these kind of frames, the refined model can be chosen as the nonlinear shell model (e.g. Brank, Perić and Damjan 1970; Brank 2005). The shell model is superior to the beam model in providing a proper local description of the strain/stress fields and the overall spread of the plasticity. This model goes even further in the domain of the representation of the local buckling of the flanges and the web, which is, in bending dominated conditions, very often the reason for localized beam failure. Considering the above, the shell model can be seen as the mesoscale model and the beam model as the macroscale model.

2.3.2.1 *Beam Element With Embedded Discontinuity*

As stated, a planar Euler–Bernoulli beam finite element is considered which can represent elastoplastic bending, including the localized softening effects associated with the strong discontinuity in the rotation. Further geometrical nonlinearity is taken into account approximately by virtual axial strains of the von Karman type, which allows this element to capture the global buckling modes.

2.3.2.1.1 Kinematics

A straight planar frame member is taken into consideration where the middle axis occupies domain $\Omega \in \mathbb{R}$. A typical 2-node finite element is presented in Figure 2.9. The following notation is used: u_i are nodal axial displacements, w_i are nodal transverse displacements, w'

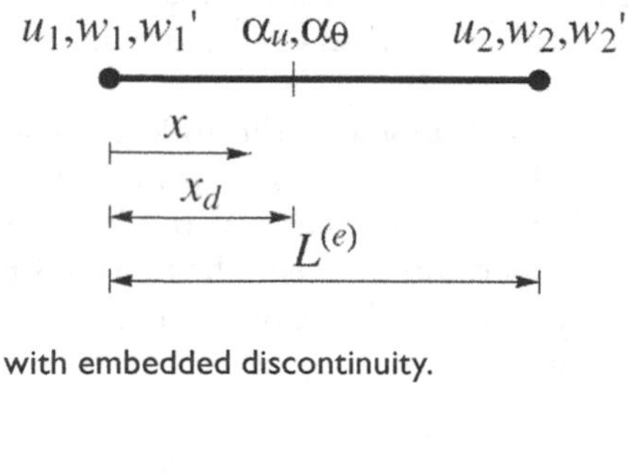

Figure 2.9 Beam finite element with embedded discontinuity.

are nodal values of the beam axis rotation and $i=1;2$ is the node number. A strong discontinuity in axial displacement is assumed α_u and beam axis rotation α_θ at $x_d \in L^{(e)}$.

Further, it is presupposed that the domain of the discontinuity influence corresponds to a single element. In respect to all these assumptions, the axial displacement is then defined as:

$$u^h(x,x_d) = \mathbf{N}^u(x)\mathbf{u} + M^u(x,x_d)\alpha_u \tag{2.61}$$

where $\mathbf{N}^u(x) = \{1 - x/L^{(e)}, x/L^{(e)}\}$, $\mathbf{u} = \{u_1, u_2\}^T$, and $M^u(x,x_d)$ is a function with zero value at the nodes with a unit jump of x_d, i.e. $M^u(0,x_d) = M^u(L^{(e)},x_d) = 0$ and $M^u(x_d^+,x_d) = M^u(x_d^-,x_d)+1$. In the same way, the transverse displacement reads:

$$w^h(x,x_d) = \mathbf{N}^w(x)\mathbf{w} + \mathbf{N}^{w'}(x)\mathbf{w}' + M^\theta(x,x_d)\alpha_\theta \tag{2.62}$$

where:

$$\mathbf{N}^w(x) = \left\{2\left(\frac{x}{L^{(e)}}\right)^3 - 3\left(\frac{x}{L^{(e)}}\right)^2 + 1, -2\left(\frac{x}{L^{(e)}}\right)^3 + 3\left(\frac{x}{L^{(e)}}\right)^2\right\}$$
$$\mathbf{w} = \{w_1, w_2\}^T \tag{2.63}$$

$$\mathbf{N}^{w'}(x) = L^{(e)}\left\{\left(\frac{x}{L^{(e)}}\right)^3 - 2\left(\frac{x}{L^{(e)}}\right)^2 + \frac{x}{L^{(e)}}, \left(\frac{x}{L^{(e)}}\right)^3 - \left(\frac{x}{L^{(e)}}\right)^2\right\}$$
$$\mathbf{w}' = \{w_1', w_2'\}^T \tag{2.64}$$

and $M^\theta(x,x_d)$ is a function with zero value at the nodes and a unit jump of its first derivative at x_d i.e. $M^\theta(0,x_d) = M^\theta(L^{(e)},x_d) = 0$ and $M^{\theta'}(x_d^+,x_d) = M^{\theta'}(x_d^-,x_d)+1$.

Making the derivative of equation (2.61) in respect to x, one obtains the beam axial strain:

$$\varepsilon(x,x_d) = \frac{\partial u^h}{\partial x} = \underbrace{\mathbf{B}^u(x)\mathbf{u} + G^u(x,x_d)\alpha_u}_{\tilde{\varepsilon}} + \underbrace{\delta_{xd}\alpha_u}_{\bar{\bar{\varepsilon}}} \tag{2.65}$$

where $\mathbf{B}^u(x) = \{-1/L^{(e)}, 1/L^{(e)}\}$, $G^u(x,x_d) = \partial M^u/\partial x$, and δ_{xd} Dirac-delta, which appears due to the discontinuous nature of axial displacement at x_d. As in equation (2.19), strain is composed of a regular part and a singular part. The beam curvature is computed as:

$$\kappa(x,x_d) = \frac{\partial^2 w^h}{\partial x^2} = \underbrace{\mathbf{B}^w(x)\mathbf{w} + \mathbf{B}^{w'}(x)\mathbf{w}' + G^\theta(x,x_d)\alpha_\theta}_{\tilde{\kappa}} + \underbrace{\delta_{xd}\alpha_\theta}_{\bar{\bar{\kappa}}} \tag{2.66}$$

where:

$$\mathbf{B}^w(x) = \left\{-\frac{6}{L^{(e)2}}\left(1 - \frac{2x}{L^{(e)}}\right), \frac{6}{L^{(e)2}}\left(1 - \frac{2x}{L^{(e)}}\right)\right\} \tag{2.67}$$

$$\mathbf{B}^{w'}(x) = \left\{-\frac{2}{L^{(e)}}\left(2 - \frac{3x}{L^{(e)}}\right), -\frac{2}{L^{(e)}}\left(1 - \frac{3x}{L^{(e)}}\right)\right\} \tag{2.68}$$

and $G^\theta(x,x_d) = \partial^2 M^\theta/\partial x^2$. The curvature κ is as well divided into a regular part $\tilde{\kappa}$ and a singular part $\bar{\bar{\kappa}}$. It is clear that $\bar{\bar{\varepsilon}}$ can be interpreted as a localized plastic axial strain on one hand and in the other $\bar{\bar{\kappa}}$ as a localized plastic curvature. In matrix notation, beam strains can be rewritten as:

$$\varepsilon = \tilde{\varepsilon} + \bar{\bar{\varepsilon}} \tag{2.69}$$

$$\tilde{\varepsilon} = \underbrace{\mathbf{Bd}}_{\bar{\tilde{\varepsilon}}} + \mathbf{G}\boldsymbol{\alpha}, \quad \bar{\bar{\varepsilon}} = \delta_{\alpha_d}\boldsymbol{\alpha} \tag{2.70}$$

Where $\varepsilon = \{\varepsilon, \kappa\}^T$, $\tilde{\varepsilon} = \{\tilde{\varepsilon}, \tilde{\kappa}\}^T$, $\bar{\tilde{\varepsilon}} = \{\bar{\tilde{\varepsilon}}, \bar{\tilde{\kappa}}\}^T$, $\bar{\bar{\varepsilon}} = \{\bar{\bar{\varepsilon}}, \bar{\bar{\kappa}}\}^T$ and

$$\mathbf{B} = \begin{Bmatrix} \mathbf{B}^u & 0 & 0 \\ 0 & \mathbf{B}^w & \mathbf{B}^{w'} \end{Bmatrix}, \mathrm{d} = \{\mathbf{u}^T, \mathbf{w}^T, \mathbf{w}'^T\}^T \tag{2.71}$$

$$\mathrm{G} = \mathrm{DIAG}\{G^u, G^\theta\}, \quad \alpha = \{\alpha_u, \alpha_\theta\}^T \tag{2.72}$$

Once derivation of **G** operator is done, the kinematic description of the element is finalized. It may be derived indirectly through the requirement that an element has to be able to describe a strain-free mode at some non-zero values of $\hat{\alpha}_u$ and $\hat{\alpha}_\theta$, (e.g. Armero and Ehrlich 2006). According to Figure 2.10, the generalized nodal displacements $\hat{\mathbf{d}}_{\mathrm{hinge}} = \{\hat{u}_1, \hat{u}_2, \hat{w}_1, \hat{w}_2, \hat{w}_1', \hat{w}_2'\}^T$ of such a strain-free mode are composed of:

$$\hat{\mathbf{d}}_{\mathrm{hinge}} = \hat{\mathbf{d}}_{\mathrm{rigid}} + \mathbf{D}_{\mathrm{hinge}}\hat{\boldsymbol{\alpha}}, \quad \mathbf{D}_{\mathrm{hinge}} = \begin{bmatrix} 0 & 1 & 0 & 0 & 0 & 0 \\ 0 & 0 & 0 & L^{(e)} - x_d & 0 & 1 \end{bmatrix}^T \tag{2.73}$$

where $\hat{\mathbf{d}}_{\mathrm{rigid}} = \{\hat{u}_1, \hat{u}_1, \hat{w}_1, \hat{w}_1 + \hat{w}_1'L^{(e)}, \hat{w}_1', \hat{w}_1'\}^T$ are generalized nodal displacements due to the rigid body motion of a complete beam, and $\mathbf{D}_{\mathrm{hinge}}\hat{\boldsymbol{\alpha}}$ are generalized nodal displacements due to the rigid body motion of one part of the beam due to imposed strong discontinuity $\hat{\boldsymbol{\alpha}} = \{\hat{\alpha}_u, \hat{\alpha}_\theta\}^T$. If we now set strains defined in (2.69) to be equal to zero for $\hat{\mathbf{d}}_{\mathrm{hinge}}$, one obtains:

$$0 = \mathbf{B}\hat{\mathbf{d}}_{\mathrm{hinge}} + \mathbf{G}\hat{\alpha} = \underbrace{\mathbf{B}\hat{\mathbf{d}}_{\mathrm{hinge}}}_{=0} + (\mathbf{G} + \mathbf{B}\mathbf{D}_{\mathrm{hinge}})\hat{\boldsymbol{\alpha}} \tag{2.74}$$

Since the above equation should hold for any $\hat{\boldsymbol{\alpha}}$, the **G** operator needs to read:

$$\mathbf{G} = -\mathbf{B}\mathbf{D}_{\mathrm{hinge}} \tag{2.75}$$

and this leads to:

$$G^u(x, x_d) = -\frac{1}{L^{(e)}} \tag{2.76}$$

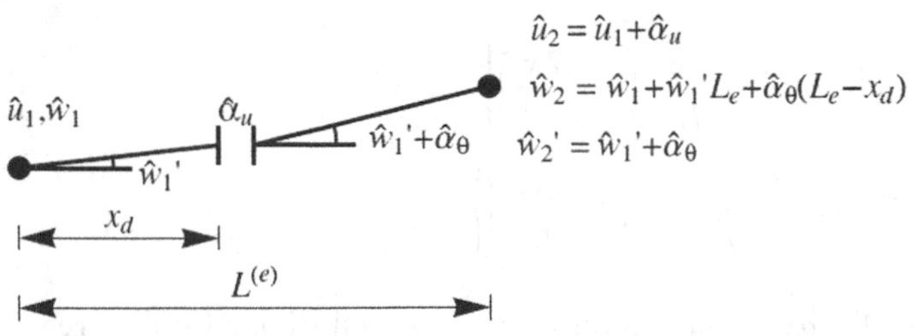

Figure 2.10 Strain-free mode of the element.

$$G^{\theta}(x,x_d) = -\frac{1+3\left(1-\frac{2x_d}{L^{(e)}}\right)\left(1-\frac{2x}{L^{(e)}}\right)}{L^{(e)}} \tag{2.77}$$

Equation (2.77) concludes the kinematic description of the geometrically linear element.

Interpolation functions M^u and M^θ can be developed by using (2.61) and (2.62) with the aim of describing the strain-free mode of Figure 2.10. Defining the values in (2.61) as $u_1 = \hat{u}_1 = 0$, $u_2 = \hat{u}_2 = \hat{\alpha}_u$, $u^h = 0$ for $x < x_d$, and $u^h = \hat{\alpha}_u$ for $x \geq x_d$, it follows that $M^u = H(x - x_d) - \mathbf{N}^u\{0,1\}$. Here, $H(x-x_d)$ is a unit-step function, which is 0 for $x < x_d$ and 1 for $x \geq x_d$. In the same sense for bending in (2.62) one gets: $M^\theta = H(x - x_d)(x - x_d) - \mathbf{N}^w \cdot \{0, L^{(e)} - x_d\} - \mathbf{N}^{w'} \cdot \{0,1\}$.

The geometrically nonlinear effects, and related global buckling, are defined by using the von Karman axial strain when computing the virtual axial strain. The real axial strain, used for computing the internal forces, will still be assumed as linear, as given in (2.65). The von Karman axial strain is defined as: $\varepsilon^{VK} = \partial u^h / \partial x + (1/2)(\partial w^h / \partial x)^2$. So, the corresponding virtual axial strain reads:

$$\delta\varepsilon^{VK} = \frac{\partial \delta u^h}{\partial x} + \frac{\partial w^h}{\partial x}\frac{\partial \delta w^h}{\partial x} \tag{2.78}$$

By interpolating $\delta u^h, w^h$ and δw^h in (2.78) as $\delta u^h = \mathbf{N}^u(x)\delta\mathbf{u} + M^u(x,x_d)\delta\alpha_u$, $w^h = \mathbf{N}^w(x)\mathbf{w} + \mathbf{N}^{w'}(x)\mathbf{w}'$, and $\delta w^h = \mathbf{N}^w(x)\delta\mathbf{w} + \mathbf{N}^{w'}(x)\delta\mathbf{w}'$, where $\delta\mathbf{u} = \{\delta u_1, \delta u_2\}^T$ is a vector of virtual nodal axial displacements, $\delta\mathbf{w} = \{\delta w_1, \delta w_2\}^T$ and $\delta\mathbf{w}' = \{\delta w_1', \delta w_2'\}^T$ are vectors of virtual nodal transverse displacements and rotations, and $\delta\alpha_u$ is a virtual discontinuity in axial displacement at x_d, the chosen interpolations direct to:

$$\delta\varepsilon^{VK} = \underbrace{\mathbf{B}^u(x)\delta\mathbf{u} + \mathbf{B}^{u,w}(x)\delta\mathbf{w} + \mathbf{B}^{u,w'}(x)\delta\mathbf{w}' + G^u(x,x_d)\delta\alpha_u}_{\delta\bar{\varepsilon}^{VK}} + \underbrace{\delta x_d \delta\alpha_u}_{\delta\bar{\bar{\varepsilon}}} \tag{2.79}$$

where:

$$\mathbf{B}^{u,w}(x) = C\frac{d\mathbf{N}^w}{dx}, \mathbf{B}^{u,w'}(x) = C\frac{d\mathbf{N}^{w'}}{dx}, C = \left(\frac{d\mathbf{N}^w}{dx}\cdot\mathbf{w} + \frac{d\mathbf{N}^{w'}}{dx}\cdot\mathbf{w}'\right) \tag{2.80}$$

When computing virtual strains $\delta\tilde{\varepsilon} = \{\delta\bar{\varepsilon}, \delta\bar{\kappa}\}^T$, the linear matrix operator $\mathbf{B}$ from (2.71) should be thus replaced with the nonlinear matrix operator $\mathbf{B}^{VK}$, meaning:

$$\begin{Bmatrix} \delta\bar{\varepsilon} = \delta\bar{\varepsilon}^{VK} \\ \delta\bar{\kappa} \end{Bmatrix} = \underbrace{\begin{Bmatrix} \mathbf{B}^u & \mathbf{B}^{u,w} & \mathbf{B}^{u,w'} \\ 0 & \mathbf{B}^w & \mathbf{B}^{w'} \end{Bmatrix}}_{\mathbf{B}^{VK}} \delta\mathbf{d} + G\delta\boldsymbol{\alpha} \tag{2.81}$$

The generalized virtual nodal displacements in (2.81) are denoted as $\delta\mathbf{d} = \{\delta\mathbf{u}^T, \delta\mathbf{w}^T, \delta\mathbf{w}'^T\}^T$ whereas virtual jumps at x_d are denoted $\delta\boldsymbol{\alpha} = \{\delta\alpha_u, \delta\alpha_\theta\}^T$.

The tangent stiffness matrix of the beam finite element with von Karman virtual axial strain has a symmetric geometric part and non-symmetric material part. Here a non-symmetric tangent stiffness matrix is used. It is important to emphasize that if von Karman definition of axial strains is used for both real and virtual strains (e.g. Reddy 2004), the element exhibits severe locking.

2.3.2.1.2 Equilibrium Equations

The principle of virtual work, as the weak form of the equilibrium equations for an element e of a chosen finite element mesh with N_{el} finite elements, can be written as:

$$\delta\Pi^{\text{int},(e)} - \delta\Pi^{\text{ext},(e)} = 0 \tag{2.82}$$

A single element contribution to the virtual work of internal forces can be defined and written, if one denotes the virtual strains as $\delta\varepsilon = \left\{\delta\varepsilon^{VK}, \delta\kappa\right\}^T$ and the virtual curvatures to be of the same form as the real curvatures as stated in (2.66):

$$\begin{aligned}\delta\Pi^{\text{int},(e)} &= \int_0^{L^{(e)}} (\delta\varepsilon)^T \boldsymbol{\sigma} dx \\ &= \underbrace{\int_0^{L^{(e)}} \delta\mathbf{d}^T \left(\mathbf{B}^{VK}\right)^T \boldsymbol{\sigma} dx}_{\text{standard}} + \underbrace{\int_0^{L^{(e)}} \delta\boldsymbol{\alpha}^T \left(\mathbf{G}^T\boldsymbol{\sigma} + \delta_{x_d}\boldsymbol{\sigma}\right) dx}_{\text{additional}}\end{aligned} \tag{2.83}$$

The vector of the internal beam forces that contains axial force N and bending moment M are defined as:

$$\boldsymbol{\sigma} = \{N, M\}^T \tag{2.84}$$

The vector of the internal nodal forces is acquired from the first intergral in equation (2.83) and reads:

$$\mathbf{f}^{\text{int},(e)} = \int_0^{L^{(e)}} \left(\mathbf{B}^{VK}\right)^T \boldsymbol{\sigma} dx \tag{2.85}$$

From the virtual work of external forces $\delta\Pi^{\text{ext},(e)}$, the vector of the element external nodal forces $\mathbf{f}^{\text{ext},(e)}$ will be obtained representing the external load applied to the element. The finite element assembly of vectors $\mathbf{f}^{\text{int},(e)}$ and $\mathbf{f}^{\text{ext},(e)}$, for all elements of the chosen mesh, leads to a set of global (i.e. mesh related) equations:

$$A_{e=1}^{N_{\text{el}}}\left(\mathbf{f}^{\text{int},(e)} - \mathbf{f}^{\text{ext},(e)}\right) = 0 \tag{2.86}$$

where A is the assembling operator.

The contribution of the second integral in equation (2.83) will be treated locally element by element. Then, in view of (2.82), the following two equations are obtained for each element of the chosen mesh:

$$\begin{aligned}\mathbf{h}^{(e)} &= \left\{h_N^{(e)}, h_M^{(e)}\right\} = \int_0^{L^{(e)}} \left(\mathbf{G}^T\boldsymbol{\sigma} + \delta_{x_d}\boldsymbol{\sigma}\right) dx = \\ &= \int_0^{L^{(e)}} \mathbf{G}^T \boldsymbol{\sigma} dx + \underbrace{\boldsymbol{\sigma}\big|_{x_d}}_{=\mathbf{t}} = \int_0^{L^{(e)}} \mathbf{G}^T \boldsymbol{\sigma} dx + \mathbf{t} = 0, \forall e \in \left[1, N_{\text{el}}\right]\end{aligned} \tag{2.87}$$

In (2.87) vector $\mathbf{t}=\boldsymbol{\sigma}|_{x_d}=\{t_N,t_M\}^T$ is defined with components t_N and t_M, which represent respectively the axial traction and moment (bending) traction at the discontinuity. The component form of (2.87) can be derived by using (2.77) and (2.84), and it reads:

$$h_N^{(e)}=\int_0^{L^{(e)}} G^u N dx + t_N = 0$$
$$h_M^{(e)}=\int_0^{L^{(e)}} G^\theta M dx + t_M = 0, \forall e \in [1, N_{\mathrm{el}}] \tag{2.88}$$

It is evident here that there is a set of global equations (2.86) and a set of equations on an element level (2.87) that need to be solved and interrelated. This will be addressed later in the chapter.

2.3.2.1.3 Constitutive Relations

It is assumed that the axial response of the beam material always remains elastic. For the bending behavior of the beam material, the following constitutive models are chosen: (i) a stress-resultant elastoplastic constitutive model with linear isotropic hardening, (ii) a stress-resultant rigid-plasticity model with linear softening at the softening plastic hinge. The basic ingredients of the chosen constitutive relations are built on classical plasticity (Ibrahimbegović, Fadi and Lotfi 1998) and are summarized in the following paragraphs.

In the same sense as in (2.70), the regular strains ε can be split into an elastic and plastic part. As the axial strain of the beam (2.65) always remains elastic, it is clear that:

$$\varepsilon = \bar{\varepsilon} = \bar{\varepsilon}^e, \bar{\bar{\varepsilon}} = 0 \Leftrightarrow \bar{\varepsilon}^p = 0, \alpha_u = 0 \tag{2.89}$$

The free energy of the beam material (before localized softening is activated) becomes:

$$\bar{\psi}\left(\bar{\varepsilon}^e, \bar{\xi}\right) := \bar{W}\left(\bar{\varepsilon}^e\right) + \bar{\Xi}\left(\bar{\xi}\right) = \frac{1}{2}\bar{\varepsilon}^{eT}\mathbf{C}\bar{\varepsilon}^e + \frac{1}{2}K_b\bar{\xi}^2 \tag{2.90}$$

where $\mathbf{C} = \mathrm{DIAG}\{EA, EI\}$, E is the elastic modulus, A and I are the area and moment of inertia of the cross-section, $\bar{\xi} \geq 0$ is the strain-like bending hardening variable and $K_b \geq 0$ is the linear bending hardening modulus.

The yield criterion for the beam material is defined in terms of the bending moment. The admissible values of the bending moment and the stress-like bending hardening variable $\bar{q}\left(\bar{\xi}\right)$ are governed by the function:

$$\bar{\phi}(M,\bar{q}) = |M| - (M_y - \bar{q}) \leq 0 \tag{2.91}$$

where $M_y > 0$ denotes the positive yield moment of the cross-section. In order to take into account the influence of the axial force N on the cross-section yielding, it is necessary to define M_y and $\bar{q}$ as functions of N. This is done via the localization (failure) criterion that activates softening in the discontinuity at x_d and is defined in terms of the bending traction t_M and the stress-like softening bending variable $\bar{\bar{q}}\left(\bar{\bar{\xi}}\right)$ (the latter is defined in terms of the bending strain-like softening variable $\bar{\bar{\xi}}$)

$$\bar{\bar{\phi}}\left(t_M, \bar{\bar{q}}\right) = |t_M| - \left(M_u - \bar{\bar{q}}\right) \leq 0 \tag{2.92}$$

where $M_u > M_y > 0$ denotes the positive ultimate (failure) moment of the cross-section. Influence of the axial force N on the cross-section failure is taken into account by defining M_u and $\bar{\bar{q}}$ as functions of N, as shown below.

The bending traction t_M at the discontinuity x_d is related to the rotation jump as shown in Figure 2.11.

$$t_M = t_M(\alpha_\theta) \tag{2.93}$$

Further thermodynamics of associative plasticity and the principle of maximum plastic dissipation are utilized for determination of the remaining ingredients of the elastoplasticity with hardening (e.g. Ibrahimbegović 2009; Lubliner 1990; Simo and Hughes 1998). In the present beam model, the elastoplasticity with hardening happens for $\boldsymbol{\alpha}=0$, which leads to $\bar{\varepsilon} = \tilde{\bar{\varepsilon}}$. By using the fact that regular strain can be decomposed into an elastic and plastic part on one hand and the free energy of the beam, as stated in (2.90), on the other hand the mechanical dissipation can be written as:

$$0 \leq \bar{\mathcal{D}} \overset{\text{def}}{=} \sigma^T \dot{\bar{\varepsilon}} - \dot{\bar{\psi}}(\bar{\varepsilon}^e, \bar{\xi}) = \left(\sigma - \frac{\partial \bar{\psi}}{\partial \bar{\varepsilon}^e}\right)^T \dot{\bar{\varepsilon}}^e + \sigma^T \dot{\bar{\varepsilon}}^p - \frac{\partial \bar{\psi}}{\partial \bar{\xi}} \dot{\bar{\xi}} \tag{2.94}$$

Once again, it is assumed that the elastic process is non-dissipative (i.e. $\bar{\mathcal{D}} = 0$), and that the plastic state variables do not change, from (2.94) one obtains:

$$\boldsymbol{\sigma} = \frac{\partial \bar{\psi}}{\partial \bar{\varepsilon}^e} = \mathbf{C}\bar{\varepsilon}^e \Rightarrow N = EA\bar{\varepsilon}, M = EI(\bar{\kappa} - \bar{\kappa}^p) \tag{2.95}$$

Taking into account the free energy of the beam material and mechanical dissipation, the hardening variable $\bar{q}$ is expressed as:

$$\bar{q} = -\frac{\partial \bar{\psi}}{\partial \bar{\xi}} = -\frac{\partial \bar{\Xi}}{\partial \bar{\xi}} = -K_h \bar{\xi} \tag{2.96}$$

With the above constitutive equations, the plastic dissipation thus reads:

$$\bar{\mathcal{D}}^p = \boldsymbol{\sigma}^T \dot{\bar{\varepsilon}}^p + \bar{\mathrm{q}} \dot{\bar{\xi}} \Rightarrow \bar{\mathcal{D}}^p = M \dot{\bar{\kappa}}^p + \bar{\mathrm{q}} \dot{\bar{\xi}} \tag{2.97}$$

Applying the principle of maximum plastic dissipation for the variables $(M, \bar{q})$ that satisfy the yield criteria $\bar{\phi}(M, \bar{q}) \leq 0$ at frozen rates $\dot{\bar{\kappa}}^p$ and $\dot{\bar{\xi}}$, the constrained optimization problem can be written:

$$\underset{M,\bar{q}}{n} \ \underset{\dot{\bar{\gamma}}}{\max}\left[\mathcal{L}^p(M, \bar{q}, \dot{\bar{\gamma}}) = -\bar{\mathcal{D}}^p(M, \bar{q}) + \dot{\bar{\gamma}}\bar{\phi}(M, \bar{q})\right] \tag{2.98}$$

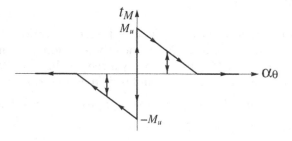

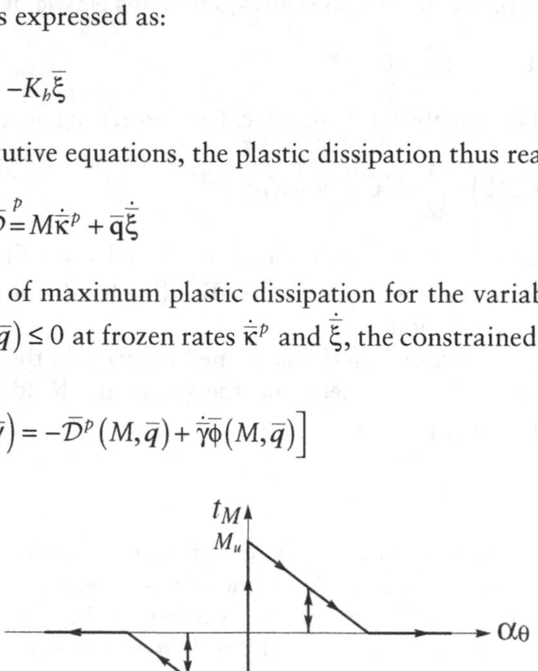

Figure 2.11 Rigid-plastic cohesive law at discontinuity.

where $\dot{\bar{\gamma}} \geq 0$ plays the role of the Lagrange multiplier. The evolution equations for internal variables is formulated by using (2.92) and (2.97):

$$\frac{\partial \bar{\mathcal{L}}^p}{\partial M} = -\dot{\bar{\kappa}}^p + \dot{\bar{\gamma}}\frac{\partial \bar{\phi}}{\partial M} = 0 \Rightarrow \dot{\bar{\kappa}}^p = \text{sign}(M)\dot{\bar{\gamma}}$$
$$\frac{\partial \bar{\mathcal{L}}^p}{\partial \bar{q}} = -\dot{\bar{\xi}} + \dot{\bar{\gamma}}\frac{\partial \bar{\phi}}{\partial \bar{q}} = 0 \Rightarrow \dot{\bar{\xi}} = \dot{\bar{\gamma}} \tag{2.99}$$

along with the Kuhn–Tucker loading/unloading conditions and the consistency condition:

$$\dot{\bar{\gamma}} \geq 0,\ \bar{\phi} \leq 0,\ \dot{\bar{\gamma}}\bar{\phi} = 0, \dot{\bar{\gamma}}\dot{\bar{\phi}} = 0 \tag{2.100}$$

It is necessary isolate the softening plastic hinge in order to find the remaining ingredients of the rigid-plastic response describing softening at the discontinuity x_d. The bending softening potential at the discontinuity is expressed as $\bar{\bar{\Xi}} = \bar{\bar{\psi}} = (1/2)K_s^2\bar{\bar{\xi}}$, where $\bar{\bar{\psi}}$ is the strain energy function due to softening. The softening potential depends on the strain-like (bending) softening variable $\bar{\bar{\xi}} \geq 0$ and the linear (bending) softening modulus $K_s \leq 0$. The dissipation at x_d can be then written as:

$$0 \leq \bar{\bar{\mathcal{D}}} \stackrel{\text{def}}{=} t_M\dot{\alpha}_\theta - \dot{\bar{\bar{\psi}}}\left(\bar{\bar{\xi}}\right) = t_M\dot{\alpha}_\theta - \frac{\partial \bar{\bar{\psi}}}{\partial \bar{\bar{\xi}}}\dot{\bar{\bar{\xi}}} \tag{2.101}$$

By defining:

$$\bar{\bar{q}} = -\frac{\partial \bar{\bar{\psi}}}{\partial \bar{\bar{\xi}}} = -\frac{\partial \bar{\bar{\Xi}}}{\partial \bar{\bar{\xi}}} = -K_s\bar{\bar{\xi}} = |K_s|\bar{\bar{\xi}} \tag{2.102}$$

the result in (2.101) can be rewritten as:

$$\bar{\bar{\mathcal{D}}} = \bar{\bar{\mathcal{D}}}^p = t_M\dot{\alpha}_\theta + \bar{\bar{q}}\dot{\bar{\bar{\xi}}} \tag{2.103}$$

The principle of maximum plastic dissipation at the rigid-plastic discontinuity can then be defined as:

$$\min_{t_M,\bar{\bar{q}}}\max_{\dot{\bar{\bar{\gamma}}}}\left[\bar{\bar{\mathcal{L}}}^p\left(t_M,\bar{\bar{q}},\dot{\bar{\bar{\gamma}}}\right) = -\bar{\bar{\mathcal{D}}}^p\left(t_M,\bar{\bar{q}}\right) + \dot{\bar{\bar{\gamma}}}\bar{\bar{\phi}}\left(t_M,\bar{\bar{q}}\right)\right] \tag{2.104}$$

and here $\dot{\bar{\bar{\gamma}}} \geq 0$ is the Lagrange multiplier. Taking into account the localization failure criterion and plastic dissipation, the following evolution equations are obtained:

$$\frac{\partial \bar{\bar{\mathcal{L}}}^p}{\partial t_M} = -\dot{\alpha}_\theta + \dot{\bar{\bar{\gamma}}}\frac{\partial \bar{\bar{\phi}}}{\partial t_M} = 0 \Rightarrow \dot{\alpha}_\theta = \text{sign}(t_M)\dot{\bar{\bar{\gamma}}}$$
$$\frac{\partial \bar{\bar{\mathcal{L}}}^p}{\partial \bar{\bar{q}}} = -\dot{\bar{\bar{\xi}}} + \dot{\bar{\bar{\gamma}}}\frac{\partial \bar{\bar{\phi}}}{\partial \bar{\bar{q}}} = 0 \Rightarrow \dot{\bar{\bar{\xi}}} = \dot{\bar{\bar{\gamma}}} \tag{2.105}$$

By noting that $\text{sign}(t_M) = \text{sign}(\alpha_\theta)$ (see (2.93) and Figure 2.11), it follows from (2.105) that:

$$\text{sign}(\alpha_\theta)\dot{\alpha}_\theta = \dot{\bar{\bar{\xi}}} \Rightarrow |\alpha_\theta| = \bar{\bar{\xi}} \tag{2.106}$$

Here, once again, the Kuhn–Tucker loading/unloading conditions and the consistency condition apply:

$$\dot{\bar{\bar{\gamma}}} \geq 0, \bar{\bar{\phi}} \leq 0, \ \dot{\bar{\bar{\gamma}}}\bar{\bar{\phi}} = 0, \dot{\bar{\bar{\gamma}}}\dot{\bar{\bar{\phi}}} = 0 \tag{2.107}$$

Finally, the total dissipation of the beam finite element, when the element is in the softening regime, can be written as:

$$\begin{aligned}
\bar{\bar{\mathcal{D}}}^{\text{tot}}_{L(e)} &= \int_0^{L(e)} \left(\boldsymbol{\sigma}^T \dot{\bar{\varepsilon}} - \dot{\bar{\psi}}\left(\bar{\varepsilon}^e, \bar{\xi}\right)\right) dx + \left(t_M \dot{\alpha}_\theta - \dot{\bar{\bar{\psi}}}\left(\bar{\bar{\xi}}\right)\right) \\
&= \int_0^{L(e)} \left(\boldsymbol{\sigma}^T \dot{\tilde{\bar{\varepsilon}}} + \boldsymbol{\sigma}^T \mathbf{G} \dot{\boldsymbol{\alpha}} - \boldsymbol{\sigma}^T \underbrace{\dot{\bar{\varepsilon}}^e}_{\tilde{\bar{\varepsilon}}^e} + \bar{q}\dot{\bar{\xi}}\right) dx + \left(t_M \dot{\alpha}_\theta + \bar{\bar{q}}\dot{\bar{\bar{\xi}}}\right) \\
&= \underbrace{\int_0^{L(e)} \left(M \dot{\tilde{\bar{\kappa}}}^p + \bar{q}\dot{\bar{\xi}}\right) dx}_{\bar{\mathcal{D}}^p} + \underbrace{\left(\int_0^{L(e)} G^\theta M dx + t_M\right)}_{=0} \dot{\alpha}_\theta + \underbrace{\bar{\bar{q}}}_{|K_s|\bar{\bar{\xi}}} \dot{\bar{\bar{\xi}}}
\end{aligned} \tag{2.108}$$

It can be seen from (2.108) that enforcing vector equations (2.88) will decouple the dissipation in the softening plastic hinge from the dissipation in the rest of the beam. In order to calculate t_M, (2.88) is employed.

The plastic work of the beam cross-section in the hardening regime is defined as:

$$\bar{W}^p = M\dot{\bar{\kappa}}^p = |M|\dot{\bar{\xi}} = \left(M_y + K_b \bar{\xi}\right)\dot{\bar{\xi}} \tag{2.109}$$

and plastic work for the beam finite element in the softening regime as:

$$\bar{\bar{W}}^p = t_M \dot{\alpha}_\theta = |t_M|\dot{\bar{\bar{\xi}}} = \left(M_u + K_s \bar{\bar{\xi}}\right)\dot{\bar{\bar{\xi}}} \tag{2.110}$$

with this, all constitutive relations are defined.

2.4 COMPUTATION OF BEAM PLASTICITY MATERIAL PARAMETERS

The material parameters that need to be known for the chosen material models are: (i) M_y and K_b for the plasticity with hardening, and (ii) M_u and K_s for the softening plastic hinge.

The yield moment M_y can be determined by considering the uniaxial yield stress of the material σ_y, the bending resistance modulus of cross-section W, the cross-section area A and the level of axial force N. One can associate the yield moment of the cross-section with the yielding of the most-stressed material fiber to get:

$$M_y(N) = W\sigma_y \left(1 - \frac{|N|}{A\sigma_y}\right) \tag{2.111}$$

If one takes material hardening into consideration, as well as the possibility of local buckling, a better estimate for M_u is obtained in contrast to the standard assumption of the elastic–perfectly-plastic response of material fibers. In this respect, a refined finite element model based on a geometrically and materially nonlinear shell element, which is able to capture local buckling and gradual spreading of plasticity over the cross-section, is formulated. The ultimate bending resistance M_u can be obtained by using results of such a shell model computation, as can be moduli K_h and K_s.

To obtain the desired results, a part of the frame member with a reference length $L^{ref} < L^{tot}$ (L^{tot} is the total length of the frame member under consideration) is: (i) modeled with shell finite elements, (ii) subjected to an external axial force $\hat{N}$ in the first loading step, and (iii) subjected to a varying external bending moment at the end cross-sections in the second loading step, while keeping $\hat{N}$ fixed (see Figure 2.12b). It is assumed that such a loading pattern will produce an approximately constant internal axial force $N = \hat{N}$ during the analysis. The computation with the shell model takes into account the geometrical and material nonlinearity which includes plasticity with hardening and strain softening, strain softening regularization and local buckling effects. The results of the shell analysis are illustrated in the diagrams presented in Figure 2.12d and g. One can associate the ultimate bending moment M_u with the peak point in the diagram in Figure 2.12d, where the applied end moment is plotted vs. the end rotation, i.e.

$$M_u(N) = M_u^{ref}(N) \tag{2.112}$$

This point can be used as well as a border-point between the hardening regime and the softening regime, where the softening can be due to material and/or geometric effects. To determine the values of the beam model hardening and softening parameters, it is assumed that the plastic work at failure should be equal for both the beam and the shell model. Meaning that the internal forces of the beam model have to produce the same amount of the plastic work as the internal forces of the shell model when considering the full complete failure of the frame element with length of L^{ref}.

Since the plastic work is done in two regimes (hardening and softening), it is necessary to assure that the amount of plastic work in each regime matches for both models, i.e.

$$E^{\bar{W}^p}(N) = E^{\bar{W}^{p,ref}}(N), E^{\bar{\bar{W}}^p}(N) = E^{\bar{\bar{W}}^{p,ref}}(N) \tag{2.113}$$

The plastic work in both regimes $E^{\bar{W}^{p,ref}}$ (hardening) and $E^{\bar{\bar{W}}^{p,ref}}$ (softening) are obtained from shell model analysis, Figure 2.12g.

The plastic work of the beam model in the hardening regime, $E^{\bar{W}^p}$, can be determined by observing that each cross-section in the frame member of length L^{ref} is approximately under the same force–moment state during the hardening regime. Integrating equation (2.109), one obtains:

$$E^{\bar{W}^p} = \int_0^{L^{ref}} \int_0^{t \text{ at } M_u} \bar{W}^p d\tau dl = L^{ref}\left(M_y\tilde{\xi} + \frac{1}{2}K_h\tilde{\xi}^2 \right) \tag{2.114}$$

In (2.114) above, $\tilde{\xi}$ is the value of hardening variable that corresponds to the bending moment M_u, Figure 2.12e. As linear hardening in the beam model is assumed (2.96), the hardening variable reads:

$$\tilde{\xi} = \frac{M_u - M_y}{K_h} \tag{2.115}$$

Hardening modulus is easily obtained by employing the three last equations:

$$K_h(N) = \frac{\left(M_u^2(N) - M_y^2(N)\right)L^{ref}}{2E^{\bar{W}^{p,ref}}(N)} \tag{2.116}$$

On the other hand, the plastic work of the beam model in the softening regime, $E^{\overline{\overline{W}}^p}$, can be determined by assuming that the softening part of the $M - \varphi$ curve in Figure 2.12d, obtained from the shell model analysis, is produced by a very localized phenomenon (in a single cross-section) related to the local buckling and/or to the localized strain softening. Integrating (2.110), one can compute the plastic work in the softening regime for the beam model as:

$$E^{\overline{\overline{W}}^p} = \int_0^t \dot{\overline{\overline{W}}}^p d\tau = \int_0^{\overline{\overline{\xi}} \text{ at } t_M = 0} \left(M_u + K_s \overline{\overline{\xi}}\right) d\overline{\overline{\xi}} = \frac{1}{2}|K_s| \overline{\overline{\overline{\xi}}}^2 \tag{2.117}$$

Here $\overline{\overline{\overline{\xi}}}$ is the value of the softening variable that corresponds to the total cross-section failure, Figure 2.12f. As linear softening is assumed in the beam model, one obtains:

$$\overline{\overline{\overline{\xi}}} = \frac{M_u}{|K_s|} \tag{2.118}$$

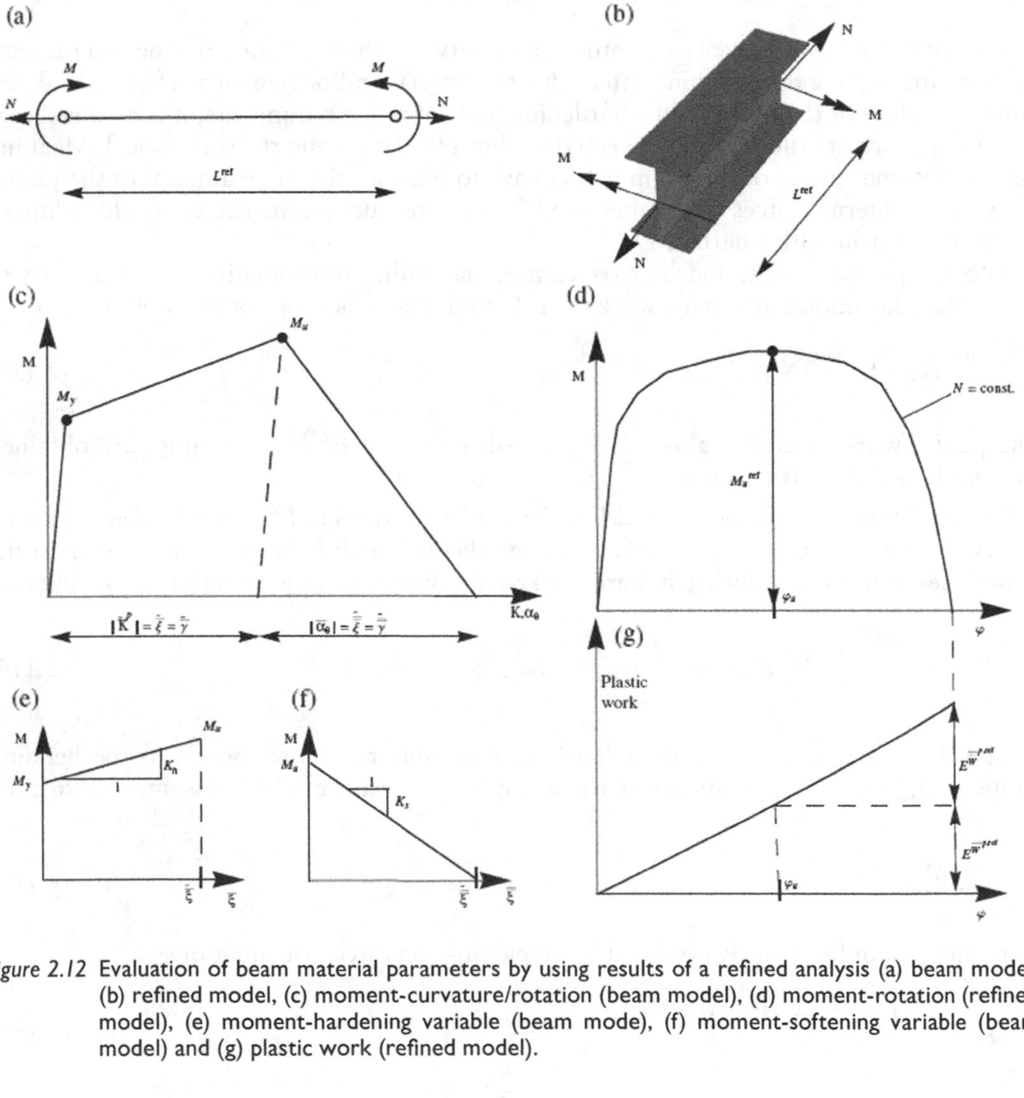

Figure 2.12 Evaluation of beam material parameters by using results of a refined analysis (a) beam model, (b) refined model, (c) moment-curvature/rotation (beam model), (d) moment-rotation (refined model), (e) moment-hardening variable (beam mode), (f) moment-softening variable (beam model) and (g) plastic work (refined model).

Taking into account the last two expressions, as well as the fact that the amount of plastic work needs to be the same in both models, the expression for softening modulus reads:

$$\left|K_s(N)\right| = \frac{M_u^2(N)}{2E^{\bar{\bar{W}}^{p,\mathrm{ref}}}(N)}, K_s \leq 0 \tag{2.119}$$

It is realized that the influence of the following values L^{ref}, K_h and K_s is very small; however, one should keep in mind that if the chosen length of L^{ref} enables large deflections then a large displacement correction of M_u has to be done.

A procedure for solving the set of global (mesh related) and the set of local (element related) nonlinear equations, generated by using the stress-resultant plasticity beam finite element with embedded discontinuity, can be found in Ibrahimbegović (2009).

2.4.1 Numerical Examples of Frames and Frame Elements

The beam model computer code was generated by using the symbolic manipulation code AceGen, and the examples were computed by using the finite element program AceFem (e.g. Korelc n.d.).

2.4.1.1 *Example 1 — Computation of Beam Plasticity Material Parameters*

As seen in the theoretical background, first of all, it is necessary to compute the beam plasticity material parameters M_u, K_h and K_s. A frame member is considered with an I-cross-section with flange width b_f=30 cm, flange thickness t_f=1.5 cm, web height b_w=40 cm and web thickness t_w=0.8 cm. The cross-section area is A=122 cm^2 and the bending resistance modulus is I=43034.2 cm^4. A part of the frame member of length $L^{\mathrm{ref}}=2L=300$ cm is modeled and it should be sufficient to capture the local softening effects due to local buckling and/or strain softening. Material properties of the frame member are presented in Table 2.2, and the uniaxial response is plotted in Figure 2.13.

The example has been computed with the finite element code ABAQUS (Hibbit, Karlsson and Sorensen n.d.) by using the shell finite element S4R with five integration points throughout the thickness. As only one half of the considered geometry was discretized (see Figures 2.14 and 2.15), the symmetry conditions $u_z = \varphi_x = \varphi_y = 0$ were employed. The mesh consists of equal squared elements. The free-end cross-section of the model was made rigid by coupling the degrees of freedom of that particular cross-section.

According to Ibrahimbegović (2009) the linear softening modulus is computed as:

$$K_s^{le} = -\frac{l_e \sigma_u^2}{2 g_s} \tag{2.120}$$

in order to minimize the mesh-dependency effect.

Table 2.2 Material Properties of the Frame Member

E	210 GPa
σ_y	0.24 GPa
σ_u	0.36 GPa
$\varepsilon_y = \sigma_y / E$	200
ε_u	0.1
ε_f	0.12778

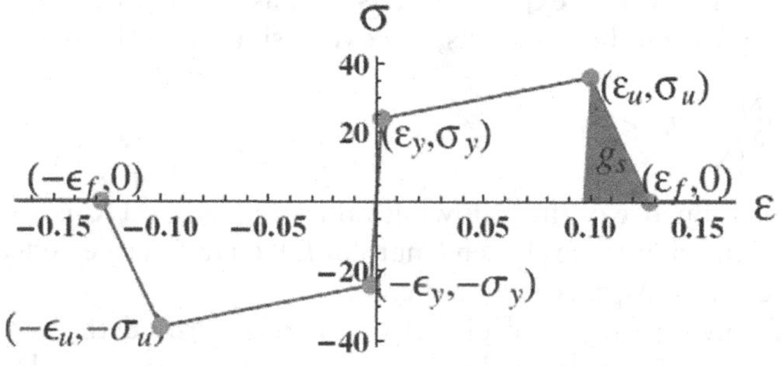

Figure 2.13 Uniaxial stress–strain curve.

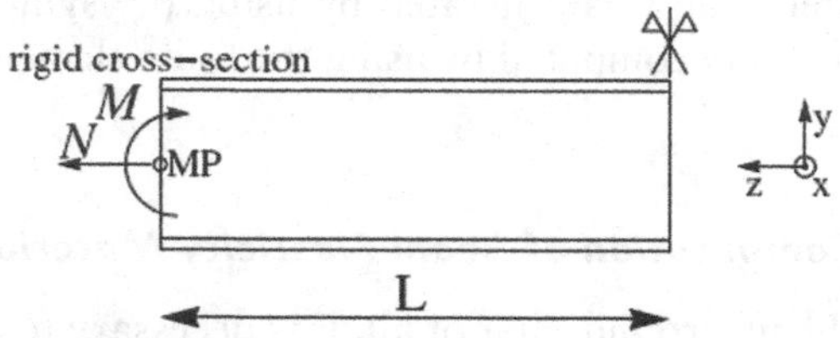

Figure 2.14 Boundary conditions for the shell model analysis.

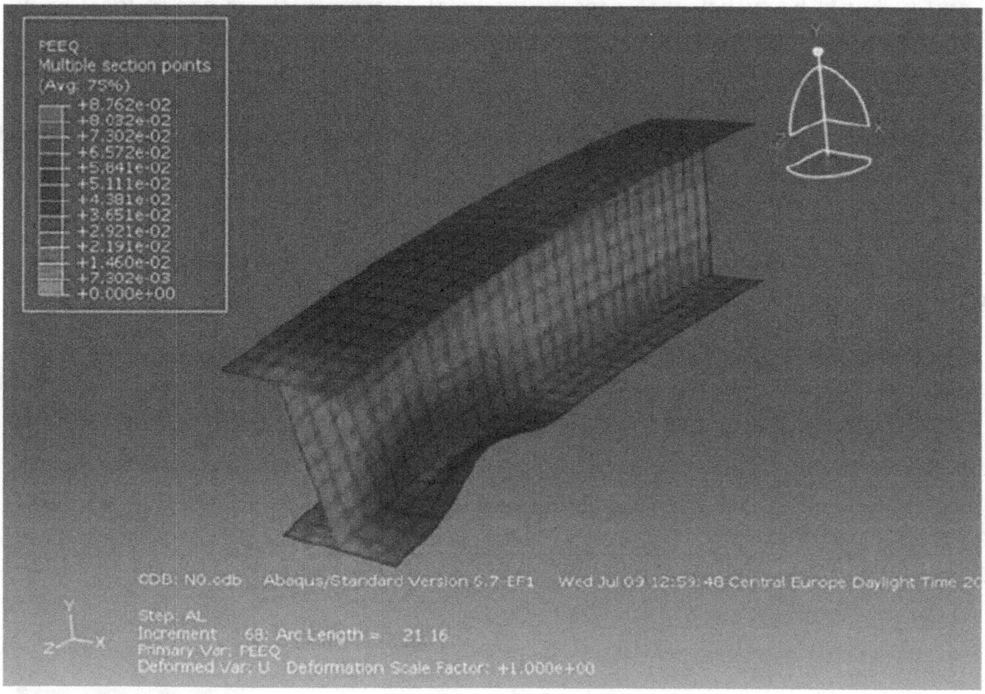

Figure 2.15 Failure mode of the representative part of the frame member as computed by the shell model.

where $l_e = 5$ cm is a characteristic dimension of the element and $g_s = 0.5\text{kN/cm}^2$ is the plastic work density (plastic work per unit volume) in the softening regime of the shell material that corresponds to the gray-colored area in Figure 2.13. The strain at failure, adjusted to the mesh, is then:

$$\varepsilon_s^{le} = \varepsilon_u + \frac{\sigma_u}{\dfrac{EK_s^{le}}{E + K_s^{le}}} = 0.10727 \tag{2.121}$$

The load was applied in two steps. For the first step, a desired level of axial force N ($N = -0.3N_y$ to $N = 0.3N_y$, $N_y = A\sigma_y$) at the mid-point MP of the rigid cross-section was applied and then the application of the bending moment M at the point MP took place. Nonlinear analysis with the path-following method was performed. The results of the analyses are presented in Figure 2.15 to Figure 2.17. The final deformed configuration of the shell model and distribution of the equivalent plastic strain are presented in Figure 2.15 for a pure bending case. It is noted that the considered part of the frame member failed by localized buckling of the bottom flange and strong localized yielding of the flange is observed which is concentrated in the neighborhood of the web. All other cases had the same failure mode. The level of axial force has a significant influence on the peak bending resistance and on the overall response (see Figure 2.16).

The value of the axial force does not influence the shape of the curve to a great extent (see Figure 2.17). The relation between the rotation and plastic work is almost linear. In Figures 2.16 and 2.17, it is assumed that points with the maximum bending moment represent a transition point from a hardening to softening regime.

The results obtained using the shell model are now used for evaluation of the beam model material parameters. Summarized results of the shell model are given in Table 2.3.

By using Table 2.3, a bi-linear approximation function for M_u^{ref} is formulated as:

$$\tilde{M}_u^{\text{ref}}(N) = \begin{cases} M_u^{\text{ref},0}\left(1.03 + 0.85\frac{N}{N_y}\right) & \text{if } N < -0.035N_y \\ M_u^{\text{ref},0} & \text{if } N \geq -0.035N_y \end{cases} \tag{2.122}$$

where $M_u^{\text{ref},0} = M_u^{\text{ref}}(N = 0)$. It is assumed that $\tilde{M}_u^{\text{ref}}(N)$ can be used to evaluate the ultimate bending moment of the beam model M_u, i.e.

$$M_u(N) = \tilde{M}_u^{\text{ref}}(N) \tag{2.123}$$

see Figure 2.18. The values for the ultimate resistance obtained with the shell analyses are marked with dots in Figure 2.18.

The beam model hardening modulus K_h can be evaluated pointwise by using (2.116), (2.111), (2.123) and the third column of Table 2.3. Average value of $K_h(N) = 1.06\times10^7\text{kN/cm}^2$ assuming that the axial force has no influence on the hardening modulus.

In the same fashion, the beam model softening modulus K_s can be evaluated pointwise by using (2.119) and (2.123) and the last column of Table 2.3. The gray-colored fields in Table 2.3 present unreliable results since for those cases the shell analysis did not converge. Again, assuming that the axial force has no influence on softening modulus the average value $K_s = -3.28\times10^5\text{kN/cm}^2$ can be adopted.

Table 2.4 gives a comparison between the shell analysis results and the corresponding beam model results. It is noted that the error in the ultimate bending moment is small, while the error in dissipated plastic work can be quite large.

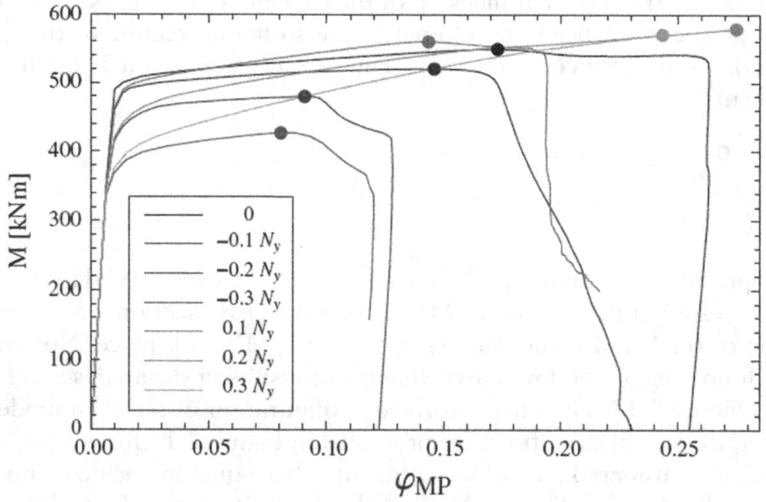

Figure 2.16 Bending moment vs. rotation curves for the end cross-section.

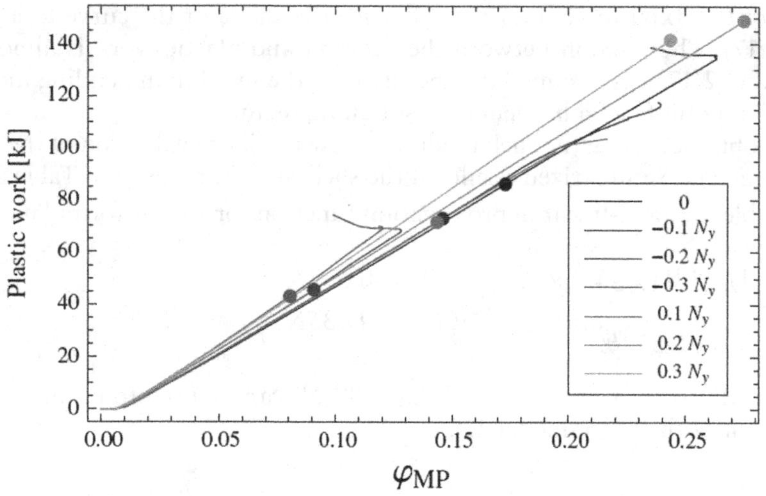

Figure 2.17 Plastic work vs. end cross-section rotation curves.

Table 2.3 Summary of Results of the Shell Model Analyses

N/N_y	M_u^{ref} (kNm)	$E^{\bar{W}^{p,ref}}$ (kJ)	$E^{\bar{\bar{W}}^{p,ref}}$ (kJ)
0	550	86	53
−0.1	521	73	45
−0.2	480	46	29
−0.3	427	43	27
0.1	560	72	32
0.2	579	149	0
0.3	571	142	0

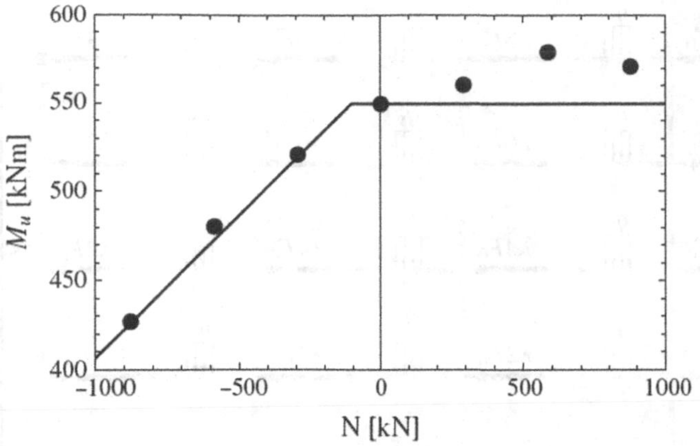

Figure 2.18 Approximation of the ultimate bending moment of the cross-section.

Table 2.4 Comparison Between Approximations and Shell Analyses Results

N/N_y	M_u(kNm)	$\left\lvert\frac{M_u - M_u^{ref}}{M_u^{ref}}\right\rvert$ (%)	$E^{\bar{W}^p}$ (kJ)	$\left\lvert\frac{E^{\bar{W}^p} - E^{\bar{W}^{p,ref}}}{E^{\bar{W}^{p,ref}}}\right\rvert$ (%)	$E^{\bar{\bar{W}}^p}$ (kJ)	$\left\lvert\frac{E^{\bar{\bar{W}}^p} - E^{\bar{\bar{W}}^{p,ref}}}{E^{\bar{\bar{W}}^{p,ref}}}\right\rvert$ (%)
0	550	0.00	50	42	46	12
−0.1	519	0.26	59	20	41	8
−0.2	473	1.54	54	17	34	19
−0.3	426	0.22	48	12	28	?
0.1	550	1.94	81	14	46	?
0.2	550	5.07	109	27	16	?
0.3	550	3.72	134	6	46	?

2.4.1.2 *Example 2 — Push-Over of an Asymmetric Frame*

An asymmetric frame is presented in Figure 2.19, where L_{B1}=600 cm, L_{B2}=500 cm, L_{B3}=400 cm, H_C=250 cm. All the other geometrical materials are the same as in the previous example. The vertical load is constant and equals q_0=0.05kN/cm. The lateral loading is presented in Figure 2.19, where F_0=1kN, is a concentrated force and λ is the load multiplier. The mesh consists of eight beam finite elements per each frame member.

The results are presented in Figures 2.20 and 2.21. The results of the geometrically linear and geometrically nonlinear analyses are nearly the same before ultimate resistance is reached at 1581.8 kN for the geometrically linear case and at 1578.4 kN for the geometrically nonlinear case. After that point, the difference between the two analyses is bigger. The final computed equilibrium configuration for the geometrically nonlinear case is at u_{top}=26.52 cm. In the next load step, one additional softening plastic hinge is activated, which results in the global failure mechanism. Since the path-following algorithm used is only governed by the increase of u_{top}, it is not possible to capture the remaining part of the load–displacement curve.

In the right part of Figure 2.20, we present the dissipated energy vs. u_{top} curves. Namely, first, there is the elastic non-dissipative phase, followed by the pure hardening dissipation phase, followed by the combined hardening and softening dissipation phase and finally the pure softening dissipation phase. In the geometrically nonlinear case, there is no final pure softening dissipation phase due to the activation of the global failure mechanism. Figure 2.21 illustrates the deformed configuration of the frame at u_{top}=26.52 cm. Locations,

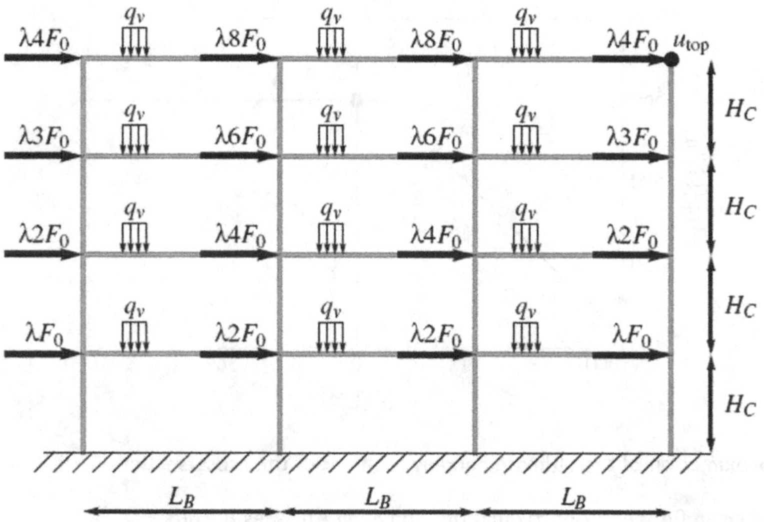

Figure 2.19 Asymmetric frame: geometry and loading.

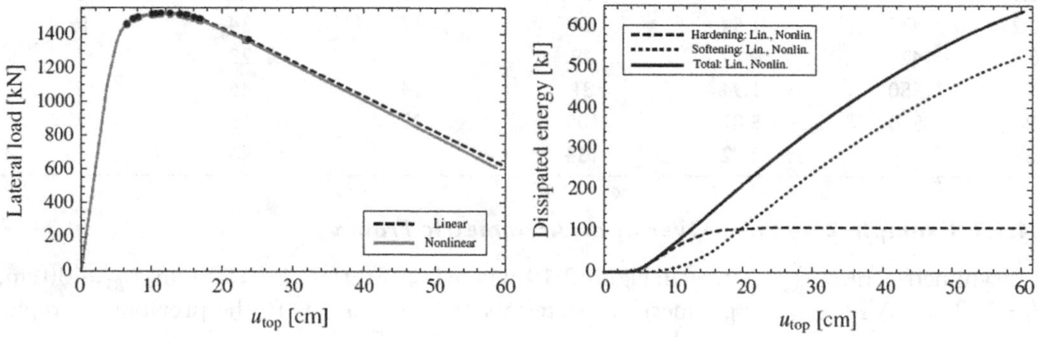

Figure 2.20 Load vs. displacement and dissipated energy vs. displacement curves.

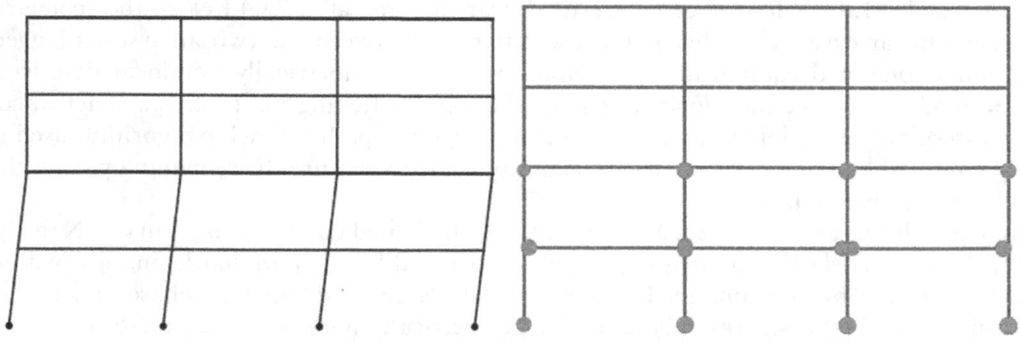

Figure 2.21 Deformed shape and locations of softening plastic hinges.

where softening plastic hinges were activated at $u_{top} \approx 26.52$ cm, are presented in the middle part of Figure 2.21 for the geometrically linear case and in the right part of the same figure for the geometrically nonlinear case.

2.4.1.3 Example 3 — The Darvall–Mendis Frame

The clamped portal frame under vertical loading first studied by Darvall and Priyantha (1985) and later examined by Armero and Ehrlich (2006) and Wackerfuß (2008) was elaborated. The geometry of the frame is presented in Figure 2.22 (Table 2.5).

The inelastic response is defined by the ultimate bending moment $M_{u,C}$ = 158.18kNm for the columns, the ultimate bending moment $M_{u,B}$ = 169.48 kNm for the beam and the softening modulus $K_s = aEI/10L$, where the values a = 0, −0.04, −0.06, −0.0718 are considered. Note, in this example, the inelastic response does not include any material hardening. It is assumed that the axial force has no influence on the ultimate bending resistance of the frame members. The mesh consists of eight geometrically linear beam finite elements (see Figure 2.22). The frame is loaded with a vertical load $\lambda F_0 (F_0 = 1kN)$ applied at node 5 (see Figure 2.22). In the numerical simulations, the load multiplier λ and the vertical displacement u_v is controlled at node 5 using the path-following method.

The vertical load vs. vertical deflection curves are presented in Figure 2.23, where the points on the curves mark configurations where the softening plastic hinge was activated in one of the elements of the mesh. In all cases, hinges form at the same locations.

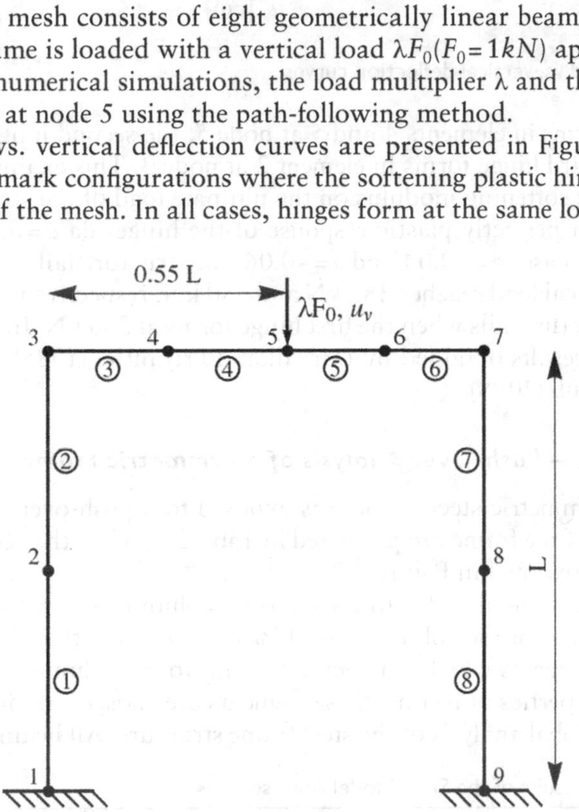

Figure 2.22 Geometry and loading of the portal frame.

Table 2.5 Geometric and Material Properties of the Frame

L_c	3.048 m
A	0.103 m²
I	0.001 m⁴
E	2.068×10⁷ kN/m²

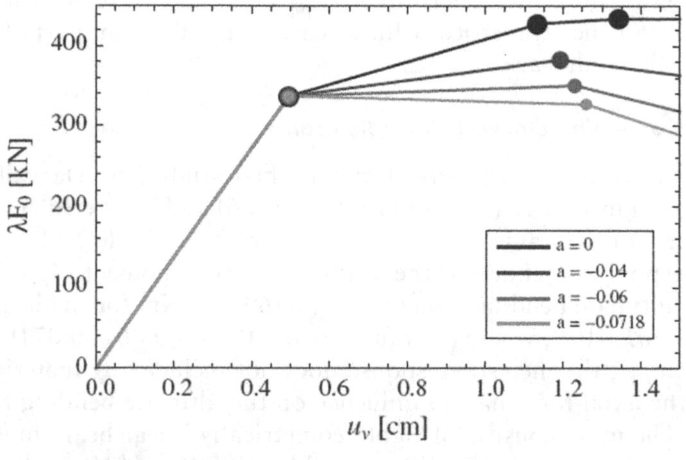

Figure 2.23 Vertical load vs. vertical deflection curves.

The first hinge forms in elements 4 and 5 at node 5, the second hinge forms in element 7 at node 7 and the third hinge forms in element 2 at node 3. This example shows the significant influence of the softening modulus on the ultimate load of the structure. The ultimate load in the case of a perfectly plastic response of the hinges đa $a=0$ is 434 kN when the third hinge forms. In cases $a=-0.04$ and $a=-0.06$, the structure fails when the second hinge forms when the vertical load reaches 383 kN and 350 kN, respectively. Finally, in the case of $a=-0.0718$, the structure fails when the first hinge forms at 336 kN. In Table 2.6, results are compared with the results obtained by Darvall and Priyantha (1985), Armero and Ehrlich (2006) and Wackerfuß (2008).

2.4.1.4 Example 4 — Push-Over Analysis of a Symmetric Frame

Additionally, the symmetric steel frame was exposed to a push-over. The geometrical and material properties of the frame are presented in Table 2.7, while the geometry and the loading of the frame is presented in Figure 2.24.

It should be noted that in order to have strong columns and a weak beak system the cross-section properties of the columns are 10% stronger than the cross-section properties of the beams. The elements which connect the beams to the columns are 10% weaker than the cross-section properties of beams; these elements are chosen to simulate the behavior of connections in the global analysis of the steel frame structure. All beams are loaded with the

Table 2.6 Summary of Results of the Shell Model Analyses

		Darvall and Mendis		*Armero and Ehrlich*		*Wackerfu*		*Present*	
a	Hinge	u_v *(cm)*	λF_0 *(kN)*	u_v *(cm)*	λF_0 *(kN)*	u_v *(cm)*	λF_0 *(kN)*	u_v *(cm)*	λF_0 *(kN)*
0	1	0.50	336	0.50	337	0.53	342	0.50	336
	2	1.14	427	1.14	428	1.13	435	1.13	427
	3	1.34	433	1.34	434	1.33	440	1.34	434
-0.04	1	0.50	336	0.50	337	0.53	349	0.50	336
	2	1.14	387	1.18	388	1.16	401	1.19	383
-0.06	1	0.50	336	0.50	337	0.52	348	0.50	336
	2	1.19	336	1.29	337	1.23	349	1.23	350
-0.0718	1	0.50	336	0.50	337			0.50	336

Table 2.7 Geometric and Material Properties of the Frame

Geometric and Material Properties of the Frame	
A	28.50 cm^2
I	1940 cm^4
E	2.10×10^4 kN/m^2
K	0.5E
Yielding and Ultimate Values	
M_y	3100 kNcm
M_u	3100 kNcm
V_y	355 kN
V_u	400 kN
Fracture Energies	
$G_{f,M}$	550 kN/cm
$G_{f,V}$	450 kN/cm

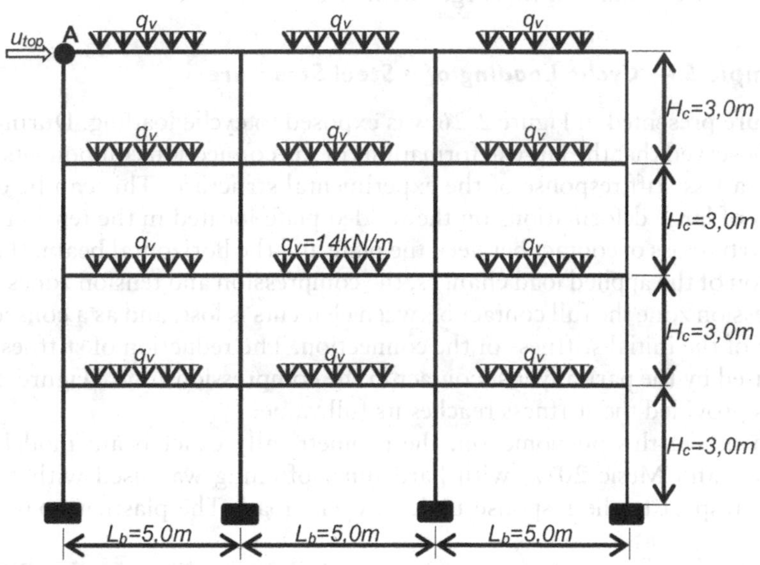

Figure 2.24 Geometrical characteristics of the frame and loading scheme.

same vertical constant uniformly distributed load simulating the dead load, while the lateral loading is applied in terms of an imposed incremental displacement (u_{top}) at the upper corner (point A, see Figure 2.24) (Imamovic, Ibrahimbegović and Mesic 2017).

The final position of the formed plastic hinges is presented in Figure 2.25 together with the obtained softening response presented in the well-known form (force–displacement diagram). The force which is taken as the adequate measure, in this case, is the reaction in the corner where the displacement is imposed (point A).

Ultimate limit analysis of any frame planar steel structure can be conducted with the application of this geometrically exact beam model with the inclusion of the second-order effects as well as different failure mechanisms. The geometrically exact beam model can perform the corresponding nonlinear analysis that allows the ultimate limit load computations with large displacement without any need for correction of the equivalent beam

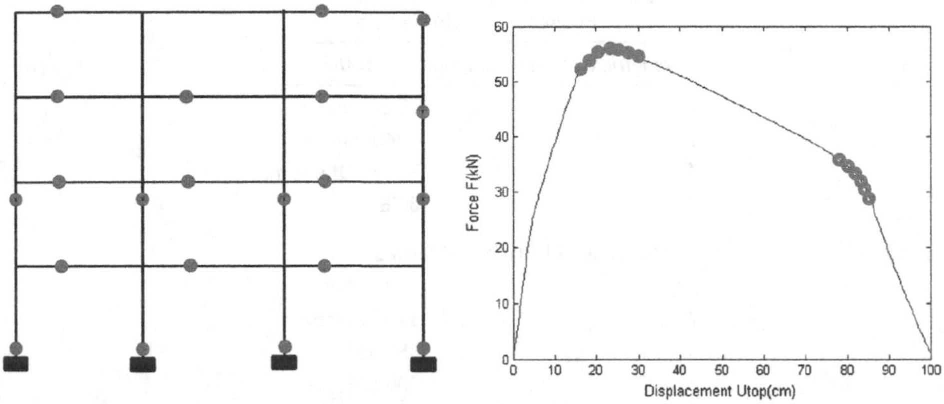

Figure 2.25 **Locations of softening plastic hinges and load vs. displacement.**

property. (Dujc, Bostjan and Ibrahimbegović 2010). This advantage is very important in a steel frame structure because of the large ductility of steel.

2.4.1.5 Example 5 — Cyclic Loading of a Steel Structure

A steel structure presented in Figure 2.26 was exposed to cyclic loading. During the experiment, it was observed that the large deformations of the connection components under cyclic loading cause a less stiff response of the experimental structure. This can be explained by the formation of large deformations on the welded plate located in the tension zone, which can cause a partial loss of contact between the plate and the horizontal beam. (Figure 2.27a). As the direction of the applied load changes, the compression and tension zones are changes. In the compression zone the full contact between elements is lost, and as a consequence there is a reduction of the initial stiffness of the connection. The reduction of stiffness in the connection is caused by the partially lost contact in the compression zone. (Figure 2.27b). Once full contact is provided the stiffness reaches its full value.

In order to model this phenomenon, the geometrically exact beam model (Imamovic, Ibrahimbegović and Mesic 2017) with hardening/softening was used with certain modifications with respect to the response under a cyclic load. The plasticity part governs the

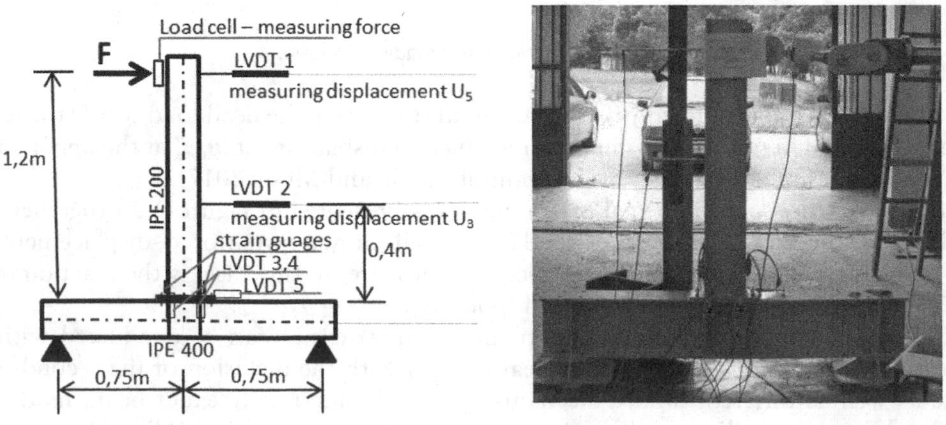

Figure 2.26 **Experimental setup and tests.**

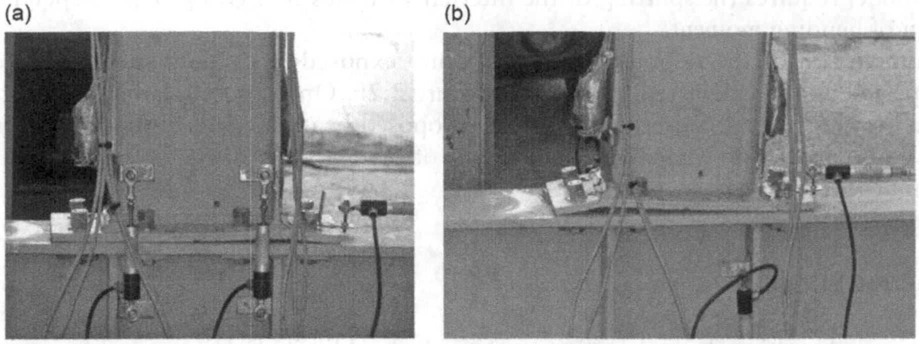

Figure 2.27 Experimental setup and tests.

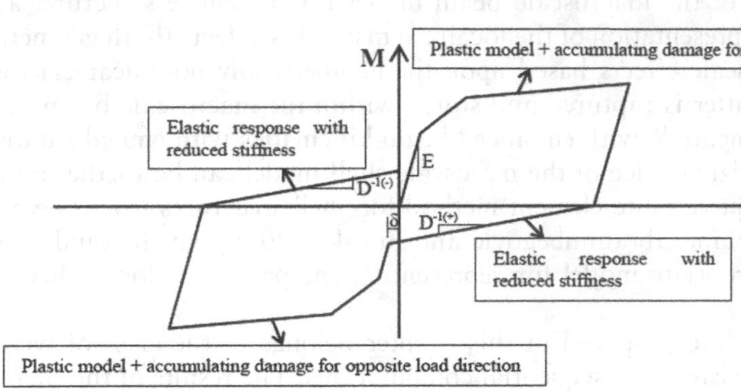

Figure 2.28 Constitutive model.

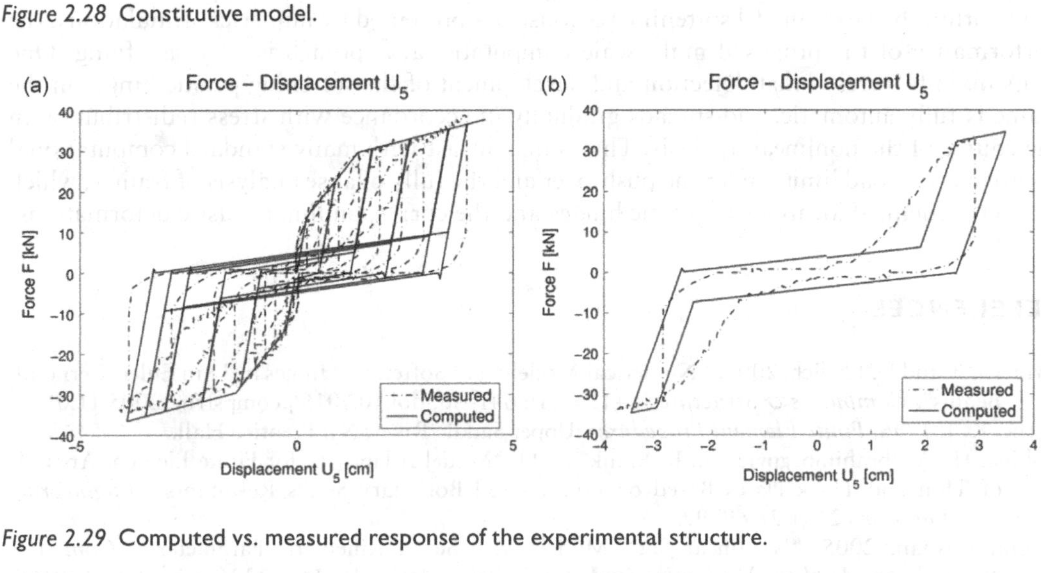

Figure 2.29 Computed vs. measured response of the experimental structure.

hardening and unloading phases, whereas the damaged part provides the reduced stiffness of the connection after the change in the sign of the bending moment: from positive to negative or vice versa (Figure 2.28). The damage model governs the connection response until full contact between the plate and horizontal beam is reached. After the full contact has been reached, the plasticity model is again activated. The gap δ corresponds to the plastic deformation in the bolts (Imamovic, Ibrahimbegović i Mesic 2018). The modification of the beam model requires the splitting of the internal variables into two groups, depending on the sign of bending moment.

A comparison of the response of the structure exposed to cyclic loading obtained by the experiment and calculated are given in Figure 2.29. One can notice that there is a difference between the responses; however, the proposed beam model significantly improved the response prediction compared to the model of the plasticity or the damage (Imamovic, Ibrahimbegović and Mesic 2018).

2.5 CONCLUSION

The presented multi-scale model for computing the load limit of planar metal frames under push-over and full collapse analysis combines the best of two worlds: on one side, the effectiveness and robustness of the macroscale beam model for the entire structure, and, on another side, a refined representation of the localized instability effects (both geometric and material) through mesoscale effects based upon the geometrically nonlinear elastoplastic shell formulation. The latter is captured and stored within the macroscale beam model in the manner which is compatible with enhanced beam kinematics with embedded discontinuity. The most appropriate choice of the mesoscale shell model can be further guided by the error-controlled adaptive finite element method for shell structures (by using a model error estimation, e.g. Bohinc, Ibrahimbegović and Brank (2009)), which could automatically find the most appropriate model for representing the particular local phenomenon under consideration.

The multi-scale procedure proposed in this chapter belongs to the class of weak coupling methods, where we carry out sequential computations. The results of the shell model computations, accounting for material and geometric localized instability, are stored to be used within the beam model softening response. As presented by numerical simulations, the performance of the proposed multi-scale computational approach is very satisfying. One of its main features is that detection and development of the softening plastic hinges in the frame is fully automatic, and spreads gradually in accordance with stress redistribution in the course of the nonlinear analysis. This is in contrast with many standard computational approaches to load limit, under the push-over and the full collapse analysis of frames, which rely on predefined locations of plastic hinges and the corresponding inelastic deformations.

REFERENCES

Armero, F. and D. Ehrlich. 2006. "Numerical Modeling of Softening Hinges in Thin Euler–Bernoulli Beams". *Computers & Structures* 84 (10–11): 641–56. doi:10.1016/j.compstruc.2005.11.010.

Bathe, K. J. 1996. *Finite Element Procedures*. Upper Saddle River, NJ: Prentice Hall.

Bohinc, U., A. Ibrahimbegović and B. Brank. 2009. "Model Adaptivity for Finite Element Analysis of Thin and Thick Plates Based on Equilibrated Boundary Stress Resultants". *Engineering Computations* 26 (1/2): 69–99.

Brank, Boštjan. 2005. "Nonlinear Shell Models with Seven Kinematic Parameters". *Computer Methods in Applied Mechanics and Engineering* 194 (21–24): 2336–62. doi:10.1016/j.cma.2004.07.036.

Brank, Boštjan, Djordje Perić and Frano B. B. Damjan. 1970. "On Large Deformations of Thin Elasto-Plastic Shells: Implementation of a Finite Rotation Model for Quadrilateral Shell Element". *International Journal for Numerical Methods in Engineering* 40 (4): 689–726. doi:10.1002/(sici)1097-0207(19970228)40:4<689::aid-nme85>3.3.co;2-z.

Darvall, Peter LePoer and Anuruddha Mend Priyantha. 1985. "Elastic-Plastic-Softening Analysis of Plane Frames". *Journal of Structural Engineering* 111 (4): 871–88. doi:10.1061/(asce)0733-9445(1985)111:4(871).

Dujc, J., B. Bostjan and A. Ibrahimbegovic. 2010. "A Multi-Scale Computational Model for Failure Analysis of Metal Frames That Includes Softening and Local Buckling". *Computational Methods in Applied Mechanics and Engineering* 1371–1385.

Hibbit, Karlsson and Sorensen. n.d. "Abaqus Manuals".

Hill, R. 1962. "Acceleration Waves in Solids". *Journal of the Mechanics and Physics of Solids* 10 (1). doi:10.1016/0022-5096(62)90024-8.

Hughes, T. J. R. 1987. *The Finite Element Methods*. Englewood-Cliffs, NJ: Prentice-Hall.

Ibrahimbegović, Adnan. 2009. *Nonlinear Solid Mechanics: Theoretical Formulations and Finite Element Solution Methods*. Springer.

Ibrahimbegović, Adnan, Gharzeddine Fadi and Chorfi Lotfi. 1998. "Classical Plasticity and Viscoplasticity Models Reformulated: Theoretical Basis and Numerical Implementation". *International Journal for Numerical Methods in Engineering* 42 (8): 1499–535. doi:10.1002/(sici)1097-0207(19980830)42:8<1499::aid-nme443>3.3.co;2-o.

Imamovic, Ismar, Adnan Ibrahimbegović and Esad Mesic. 2018. "Coupled Testing-Modeling Approach to Ultimate State Computation of Steel Structure with Connections for Statics and Dynamics". *Couples System Mechanics* 7 (5): 555–81.

Imamovic, Ismar, Adnan Ibrahimbegović and Esad Mesic. 2017. "Nonlinear Kinematics Reissner's Beam with Combined". *Computer and Structures* 12–20.

Korelc, J. n.d. "Fakulteta za gradbeništvo in geodezijo, AceGen, AceFem". http://www.fgg.uni-lj.si/Symech.

Lubliner, Jacob. 1990. *Plasticity Theory*. New York: Macmillan.

Luenberger, David G. 1984. *Linear and Non-Linear Programming*. 2nd ed. Reading, MA: Addison-Wesley.

Mandel, J. 1996. "Conditions de Stabilite´ et postulat de drucker". *Rheology and Soil Mechanics*. In: Kravtchenko J. and Sirieys P. M. (eds). Grenoble 1964: IUTAM Symposium.

Maugin, Gerard A. 1992. *The Thermomechanics of Plasticity and Fracture*. Cambridge: Cambridge university press.

Nagtegaal, J. C., D. M. Parks and J. R. Rice. 1974. "On Numerically Accurate Finite Element Solutions in the Fully Plastic Range". *Computer Methods in Applied Mechanic and Engineering* 4: 153–78.

Reddy, J. N. 2004. *An Introduction to Nonlinear Finite Element Analysis*. Oxford: Oxford University Press.

Rice, J. R. 1976. "The Localization of Plastic Deformation. In: Theory and Applied Mechanics". *14th IUTAM Congress*. Delft, North-Holland, Amsterdam, 207–20.

Simo, J. C. and T. J. R. Hughes. 1998. *Computational Inelasticity*. Springer.

Stakgold, Ivar. 1998. *Green's Functions and Boundary Value Problems*. New York: John Wiley & Sons.

Strang, G. 1986. *Introduction to Applied Mathematics*. Cambridge: Wellesley- Cambridge Press.

Truesdell, C. and W. Noll. 1965. *The Non-Linear Field Theories*. Berlin: Springer.

Wackerfuß, Jens. 2008. "Efficient Finite Element Formulation for the Analysis of Localized Failure in Beam Structures". *International Journal for Numerical Methods in Engineering* 73 (9): 1217–50. doi:10.1002/nme.2116.

Zienkiewicz, O. C. and R. L. Taylor. 2000. *The Finite Element Method*. 5th ed. Oxford: Butterworth-Heinemann.

Chapter 3

Reinforced concrete structures

The modeling of reinforced concrete (RC) structures in order to ensure the optional design with respect to structural integrity has a long tradition, which has been summarized in several reference works (e.g. Chen 1982; Hofstetter and Mang 1995). The finite element models are used extensively in this respect (e.g. Stein and Tihomirov 1999; Bažant 1986) where the equivalent properties of the RC composite are obtained by smearing the reinforcement within the concrete layers to provide a corresponding match of the total resistance. This can be acceptable with respect to the main design goal pertaining to ensuring the structural integrity, but not taking into consideration the accuracy of the behavior of each component of the RC composite. It is well known that the most efficient approach of this kind can be developed for RC structures in terms of stress resultants, i.e. moments, shear and membrane forces (e.g. Ibrahimbegović and Frey 1993a,b; Armero 1997). It is clear that any such stress-resultant RC model cannot provide reliable information on the detailed sequence of events pertaining to different failure mechanism creations, but only the corresponding bounds with respect to the structure or structural component resistance and integrity.

However, the current engineering design practice of RC structures no longer only requires proof of structural integrity, but also structural durability, either under long term, exploitation loading (such as aging problems) or under short term, extreme loading (such as fire exposure or aggressive agent exposure). For this, detailed information on concrete cracking, with the results on crack-spacing and opening or the bond-slip, is needed and cannot be provided by the smeared RC models. In this respect, a more detailed modeling of each of the RC composite ingredients, capable of describing the cracking of concrete, the bond-slip as well as the plastic yielding of steel is required.

One such model is proposed herein, combining the concrete finite element model capable of representing concrete cracking, both in terms of micro-cracks in the fracture process zone (FPZ) and macro-cracks (e.g. Brancherie and Ibrahimbegović 2009), with the elastoplastic model for the yielding and localized failure of reinforcement (e.g. Ibrahimbegović and Brancherie 2003; Ibrahimbegović 2009) coupled together with a simple finite element model for bond-slip based upon Drucker–Prager plasticity.

The goal is to compute crack-spacing and opening in a case when the external load is applied upon the concrete, and where only the cracking of the concrete will lead toward bond-slip activation. It is the crack-spacing and opening that pertains to the structural durability.

For bond-slip, a simple Drucker–Prager plasticity model is proposed as it has a reduced number of parameters with a clear physical meaning in respect to the essential features of bond-slip within the context of the RC composite. The simple bond-slip model, similar to the non-associative perfect plasticity, sensitive to confining pressure, is shown to be quite sufficient for the problems of interest where the bond-slip is not excessive. The failure mechanism with pull-out of the reinforcement is not considered herein, which allows for limitation of the analysis to the linear kinematics framework.

The new approach proposed herein combines all the ingredients of the RC composite into a single finite element model, which has practically the same outside appearance as the standard finite element model in terms of the macro finite element, but provides an inside detailed description of the strain field for each ingredient, concrete, steel or bond-slip. The latter is achieved with the corresponding enhancements of the macro element strain field, where the incompatible mode method (e.g. Ibrahimbegović and Wilson 1991) is used to handle the macro-crack appearance in the concrete, and the X-FEM method (e.g. Belytschko et al. 2001; Belytschko et al. 2002; Belytschko et al. 2002; Belytschko and Black 1999) is used for representing the bond-slip along a particular reinforcement bar.

The suitable form of the operator split procedure is then constructed in order to separate the main update procedure for the concrete displacement, which also implies the corresponding steel displacement with the frozen value of bond-slip, from the new slip computation. The latter is not equivalent to the local (Gauss quadrature point) plastic deformation, since it is computed from the corresponding slip equilibrium equations assembled from all the bond-slip contributions along the particular reinforcement bar. The key ingredient that allows for this kind of operator split solution procedure, separating the bond-slip computation from the RC displacement (with the value of slip kept fixed), pertains to describing the bond-slip with relative degrees of freedom. This is in sharp contrast with the standard parametrization of interface discontinuities (e.g. Belytschko et al. 2001; Jirásek 2002), which (as noted by Ibrahimbegović and Melnyk (2007)) would not allow for the use of the same operator split procedure.

3.1 CONCRETE MODELING WITH A DAMAGE MODEL

The concrete model has been developed in view of failure models for massive structures (e.g. Ibrahimbegović and Brancherie 2003) where the elastic behavior is followed by the creation of the FPZ with a large number of micro-cracks and subsequent final failure mode in terms of the macro-cracks. The FPZ is represented by the isotropic continuum damage model since the number of micro-cracks is considered sufficiently large and their orientation random. The macro-crack is represented by the displacement discontinuity across the FPZ, as done for a number of damage models for localized failure (e.g. Armero 1997; Jirásek and Zimmermann 2001; Oliver 2000; Oliver, Huespe and Sánchez 2006). The original feature of this concrete cracking model, with respect to those proposed previously, is its ability to combine two failure modes: FPZ and macro-cracks. Therefore, the proposed model not only accounts for damage-induced anisotropy, but also provides the most appropriate representation of failure mechanisms for massive structures, and allows for the reproduction of the proper sequence of failure modes creation and the induced stress redistribution leading to a better prediction of macro-cracks orientation (e.g. Ibrahimbegović and Brancherie 2003; Kucerova et al. 2009). The concrete failure model of this kind is built from three main ingredients: a non-homogeneous representation of the strain field, the strain energy in a quadratic form and, finally, a corresponding damage criterion indicating the creation of the FPZ and the creation of the failure-inducing macro-cracks.

(i) A non-homogeneous representation of the strain field is proposed in order to capture the strain localization at the onset of macro-cracks creation, leading to the softening phenomena. The total strain in the concrete $\varepsilon^c(\mathbf{x},t)$ can be written as the sum of the smooth part $\bar{\varepsilon}^c(\mathbf{x},t)$ and the singular part $\tilde{\mathbf{G}}\boldsymbol{\alpha}^c + \delta_\Gamma\boldsymbol{\alpha}^c$ corresponding to the displacement discontinuity contribution of the macro-cracks described by the surface Γ.

$$\varepsilon^c(\mathbf{x},t) = \overbrace{\underbrace{\bar{\varepsilon}^c(\mathbf{x},t)}_{\nabla^s \mathbf{u}} + \tilde{\mathbf{G}}(\mathbf{x})\boldsymbol{\alpha}^c(t)}^{\tilde{\varepsilon}^c} + \delta_\Gamma(\mathbf{x})\left(\boldsymbol{\alpha}^c(t)\otimes\mathbf{n}\right)^s \tag{3.1}$$

where Γ denotes the surface of discontinuity of the displacement field, δ_Γ is the Dirac-Delta function on Γ and $\boldsymbol{\alpha}^C$ denotes the displacement jump on Γ. $\tilde{\mathbf{G}}(\mathbf{x})$ corresponds to the influence function and is chosen in order to limit the macro-cracks influence on the FPZ; in the finite element setting, the representation in terms of the enhanced constant strain triangle (CST) element is chosen (e.g. Zienkiewics and Taylor 2005) with embedded discontinuity (see Figure 3.1).

Such a choice for the embedded discontinuity permits the formulation of the corresponding approximation for $\tilde{\mathbf{G}}(\mathbf{x})$ by using:

$$\tilde{\mathbf{G}} = -\sum_{i\in\Omega^{e+}} \nabla^s N_i \tag{3.2}$$

The virtual strain field $\gamma^c(\mathbf{x},t)$ is typically constructed in a slightly modified form (Ibrahimbegović and Brancherie 2003; Ibrahimbegović and Wilson 1991), which guarantees that the patch test condition for convergence (e.g. Zienkiewics and Taylor 2005; Ibrahimbegović 2009) is satisfied:

$$\gamma^c(\mathbf{x},t) = \bar{\gamma}^c(\mathbf{x},t) + \hat{\mathbf{G}}(\mathbf{x})\beta^c(t) + \delta_{\Gamma^e}(\mathbf{x})\left(\beta^c(t)\otimes\mathbf{n}\right)^s \tag{3.3}$$

where $\hat{\mathbf{G}}(\mathbf{x})$ is the modified version of the function $\tilde{\mathbf{G}}(\mathbf{x})$ and β^c denotes the virtual displacement discontinuity.

(ii) The strain energy density of the concrete model $\psi^c(\mathbf{x},t)$ is constructed by taking into account both the smooth field and the discontinuity contributions:

$$\psi^c(\mathbf{x},t) = \boldsymbol{\sigma}\cdot\varepsilon^c - \frac{1}{2}\boldsymbol{\sigma}\cdot D\boldsymbol{\sigma} + \frac{1}{2}\bar{K}\bar{\zeta}^2 + \frac{1}{2}\bar{\bar{K}}\bar{\bar{\zeta}}^2 \tag{3.4}$$

where $\boldsymbol{\sigma}$ is the Cauchy stress tensor, D is the damage compliance tensor, $\bar{K}$ and $\bar{\bar{K}}$ are, respectively, the hardening and softening moduli for the FPZ and macro-crack-induced discontinuity, whereas $\bar{\zeta}$ and $\bar{\bar{\zeta}}$ are the corresponding internal variables governing the evolution of these two mechanisms.

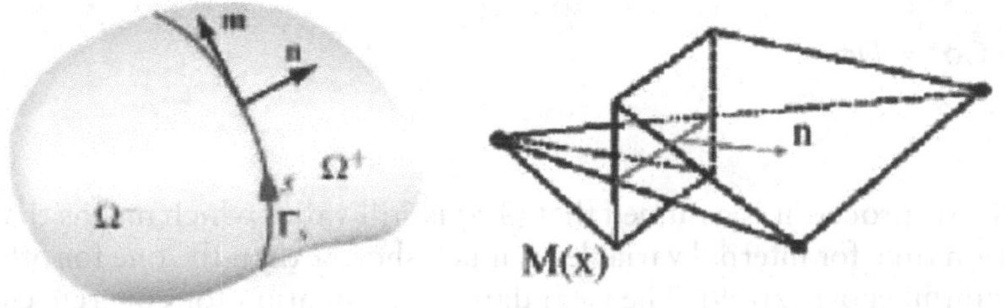

Figure 3.1 Macro-cracks with the fracture process zone and its finite element representation.

(iii) The damage criteria indicating the creation of the FPZ and the macro-crack are written, respectively, as:

$$\bar{\phi}^c(\boldsymbol{\sigma},\bar{q}) = \sqrt{\boldsymbol{\sigma}\cdot D^e\boldsymbol{\sigma}}^{\,c} - \frac{1}{\sqrt{E}}(\bar{\sigma}_f - \bar{q}) \le 0$$

$$\bar{\bar{\phi}}_n{}^c(\boldsymbol{\sigma},\bar{\bar{q}}) = \mathbf{t}_\Gamma\cdot\mathbf{n} - (\bar{\bar{\sigma}}_f - \bar{\bar{q}}) \le 0 \tag{3.5}$$

$$\bar{\bar{\phi}}_s{}^c(\boldsymbol{\sigma},\bar{\bar{q}}) = |\mathbf{t}_\Gamma\cdot\mathbf{m}| - \left(\bar{\bar{\sigma}}_s - \frac{\bar{\bar{\sigma}}_s}{\bar{\bar{\sigma}}_f}\bar{\bar{q}}\right) \le 0$$

where:

$D^e = C^{-1}$ is the elastic damage compliance tensor (equal to the inverse of the elasticity tensor),
E is Young's modulus,
$\mathbf{t}_\Gamma = \boldsymbol{\sigma}\mathbf{n}$ is the driving traction at the discontinuity,
$\bar{\sigma}_f$ is the elasticity limit indicating the creation of the FPZ,
$\bar{\bar{\sigma}}_f$ and $\bar{\bar{\sigma}}_s$ are the threshold values indicating the beginning of the localized failure in traction (mode I) and in shear (mode II); whereas, $\bar{q}$ and $\bar{\bar{q}}$ are the stress-like variables governing the evolution of the stress threshold.

The remaining governing equations can be obtained using the standard thermodynamics arguments based on the second principle defining dissipation in concrete $\mathcal{D}^c$(e.g. Ibrahimbegović 2009) according to:

$$0 \le \mathcal{D}^c = \sigma\cdot\dot{\varepsilon}^c - \frac{d}{dt}\psi^c = \dot{\sigma}\cdot(D\sigma - \varepsilon^c) + \frac{1}{2}\sigma\cdot\dot{D}\sigma + \bar{q}\dot{\bar{\zeta}} + \bar{\bar{q}}\dot{\bar{\bar{\zeta}}}\delta_\Gamma \tag{3.6}$$

where $\bar{q} = -\bar{K}\bar{\zeta}$ and $\bar{\bar{q}} = -\bar{\bar{K}}\bar{\bar{\zeta}}$. The last equation applies to the linear softening law, which is often replaced by exponential softening (Ibrahimbegović and Brancherie 2003) in order to ensure the robustness of the computations. However, either case of softening law can be characterized by physically sound parameters, such as the total fracture energy G_f (Ibrahimbegović 2009). For the elastic process, as seen in (3.6), where internal variables remain frozen with $\dot{D} = 0, \dot{\bar{\zeta}} = 0, \dot{\bar{\bar{\zeta}}} = 0$:

$$\varepsilon^c = D\sigma \tag{3.7}$$

From (3.1), it can be worked out that:

$$D = \bar{D} + \bar{\bar{D}}\delta_\Gamma$$

$$\tilde{\varepsilon}^c := \bar{\varepsilon}^c + \tilde{G}\alpha^c = \bar{D}\boldsymbol{\sigma} \tag{3.8}$$

$$\alpha^c = \bar{\bar{D}}\mathbf{t}_\Gamma$$

For the inelastic process it is assumed that (3.8) is still valid, which means that the characteristic evolution time for internal variables is much shorter than the one for other state variables (e.g. Ibrahimbegović 2009). The total damage dissipation for concrete can be defined from (3.6):

$$0 \le \mathcal{D}^c = \underbrace{\frac{1}{2}\sigma \cdot \dot{\bar{D}}\sigma + \bar{q}\dot{\bar{\zeta}}}_{\bar{\mathcal{D}}^c} + \underbrace{\left(\frac{1}{2}\mathbf{t}_\Gamma \cdot \dot{\bar{\bar{D}}}\,\mathbf{t}_\Gamma + \bar{\bar{q}}\dot{\bar{\bar{\zeta}}}\right)\delta_\Gamma}_{\bar{\bar{\mathcal{D}}}^c} \tag{3.9}$$

Applying the principle of maximum dissipation, the corresponding evolution equations for damage variables for concrete cracking can be obtained according to:

$$\max_{\bar{\phi}^c=0} \bar{\mathcal{D}}^c \Rightarrow \begin{cases} \dot{\bar{D}}\boldsymbol{\sigma} = \dot{\bar{\gamma}}\dfrac{\partial\bar{\phi}^c}{\partial\boldsymbol{\sigma}} \\ \dot{\bar{\zeta}} = \dot{\bar{\gamma}}\dfrac{\partial\bar{\phi}^c}{\partial\bar{q}} \\ \dot{\bar{\gamma}} \ge 0, \bar{\phi}^c \le 0, \dot{\bar{\gamma}}\bar{\phi}^c = 0 \end{cases} \tag{3.10}$$

and

$$\max_{\bar{\bar{\phi}}_n{}^c=0, \bar{\bar{\phi}}_m{}^c=0} \bar{\bar{\mathcal{D}}}^c \Rightarrow \begin{cases} \dot{\bar{\bar{D}}}\mathbf{t}_\Gamma = \displaystyle\sum_{a=m,n} \dot{\bar{\bar{\gamma}}}_a \frac{\partial\bar{\bar{\phi}}_a^c}{\partial\mathbf{t}_\Gamma} \\ \dot{\bar{\bar{\zeta}}} = \displaystyle\sum_{a=m,n} \dot{\bar{\bar{\gamma}}}_a \frac{\partial\bar{\bar{\phi}}_a{}^c}{\partial\bar{\bar{q}}} \end{cases} \tag{3.11}$$

In the finite element implementation based on the enhanced CST element (Figure 3.1), the internal variables for the FPZ are computed with a single quadrature point. A single value per element is also used for each discontinuity inducing macro-crack in both mode I and mode II. For a typical time step $\Delta t = t_{n+1} - t_n$. The computation is started by assuming the trial elastic step:

$$\begin{aligned} &\bar{\gamma}_{n+1}^{c,\text{trial}} = 0 \Rightarrow \bar{D}_{n+1}^{\text{trial}} = \bar{D}_n, \bar{\zeta}_{n+1}^{\text{trial}} = \bar{\zeta}_n \\ &\boldsymbol{\sigma}_{n+1}^{\text{trial}} = \bar{D}_n^{-1}\tilde{\varepsilon}_{n+1}^{c,(i)}, \quad \bar{q}_{n+1}^{\text{trial}} = -\bar{K}\bar{\zeta}_n = \bar{q}_n \\ &\bar{\phi}_{n+1}^{c,\text{trial}} = \sqrt{\boldsymbol{\sigma}_{n+1}^{\text{trial}} \cdot D^e \boldsymbol{\sigma}_{n+1}^{\text{trial}}} - \frac{1}{\sqrt{E}}(\bar{\sigma}_f - \bar{q}_n) \end{aligned} \tag{3.12}$$

If $\bar{\gamma}_{n+1}^{c,\text{trial}} \le 0$, the step is indeed elastic and no change of the FPZ will result. However, if $\bar{\gamma}_{n+1}^{c,\text{trial}} \ge 0$, a correct (positive) value has to be obtained $\bar{\gamma}_{n+1}^c$ as:

$$\bar{\gamma}_{n+1}^c = \frac{\bar{\phi}_{n+1}^{c,\text{trial}}}{\dfrac{\partial\bar{\phi}_{n+1}^{c,\text{trial}}}{\partial\boldsymbol{\sigma}} \cdot \bar{D}_{n+1}^{-1} \dfrac{\partial\bar{\phi}_{n+1}^{c,\text{trial}}}{\partial\boldsymbol{\sigma}} + \bar{K}\left(\dfrac{\partial\bar{\phi}_{n+1}^{c,\text{trial}}}{\partial\bar{q}}\right)^2} \tag{3.13}$$

and the corresponding change of smooth damage variables according to:

$$\begin{aligned} &\bar{D}_{n+1}\boldsymbol{\sigma}_{n+1} = \bar{D}_n\boldsymbol{\sigma}_{n+1} + \bar{\gamma}_{n+1}^c \frac{\partial\bar{\phi}_{n+1}}{\partial\boldsymbol{\sigma}_{n+1}} \\ &\bar{\zeta}_{n+1} = \bar{\zeta}_n + \bar{\gamma}_{n+1}^c \frac{\partial\bar{\phi}_{n+1}}{\partial\bar{q}_{n+1}} \end{aligned} \tag{3.14}$$

The final stress from the obtained results can be then obtained as:

$$\boldsymbol{\sigma}_{n+1}^{c} = \bar{D}_{n+1}^{-1}\tilde{\varepsilon}_{n+1}^{c,(i)} \tag{3.15}$$

It is from (3.15) that one can test for the eventual evolution of the macro-cracks. In this sense, one needs to first compute the trial value of traction.

$$\mathbf{t}_{n+1}^{\text{trial}} = -\int_{\Omega^e} \hat{\mathbf{G}}^T \boldsymbol{\sigma}_{n+1}^{c} dV \tag{3.16}$$

and the corresponding trial values of the discontinuity damage criteria by assuming that no change in the macro-crack would occur:

$$\bar{\bar{\phi}}_n^{c,\text{trial}}\left(\boldsymbol{\sigma},\bar{\bar{q}}\right) = \mathbf{t}_{n+1}^{\text{trial}} \cdot \mathbf{n} - \left(\bar{\bar{\sigma}}_f - \bar{\bar{q}}_n\right) \tag{3.17a}$$

$$\bar{\bar{\phi}}_n^{c,\text{trial}}\left(\boldsymbol{\sigma},\bar{\bar{q}}\right) = \left|\mathbf{t}_{n+1}^{trial} \cdot \mathbf{m}\right| - \left(\bar{\bar{\sigma}}_s - \frac{\bar{\bar{\sigma}}_s}{\bar{\bar{\sigma}}_f}\bar{\bar{q}}_n\right) \tag{3.17b}$$

If these trial values remain negative, no change will occur for any of the fracture modes (mode I or mode II). At the opposite end, one has to compute the corresponding values of the damage multipliers $\bar{\gamma}_{n+1}^{c}$ and $\bar{\bar{\gamma}}_{n+1}^{c}$ giving the final contribution of each mode in the current step. The final result of such a computation concerns the potential evolution of the discontinuity parameters, $\boldsymbol{\alpha}_{n+1}^{c}$, due to the macro-crack evolution, which in return will also modify the total deformation field $\varepsilon^c(\mathbf{x},t) = \bar{\varepsilon}^c + \tilde{\mathbf{G}}\boldsymbol{\alpha}^c$. Now, the concrete part contribution to the RC residual $\mathbf{r}_{n+1}^{c}$can be computed according to:

$$\mathbf{r}_{n+1}^{c} = \int_{\Omega^e} \mathbf{N}^T \mathbf{b}_{n+1} dV - \int_{\Omega^e} \mathbf{B}^T \boldsymbol{\sigma}_{n+1}^{c} dV \tag{3.18}$$

where $\boldsymbol{\sigma}_{n+1}^{c}$ is the final admissible value of stress from the standpoint of the all damage criteria $\bar{\phi}^c, \bar{\bar{\phi}}_n^c, \bar{\bar{\phi}}_m^c$. The concrete part also provides a corresponding contribution to the tangent moduli of the RC composite, which can be obtained upon the static condensation of the macro-crack-induced displacement discontinuity parameters according to:

$$\tilde{\mathbf{K}}_{n+1}^{c} = \mathbf{K}_{n+1}^{c} - \bar{\mathbf{F}}_{n+1}^{T}\mathbf{H}_{n+1}^{-1}\mathbf{F}_{n+1} \tag{3.19}$$

with:

$$\begin{aligned} \mathbf{K}_{n+1}^{c} &= \int_{\Omega^e} \mathbf{B}^T \bar{\mathscr{C}}_{n+1}^{d} \mathbf{B} dV, \quad \bar{\mathbf{F}}_{n+1}^{T} = \int_{\Omega^e} \mathbf{B}^T \bar{\mathscr{C}}_{n+1}^{d} \tilde{\mathbf{G}} dV \\ \mathbf{H}_{n+1} &= \int_{\Omega^e} \mathbf{G}^T \bar{\mathscr{C}}_{n+1}^{d} \tilde{\mathbf{G}} dV + \bar{\bar{\mathrm{D}}}_{n+1}^{-1}, \quad \mathbf{F}_{n+1} = \int_{\Omega^e} \mathbf{G}^T \bar{\mathscr{C}}_{n+1}^{d} \mathbf{B} dV \end{aligned} \tag{3.20}$$

where $\bar{\mathscr{C}}_{n+1}^{d}$ denotes the consistent tangent modulus of the bulk material (outside the discontinuity). As $\tilde{\mathbf{K}}_{n+1}^{c}$, the tangent stiffness of the concrete part, is non-symmetric it can provide detailed information on different concrete fracture modes and increased computational time is here of secondary importance.

3.1.1 A Numerical Example of Small Experiments

A simple example of concrete specimen failure in a simple tension test illustrates the main advantages of such a concrete model for a RC structure in providing crack-spacing information and a decrease in resistance with an increase in crack-opening is shown. A square specimen of length and width equal to 200 mm, with a unit thickness, is chosen having the mechanical properties as stated in Table 3.1.

Owing to symmetry, only half of the specimen is considered in the computations (see Figure 3.2). The computations are performed by incrementing the imposed displacement at the free-end of the specimen in order to simulate a displacement-controlled tension test. For the loading program that includes both loading and unloading, the force–displacement response computed for coarse and irregular vs. fine and regular mesh remains practically the same (see Figure 3.3).

The only difference concerns the computed total crack lengths for both a coarse and fine finite element mesh (see Figure 3.2). This crack length difference appears as the consequence of the adopted strategy to allow for crack creation only at the element centers. The presented concrete model (with additional inelastic mechanism representing the FPZ) remains capable of providing an invariant computed response (see Figure 3.3).

3.2 BOND-SLIP MODEL

In the spirit of the bond-slip role for the RC composite, several features have to be adopted. First, the bond-slip element should be reduced to the concrete–steel interface and should be thus without thickness. The bond-slip should not be activated as long as the dilatations in the adjacent steel and concrete remain the same. The bond-slip is represented through the shear strain, and its resistance can be increased by the normal pressure-induced confinement, as well as decreased through transverse tension. This kind of interface model is developed by combining the finite element interpolation for contact problems (Ibrahimbegović and

Table 3.1 Material Properties for Simple Tension Test

Continuum model		*Localized model*	
Young's modulus	38 GPa	$\bar{\bar{\sigma}}_f$	2.55 MPa
Poisson ratio	0.18		2.5 MPa (weakened element)
$\bar{\sigma}_f$	2 MPa	$\bar{\bar{\sigma}}_s/\bar{\bar{\sigma}}_f$	0.3
$\bar{K}$	1000 MPa	Fracture energy G_f	1.275 N/mm

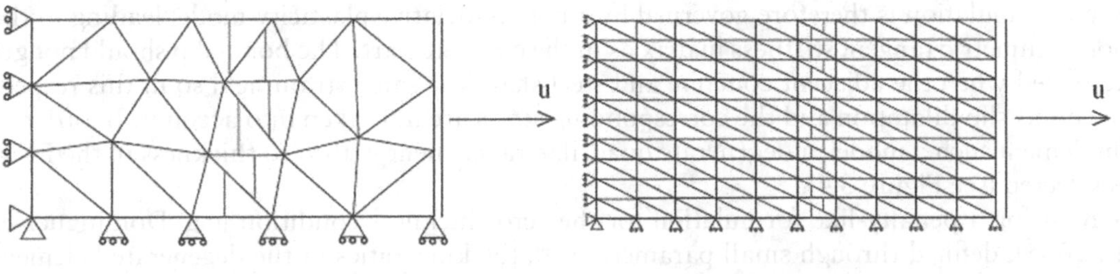

Figure 3.2 Finite element model and computed macro-crack representation for coarse irregular mesh and fine regular mesh.

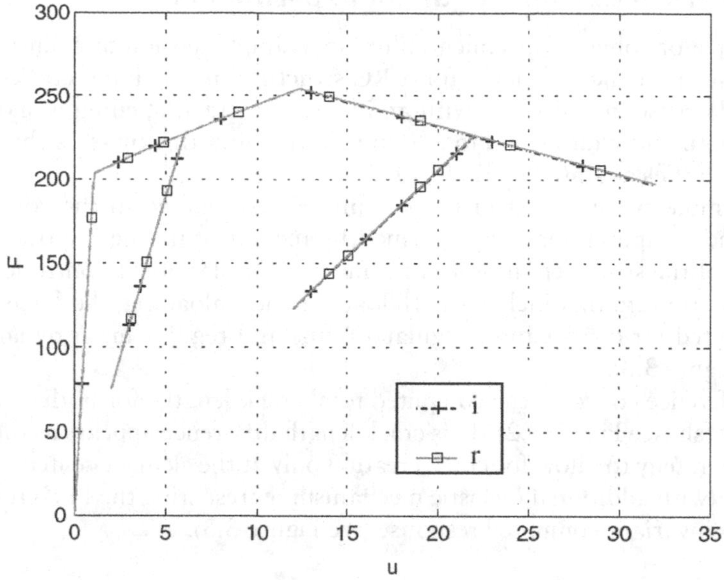

Figure 3.3 Load–displacement diagram for regular and irregular meshes.

Wilson 1992) with the thermodynamics-based formulation of the inelastic slip mechanism. The latter and essential features are here represented by the Drucker–Prager non-associative plasticity model (Ibrahimbegović 2009). The equation for the bond-slip model then reads:

$$\bar{\phi}^{bs}(\boldsymbol{\sigma}) = \sqrt{\frac{3}{2}}\left\|\operatorname{dev}[\boldsymbol{\sigma}]\right\| - \kappa(p) \le 0, \kappa(p) = \sigma_c - pf, p = \frac{1}{3}\operatorname{tr}[\boldsymbol{\sigma}] \tag{3.21}$$

where $\operatorname{dev}[\boldsymbol{\sigma}]$ and $\operatorname{tr}[\boldsymbol{\sigma}]$ denote, respectively, the deviatoric and spherical parts of the stress tensor, and σ_c is the sliding resistance at zero pressure, with no confinement. It is evident that with the increase of pressure, the sliding resistance with a factor proportional to the model parameter f will also increase. On the contrary, the tension stress across the interface can actually reduce the lateral resistance (see Figure 3.5). Once the slip criterion is satisfied, the slip is activated. This kind of slip is considered as the internal variable of the model, and its evolution is governed by the von Mises plasticity potential with:

$$\dot{\varepsilon}^p = \dot{\gamma}\frac{\partial \bar{\phi}^{vM}}{\partial \boldsymbol{\sigma}}, \bar{\phi}^{vM}(\boldsymbol{\sigma}) = \sqrt{\frac{3}{2}}\left\|\operatorname{dev}[\boldsymbol{\sigma}]\right\| - \sigma_y \tag{3.22}$$

Slip accumulation is therefore governed by a non-associative plasticity model leading to the non-symmetric tangent stiffness matrix as in the concrete part. The bond-slip should not get activated when the adjacent concrete and steel share the same strain field so in this respect the bond-slip dilatation and the corresponding stress are not taken into account. In order to implement such a model, a degenerate quadrilateral element with zero thickness in the FEM is selected (see Figure 3.4).

By using a penalty-like formulation for the zero-thickness condition (e.g. Dominguez et al. 2005), defined through small parameter h^{pen}, the kinematics of the degenerated element can be constructed by following the standard procedure for isoparametric quadrilateral elements (e.g. Zienkiewics and Taylor 2005; Ibrahimbegović 2009)). The shape function derivatives which are required for strain computations, can be formulated according to:

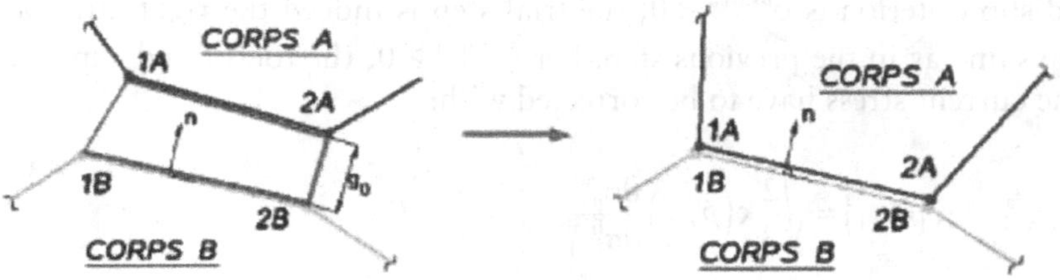

Figure 3.4 Degenerated quadrilateral bond-slip interface element.

$$\frac{\partial N_i}{\partial x} = \left[\left(\frac{\partial N_i}{\partial \zeta}\right)\left(\frac{\partial y}{\partial \eta}\right) - \left(\frac{\partial N_i}{\partial \eta}\right)\left(\frac{\partial y}{\partial \zeta}\right)\right] / j$$

$$\frac{\partial N_i}{\partial y} = \left[\left(\frac{\partial N_i}{\partial \zeta}\right)\left(\frac{\partial x}{\partial \eta}\right) - \left(\frac{\partial N_i}{\partial \eta}\right)\left(\frac{\partial x}{\partial \zeta}\right)\right] / j \tag{3.23}$$

$$j = \left(\frac{\partial x}{\partial y}\right)\left(\frac{\partial y}{\partial \eta}\right) - \left(\frac{\partial x}{\partial \eta}\right)\left(\frac{\partial y}{\partial \zeta}\right)$$

where we obtain the derivatives with respect to the natural coordinates:

$$\frac{\partial y}{\partial \zeta} = \left[-y_1 + y_2 + \left(y_3 + h^{\text{pen}}\right) - \left(y_4 + h^{\text{pen}}\right)\right] / 4 + y_0 \eta$$

$$\frac{\partial y}{\partial \eta} = \left[-y_1 - y_2 + \left(y_3 + h^{\text{pen}}\right) + \left(y_4 + h^{\text{pen}}\right)\right] / 4 + y_0 \zeta \tag{3.24}$$

$$y_0 = \left[-y_1 - y_2 + \left(y_3 + h^{\text{pen}}\right) - \left(y_4 + h^{\text{pen}}\right)\right] / 4$$

As h^{pen} allows for one to carry out this kind of computation even for zero thickness of the bond-slip element (nodal coordinates coincide for the corresponding nodes pairs), from the given current iterative displacements of the four nodes of the bond-slip element $\mathbf{d}_{i,n+1}^{(i)}, i = 1,2,3,4$, the corresponding best iterative guess of the total strain field for this element can be computed as:

$$\begin{pmatrix} \varepsilon_{yy}^{bs} \\ \gamma_{xy}^{bs} \end{pmatrix} =: \varepsilon_{n+1}^{bs,(i)} = \sum_{i=1}^{4} \mathbf{B}_i^{bs} d_{i,n+1}^{(i)}, \mathbf{B}_i^{bs} = \begin{bmatrix} 0 & \dfrac{\partial N_i}{\partial y} \\ \dfrac{\partial N_i}{\partial y} & \dfrac{\partial N_i}{\partial x} \end{bmatrix}, d_{i,n+1}^{(i)} : \left\{d_1^c, d_1^s, d_2^c, d_2^s\right\} \tag{3.25}$$

where $\mathbf{B}_i^{bs}$ is the appropriate form of the strain–displacement matrix for the chosen ordering of concrete and steel nodal displacements. The trial value of the bond-slip stress field is then computed by assuming that the slip remains frozen with:

$$\boldsymbol{\sigma}_{n+1}^{\text{trial}} = C^{bs}\left(\varepsilon_{n+1}^{bs,(i)} - \varepsilon_n^{bs,p}\right)$$

$$\Leftrightarrow p_{n+1}^{\text{trial}} = Ktr\left[\varepsilon_{n+1}^{bs,(i)}\right], \quad \text{dev}\left[\boldsymbol{\sigma}_{n+1}^{\text{trial}}\right] = \mu\left(\text{dev}\left[\varepsilon_{n+1}^{bs,(i)}\right] - \varepsilon_n^{bs,p}\right) \tag{3.26}$$

$$\phi^{bs,\text{trial}} = \sqrt{\frac{3}{2}}\left\|\text{dev}\left[\boldsymbol{\sigma}_{n+1}^{\text{trial}}\right]\right\| - \kappa\left(p_{n+1}^{\text{trial}}\right)$$

If the bond-slip criterion is $\phi^{bs,\text{trial}} < 0$, the trial step is indeed the right guess and the slip remains the same as in the previous step. For $\phi^{bs,\text{trial}} \geq 0$, the total bond-slip in the current step and the current stress have to be corrected with:

$$p_{n+1} = p_{n+1}^{\text{trial}}, \text{dev}\left[\boldsymbol{\sigma}_{n+1}\right] = \sqrt{\frac{2}{3}}\kappa\left(p_{n+1}^{\text{trial}}\right)\frac{\partial\bar{\phi}_{n+1}^{vM}}{\partial\boldsymbol{\sigma}_{n+1}^{\text{trial}}}$$
$$\boldsymbol{\sigma}_{n+1}^{bs} = p_{n+1}\mathbf{1} + \text{dev}\left[\boldsymbol{\sigma}_{n+1}\right] \tag{3.27}$$

The consistent tangent can then be computed for the active slip case according to:

$$\mathbf{K}_{n+1}^{bs} = \int_{\Gamma^e} \mathbf{B}^{bs,T}\bar{C}_{n+1}^{ep}\mathbf{B}^{bs}dA,$$
$$\bar{C}_{n+1}^{ep} = K\mathbf{1}\otimes\mathbf{1} + 2G\left[\mathbf{I} - \frac{1}{3}\mathbf{1}\otimes\mathbf{1} - \frac{\partial\bar{\phi}_{n+1}^{vM}}{\partial\boldsymbol{\sigma}_{n+1}^{\text{trial}}}\otimes\frac{\partial\bar{\phi}_{n+1}^{vM}}{\partial\boldsymbol{\sigma}_{n+1}^{\text{trial}}}\right] + K\sqrt{\frac{2}{3}}\kappa' p_{n+1}\frac{\partial\bar{\phi}_{n+1}^{vM}}{\partial\boldsymbol{\sigma}_{n+1}^{\text{trial}}}\otimes\mathbf{1} \tag{3.28}$$

3.3 STEEL MODEL WITH CLASSICAL ELASTOPLASTICITY

The standard elastoplastic model with isotropic hardening, where the stress is defined from the total and plastic strain, is chosen for steel making sure that the plastic admissibility of stress is guaranteed (Figure 3.5).

$$\sigma_{n+1}^{s} = E\left(\varepsilon_{n+1}^{s} - \varepsilon_{n+1}^{s,p}\right)$$
$$\phi^{s}(\sigma) = |\sigma| - \left(\sigma_y + K\zeta\right) \leq 0, \gamma_{n+1} \geq 0, \gamma_{n+1}\phi^{s} = 0 \tag{3.29}$$

The value of this total strain is computed from the simplest two-node displacement field representation for the bar with:

$$\varepsilon_{n+1}^{s} = \frac{d_{2,n+1}^{s} - d_{1,n+1}^{s}}{l^{e}} \tag{3.30}$$

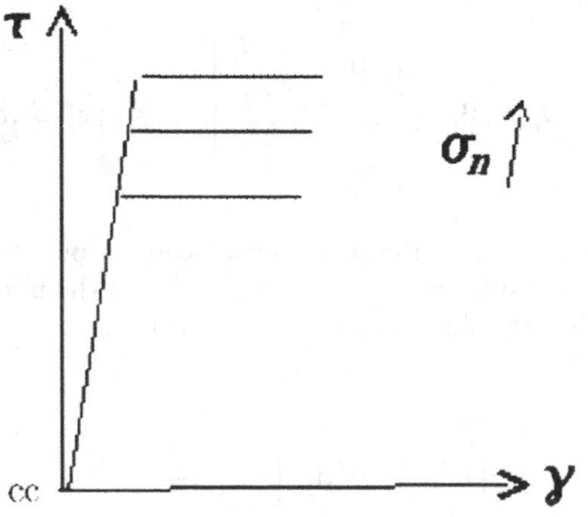

Figure 3.5 Elastoplastic bond-slip law accounting for transverse pressure confinement.

The plastic strain is computed from the standard procedure (e.g. Ibrahimbegović 2009) for the model of this kind:

$$\begin{aligned}
\sigma_{n+1}^{s,\text{trial}} &= E\left(\varepsilon_{n+1}^{s} - \varepsilon_{n}^{s,p}\right) \\
\phi_{n+1}^{s,\text{trial}} &= \left|\sigma_{n+1}^{s,\text{trial}}\right| - \left(\sigma_y + K\zeta_n\right) \\
\gamma_{n+1} &= \frac{\phi_{n+1}^{s,\text{trial}}}{K+E} \geq 0 \\
\varepsilon_{n+1}^{s,p} &= \varepsilon_{n}^{s,p} + \gamma_{n+1}\text{sign}\left(\sigma^{s}\right)
\end{aligned} \tag{3.31}$$

The steel bar stiffness matrix pertaining to steel displacement d_i^s can then be written as:

$$K_{n+1}^{s} = \frac{C_{n+1}^{\text{ep}}A}{l}\begin{bmatrix} I & -I \\ -I & I \end{bmatrix} \tag{3.32}$$

where l and A are, respectively, the reinforcement bar length and cross-section and C^{ep} is the corresponding value of the elastoplastic tangent modulus for elastic or plastic step, respectively:

$$C_{n+1}^{\text{ep}} = \begin{cases} E & \text{if } \gamma_{n+1} = 0 \\ \dfrac{EK}{E+K} & \text{if } \gamma_{n+1} > 0 \end{cases} \tag{3.33}$$

We can further study the failure mode related to reinforcement bar failure if we consider the post-peak softening response for the steel bar; the details of such a model enhancement are very much similar to the corresponding modification of the concrete softening model, and are thus omitted (e.g. Ibrahimbegović and Brancherie (2003) for a detailed discussion).

3.4 REINFORCED CONCRETE MODEL

It is important to make a correct assembly using the corresponding contributions from each of the presented ingredients of the RC model. Such an assembly should be able to provide the correct representation for a number of possible damage modes along with the right order of activation between the concrete cracking, bond-slip and steel yielding. It is anticipated that it will capture the complete failure mode for the RC, which is caused either by the failure of the concrete with the creation of a macro-crack or a failure of the steel with the rupture of the reinforcement. A potential failure mode of RC related to the eventual pull-out of the reinforcement bar is not considered, assuming either that special provisions are made to ensure steel reinforcement anchorage or that the concrete pressure confinement at the end of each steel bar is sufficient to prevent this kind of extreme bond-slip. Such a hypothesis implies that the slips remain equal to zero at both ends of the bar and sufficiently small in-between so that the analysis of the failure mode can be carried out within the framework of the small displacement gradient theory or linearized kinematics. It is important to note that even such a small amount of slip along the inner portion of steel reinforcement bar can lead to significant stress redistribution between the concrete and the steel, and, more importantly, provide the reliable information on final crack-spacing and opening, which will be very valuable for durability studies.

3.4.1 Kinematics of Bond-Slip Along Reinforcement Bar and its X-FEM Representation

A typical finite element for the composite RC model (see Figure 3.6) is considered, represented by structured finite element mesh.

The black nodes of the finite element mesh provide the corresponding description of the concrete displacement field, whereas the white nodes are used to describe the displacement field of the steel bar. It is considered that the bond-slip element possesses two black nodes attached to concrete and two white nodes attached to steel. It is important to note that the bond-slip element has a zero thickness, and therefore the black and white nodes share the same sets of coordinates. The corresponding nodal displacements, however, are different due to a bond-slip between the concrete and the steel. Such a displacement field can conveniently be represented by using X-FEM (see Belytschko et al. 2001; Belytschko et al. 2002; Belytschko and Black 1999) where the global bond-slip along a particular reinforcement bar is represented by the extended finite element interpolation; for example, with the horizontal steel bar placed parallel to x-axis at the distance $\bar{y}$, the X-FEM representation of the displacement covering one entire composite element for RC can be written as:

$$\mathbf{u}(\mathbf{x})\big|_{\Omega^e} = \sum_{i=1}^{4} N_i(\mathbf{x})\left(\mathbf{d}_i^c + H_{\bar{y}}(y)\boldsymbol{\alpha}(\mathbf{x})\right), H_{\bar{y}}(y) = \begin{cases} 1, y < \bar{y} \\ 0, y \geq \bar{y} \end{cases} \tag{3.34}$$

where $\mathbf{N}_i(\mathbf{x})$ represents the standard isoparametric shape functions of the corresponding element used for concrete, $\mathbf{d}_i^c$ are the nodal values of the concrete displacement, $H_{\bar{y}}(y)$ is the Heaviside function and $\boldsymbol{\alpha}(x)$ is the distribution of slip along the part of the reinforcement bar that belongs to a particulate element. The Heaviside function $H_{\bar{y}}(y)$ in (3.34) introduces a displacement discontinuity in y direction due to bond-slip, whereas $\boldsymbol{\alpha}(x)$ defines the corresponding slip variation in x direction. It is important to emphasize that $\boldsymbol{\alpha}(x)$ is a global field describing the complete variation of the slip along a particular steel bar. The expression in (3.34) represents the part of that field that belongs within a chosen element Ω^e, which is constructed by appealing to the partition of unity properties (e.g. Melenk and Babuška 1996) of the isoparametric shape functions $N_i(\mathbf{x})$. Therefore, the appropriate local representation of the global slip field can be obtained, and thus the standard finite element method for the computations related to the slip field is exploited.

Such a discontinuity of the displacement field to account for bond-slip will impose the corresponding strain discontinuity, which can be written as follows:

$$\varepsilon_{xx}(x,y)\big|_{\Omega^e} = \frac{\partial \bar{u}(x,y)}{\partial x} + H_{\bar{y}}(y)\frac{\partial \alpha(x)}{\partial x} \tag{3.35}$$

$$\varepsilon_{yy}(x,y)\big|_{\Omega^e} = \frac{\partial \bar{v}(x,y)}{\partial y} + \delta_{\bar{y}}(y)\beta(x) \tag{3.36}$$

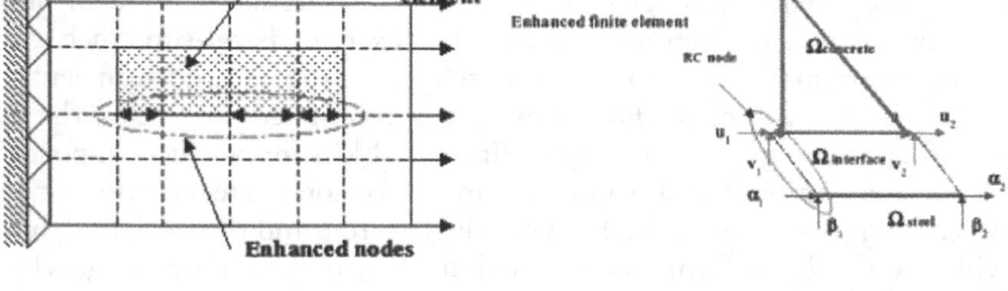

Figure 3.6 Finite element model of reinforced concrete constructed by a structured mesh, with nodes that belong to concrete and nodes that belong to steel.

$$\gamma_{xy}(x,y)\Big|_{\Omega^e} = \frac{\partial \bar{u}(x,y)}{\partial y} + \frac{\partial \bar{v}(x,y)}{\partial x} + \delta_{\bar{y}}(y)\alpha(x,\bar{y}) + H_{\bar{y}}(y)\frac{\partial \beta(x,\bar{y})}{\partial x} \tag{3.37}$$

where $\delta_{\bar{y}}(y)$ is the Dirac function obtained as the corresponding derivative of the Heaviside function. The main advantage of the X-FEM representation of the bond-slip field is the ability to provide the appropriate coupling between all the bond-slip elements along a particular steel bar. This is shown for the simplest case (chosen for clarity, but without any essential restriction), where the bond-slip coincides with an element edge with $\bar{y} = 0$. In this particular case, the slip field function over a single element can be represented by a linear polynomial expression, which is uniquely defined by two sets of nodal values $\boldsymbol{\alpha}_j, j = 1,2$:

$$\boldsymbol{\alpha}(x,y)\Big|_{\Gamma^e_{bs}} = \sum_{j=1}^{2} M_j(x)\boldsymbol{\alpha}_j, M_j(x) = N_j(x,0) \tag{3.38}$$

where Γ^e_{bs} is the corresponding element edge where the bond-slip element is placed. The finite element interpolation in (3.31) can be written as:

$$\mathbf{u}(\mathbf{x})\Big|_{\Omega^e} = \sum_{i=1}^{4} N_i(\mathbf{x})\left(\mathbf{d}^c_i + \sum_{j=1}^{2} M_j(x)\boldsymbol{\alpha}_j\right) \tag{3.39}$$

In such a case, $\boldsymbol{\alpha}_j$ can be defined as the nodal values of bond-slip displacements relative to the concrete nodal displacement $\mathbf{d}^c_i$. The same displacement field can also be represented by separating the description of the concrete vs. steel displacements according to:

$$\begin{aligned} \mathbf{u}^c(\mathbf{x})\Big|_{\Omega^e} &= \sum_{i=1}^{4} N_i(\mathbf{x})\mathbf{d}^c_i \\ \mathbf{u}^s(\mathbf{x})\Big|_{\Gamma^e_{bs}} &= \sum_{j=1}^{2} M_j\mathbf{d}^s_j(x), \mathbf{d}^s_j = \mathbf{d}^c_j + \boldsymbol{\alpha}_j \end{aligned} \tag{3.40}$$

where $\mathbf{d}^s_j$ are now the absolute nodal displacements of steel bar expressed in the fixed reference frame. While these two descriptions of the RC composite indeed represent the same displacement field for the concrete and the steel components, the crucial difference between them concerns the description of the bond-slip. In the first case in (3.39), the bond-slip is measured in a relative sense in respect to the concrete displacement, which further implies that one part of the steel bar displacement will accompany a given concrete displacement as defined before by the current value of slip. This includes the starting value of zero slip, in which case both the concrete and the steel undergo the same displacement. The inelastic value of slip is what remains as the permanent differential displacement between the concrete and the steel, and plays the role of the inelastic deformation localized at the interface between the concrete and the steel, with a given non-local distribution of bond-slip along the steel bar. In the second case in (3.40), the bond-slip is measured in the absolute sense, produced either by a given displacement of concrete (while keeping the steel fixed) or by a given displacement of steel (while keeping the concrete fixed). The two displacement field descriptions in (3.39) and (3.40) lead to very different solution procedures.

Here only the solution procedure for the relative displacement of the bond-slip element (3.40) will be presented as the procedure for the absolute displacement of the bond-slip element (3.39) is not robust enough to make this approach plausible. The main reason for that lack of robustness is in the very unfavorable sequence of bond-slip computations, where the motion of concrete nodes, keeping the steel nodes fixed, does not really correspond to the real behavior of the RC assembly. An alternative manner to achieve the same goal of simplifying the system while keeping the robustness is presented in the next section.

3.4.2 Solution Procedure for Relative Displacements of Bond-Slip Element

The bond-slip phenomenon is considered in terms of relative motion in respect to the concrete. More precisely, the total motion of the RC composite is split between the motion of RC with a fixed slip (including the zero slip value initially) imposing the mutually constrained displacement of concrete and steel nodes, followed by the residual bond-slip that occurs due to the redistribution of the internal force upon the concrete cracking. This kind of split is more closely related to the more physical phenomena, which take place during the cracking of concrete and the related stress redistribution within the RC composite. Additionally, this kind procedure provides more robust computations, where the operator split procedure can be used to separate the global slip of RC motions computation (with a fixed slip) from the local computation of slip along a particular bar. It is obvious that the latter plays the same role as the local damage deformation in concrete or local plastic deformation in steel, but it is related to the complete steel bar. The key ingredient that allows one to integrate this kind of non-local field within standard finite element computational procedure is related to the special representation of the bond-slip field in terms of the X-FEM approximations (e.g. Belytschko et al. 2001; Belytschko et al. 2002; Belytschko and Black 1999). To that end the virtual displacement field will be constructed in the same manner as the real displacement field defined in (3.39), but with nodal value of real displacement and relative bond-slip $\mathbf{d}_i^c$ and $\boldsymbol{\alpha}_j$ replaced by the corresponding virtual ones, denoted as $\mathbf{w}_i^c$ and $\boldsymbol{\eta}_j$, respectively; this gives the corresponding representation of the virtual displacement field over a typical element Ω^e of RC composite with the bond-slip, which can be written as:

$$\mathbf{w}(\mathbf{x})\big|_{\Omega^e} = \sum_{i=1}^{4} N_i(\mathbf{x}) \left(\mathbf{w}_i^c + \sum_{j=1}^{2} M_j(x) \boldsymbol{\eta}_j \right) \tag{3.41}$$

By assuming that such a virtual displacement field is applied at the deformed configuration at pseudo-time t_{n+1} of a particular loading program, the weak form of the equilibrium equations can be obtained. The latter contains the first global term where the contribution of all concrete elements is accounted for along with the corresponding contribution of the steel bar, which is constrained to move along with concrete as defined by the given value of bond-slip. The corresponding residual then reads:

$$\begin{aligned}
&\mathbf{r}^{cs} := \mathop{\mathbb{A}}_{e=1}^{n_{el}} f^{c,\text{int},e}\left(\mathbf{d}_{n+1}^c\right) + \mathop{\mathbb{A}}_{e=1}^{n_{el}^{bs}} \mathbf{f}^{s,\text{int},e}\left(\mathbf{d}_{n+1}^c, \boldsymbol{\alpha}_n\right) - \mathbf{f}_{n+1}^{\text{ext}} = 0 \\
&\mathbf{f}_i^{c,\text{int},e}\left(\mathbf{d}_{n+1}^c\right) = \int_{\Omega^e} \left(\nabla^s N_i\right)^T \boldsymbol{\sigma}_{n+1}^c\left(\mathbf{d}_{n+1}^c\right) dV \\
&\mathbf{f}_j^{c,\text{int},e}\left(\mathbf{d}_{n+1}^c, \boldsymbol{\alpha}_n\right) = \int_{\Gamma_{bs}^e} \left(\frac{dM_j}{dx}\right)^T \sigma_{n+1}^s\left(\mathbf{d}_{n+1}^c, \boldsymbol{\alpha}_n\right) dA \\
&\mathbf{f}_{i,n+1}^{\text{ext}} = \int_{\Omega^e} \mathbf{N}_i^T \mathbf{b} dV
\end{aligned} \tag{3.42}$$

This global equation is accompanied by the local equilibrium equation, which concerns redistribution of the slip with:

$$\mathbf{r}^{\text{bss}} := \mathop{\mathbb{A}}_{e=1}^{n_{\text{el}}}\left[\mathbf{f}^{s,\text{int},e}\left(\mathbf{d}_{n+1}^{c},\boldsymbol{\alpha}_{n+1}\right)+\mathbf{f}^{bs,\text{int},e}\left(\boldsymbol{\alpha}_{n+1}\right)\right]=0$$

$$\mathbf{f}_{j}^{s,\text{int},e}\left(\mathbf{d}_{n+1}^{c},\boldsymbol{\alpha}_{n+1}\right)=\int_{\Gamma_{bs}^{e}}\left(\frac{dM_j}{dx}\right)^{T}\sigma_{n+1}^{s}\left(\mathbf{d}_{n+1}^{c},\boldsymbol{\alpha}_{n}\right)dA \tag{3.43}$$

$$\mathbf{f}_{j}^{bs,\text{int},e}\left(\boldsymbol{\alpha}_{n+1}\right)=\int_{\Gamma_{bs}^{e}}\mathbf{B}^{bs,T}\boldsymbol{\sigma}_{n+1}^{sb}\left(\boldsymbol{\alpha}_{n+1}\right)dA$$

The computation of the stress values for each component of the RC is again carried out as described previously. From equations (3.42) and (3.43) it is evident that only the steel bar contributes to both sets of equations, whereas the bond-slip computation is limited only to the last equation. In fact, such an equation is very much like the one used for computing the internal variables for representing concrete cracking and steel yielding; only it concerns more than one element at the time since connecting all the bond-slip elements along a particular bar.

For that reason, the operator split methods, where the sets of in (3.42) and (3.43) are solved sequentially, provide the most appropriate approach. The benefit of this approach is dual; firstly, one is not to deal simultaneously with two sets of equations of different sizes, as well as the guarantee of the computational robustness where the corresponding slip evolution within a particular time step is always computed by starting from the converged value obtained in the previous step. The latter allows avoiding eventual spurious loading–unloading slip cycle, which can occur when advancing the slip and the concrete displacement computations simultaneously.

For a typical time step between t_n and t_{n+1} we can thus write:

- For n=0,1,2...
- Given: $\mathbf{d}_n^c, \boldsymbol{\alpha}_n$dc n, an, internal variables at t_n
- Find: $\mathbf{d}_{n+1}^c, \boldsymbol{\alpha}_{n+1}$, internal variables at t_{n+1}
- Iterate: (i)=1,2...
- Iterate: (j)=1,2...

compute internal variable evolution for given $\mathbf{d}_{n+1}^{c,(i)}, \boldsymbol{\alpha}_{n+1}^{(j)}$.

$$\frac{\partial \mathbf{r}^{\text{sbs},(i)}}{\partial \boldsymbol{\alpha}_{n+1}}\left({\boldsymbol{\alpha}_{n+1}}^{(j+1)}-{\boldsymbol{\alpha}_{n}}^{(j+1)}\right)=-\mathbf{r}^{\text{sbs}}\left(\mathbf{d}_{n+1}^{c,(i)},{\boldsymbol{\alpha}_{n+1}}^{(j)}\right) \tag{3.44}$$

Test convergence locally:

$$\text{IF}\left\|\mathbf{r}^{\text{sbs}}\left(\mathbf{d}_{n+1}^{c,(i)},{\boldsymbol{\alpha}_{n+1}}^{(j)}\right)\right\|>\text{tol NEXT }(j)$$

$$\text{ELSE}\left\|\mathbf{r}^{\text{sbs}}\left(\mathbf{d}_{n+1}^{c,(i)},{\boldsymbol{\alpha}_{n+1}}^{(j)}\right)\right\|<\text{tol}\Rightarrow{\boldsymbol{\alpha}_{n+1}}^{(i+1)}={\boldsymbol{\alpha}_{n+1}}^{(j+1)}\text{ NEXT }(i)$$

$$\left[\frac{\partial \mathbf{r}^{cs}}{\mathbf{d}_{n+1}^{c,(i)}}-\frac{\partial \mathbf{r}^{cs}}{{\boldsymbol{\alpha}_{n+1}}^{(i)}}\left(\frac{\partial \mathbf{r}^{\text{sbs}}}{{\boldsymbol{\alpha}_{n+1}}^{(i)}}\right)^{-1}\frac{\partial \mathbf{r}^{\text{sbs}}}{{\boldsymbol{\alpha}_{n+1}}^{(i)}}\right]\left(\mathbf{d}_{n+1}^{c,(i+1)}-\mathbf{d}_{n+1}^{c,(i)}\right)=-\mathbf{r}_{n+1}^{cs,(i)} \tag{3.45}$$

Test convergence globally:

$$\text{IF } \left\| \mathbf{r}^{cs}\left(\mathbf{d}_{n+1}^{c,(i+1)}, \boldsymbol{\alpha}_{n+1}^{(i+1)}\right) \right\| > \text{tol NEXT } (i)$$
$$\text{ELSE } \left\| \mathbf{r}^{cs}\left(\mathbf{d}_{n+1}^{c,(i+1)}, \alpha_{n+1}^{(i+1)}\right) \right\| < \text{tol} \Rightarrow \boldsymbol{\alpha}_n \leftarrow \boldsymbol{\alpha}_{n+1}^{(i+1)}, \mathbf{d}_n \leftarrow \mathbf{d}_{n+1}^{c,(i+1)} \text{NEXT } (n)$$

From (3.44) and (3.45) it is evident that the computation of the bond-slip parameters is carried out for the given nodal values of concrete displacements. Hence, a smaller set of equations governing bond-slip is forced to converge for a given iterative value of concrete displacement. This could be interpreted as imposing the admissibility of the bond-slip forces in the given slip criterion, which is fully equivalent to imposing the admissibility of the stress field for each component of the RC composite in the sense of the given damage criterion for concrete and the plasticity criterion for steel. It is important to note that such a computation is also carried out for each iterative sweep (either (i) or (j)) in order to provide the corresponding values of the internal variables governing the cracking of concrete or the plastic yielding of steel. With these values in hand, the corresponding form of the tangent moduli for concrete can also be obtained, steel and bond-slip, which allows one to compute accordingly the tangent operators needed in (3.45). Hence, explicitly it can be written as:

$$\begin{aligned} H_{ij}^{\text{sbs},(j)} &= \left.\frac{\partial \mathbf{r}_i^{\text{sbs}}}{\partial \boldsymbol{\alpha}_{j,n+1}^{(j)}}\right|_{\Gamma_{\text{bs}}^e} = \int_{\Gamma_{\text{bs}}^e} \frac{dM_i}{dx} C_{n+1}^{\text{ep},s}\left(\mathbf{d}_{n+1}^{c,(i)}, \boldsymbol{\alpha}_{n+1}^{c,(i)}\right) \frac{dM_j}{dx} dA \\ &+ \int_{\Gamma_{bs}^e} \overline{\mathbf{B}}_i^{\text{bs},T} C_{n+1}^{\text{ep,bs}}\left(\boldsymbol{\alpha}_{n+1}^{(i)}\right) \overline{\mathbf{B}}_j^{\text{bs}} dA \end{aligned} \tag{3.46}$$

where $C_{n+1}^{\text{ep},s}$ and $C_{n+1}^{\text{ep,bs}}$ are given in (3.33) and (3.28), respectively. In (3.46) above, $\overline{\mathbf{B}}^{\text{bs}}$ is the strain–displacement matrix describing the kinematics of the bond-slip element with relative degrees of freedom. Namely, by accounting for the relation between the absolute and the relative displacement given in (3.40) it can be stated that:

$$\overline{\mathbf{B}}^{\text{bs}} = \mathbf{B}^{\text{bs}}\mathbf{T}, \mathbf{T} = \begin{bmatrix} \mathbf{I} & 0 \\ \mathbf{I} & \mathbf{I} \end{bmatrix} \tag{3.47}$$

The explicit form of the tangent stiffness for concrete will not change with respect to the result in (3.19) and if Newton's iterative method is used, the tangent operators can be computed as:

$$\begin{aligned} \mathbf{K}_{n+1}^{\text{cbsc},(i)} &= \frac{\partial \mathbf{r}^{\text{cbs}}}{\partial \mathbf{d}_{n+1}^{c,(i)}}, \mathbf{K}_{n+1}^{\text{cbss},(i)} = \frac{\partial \mathbf{r}^{\text{cbs}}}{\partial \mathbf{d}_{n+1}^{s}} \\ \mathbf{K}_{n+1}^{\text{sbsc},(i)} &= \frac{\partial \mathbf{r}^{\text{sbs}}}{\partial \mathbf{d}_{n+1}^{c,(i)}}, \mathbf{K}_{n+1}^{\text{sbss},(i)} = \frac{\partial \mathbf{r}^{\text{sbs}}}{\partial \mathbf{d}_{n+1}^{s}} \end{aligned} \tag{3.48}$$

Finally, the coupling term in the global tangent stiffness matrix is provided by the steel component and can be written as:

$$\begin{aligned} \mathbf{F}_{ij,n+1}^{\text{sbs},e,(i)} &:= \frac{\partial \underline{\mathbf{r}}^{\text{sbs},(i),e}}{\partial \mathbf{d}_{n+1}^{c}} \int_{\Gamma_{\text{bs}}^e} = \frac{dN_i}{dx} C_{n+1}^{\text{ep},s} \frac{dN_j}{dx} dA \\ \mathbf{F}_{ij,n+1}^{\text{bss},e,(i)} &:= \frac{\partial \mathbf{r}_i^{\text{cs},(i),e}}{\partial \boldsymbol{\alpha}_{n+1}} \int_{\Gamma_{\text{bs}}^e} = \frac{dM_a}{dx} C_{n+1}^{\text{ep},s} \frac{dN_b}{dx} dA \end{aligned} \tag{3.49}$$

where $C^{ep,s}$ is given in (i), for the corresponding case of internal variables' computations.

3.4.3 Numerical Examples

In this section, the results of several numerical simulations are validated, which are given in order to illustrate the corresponding advantages of the proposed approach. All the computations are carried out by a research version of the computer program FEAP developed by Prof. R.L. Taylor at UC Berkeley (e.g. Zienkiewics and Taylor 2005).

3.4.3.1 Traction Test and Proposed RC Model Validation

The first example considers the traction test on an RC beam, which was carried out by Mivelaz (1996). The experimental results presented in Mivelaz (1996) pertain both to a global response, such as the resulting force–displacement diagram, and to a local response, such as the observed cracks' distribution and measured cracks' width. For that reason, this particular test was used to provide complete validation of the proposed RC model. In the first phase formation of a macro-crack (visible to the naked eye) is observed once the ultimate concrete resistance is reached. In the second phase, once the bond-slip mechanism is activated redistribution of the internal force between the steel and the cracked concrete entering the softening response phase is observed and this is tested as well. It is the latter that decides, to a large extent, not only the global response but also the details of the local response with resulting crack-spacing and opening.

The selected specimen geometry and loading conditions used in the test are presented in Figure 3.7. The test specimen is in a large RC panel with the dimensions: 3000×420×1000 mm, reinforced with three layers of steel bars. Each layer contains six steel bars whose diameter is equal to 16 mm. The material properties of concrete, steel and the bond-slip are given in Table 3.2.

The finite element mesh is constructed with CST elements (Brancherie and Ibrahimbegović 2009), as shown in (Figure 3.8), by using 200×28 elements. Three layers of enhanced RC finite elements presented herein, which are used to provide the proper representation of the

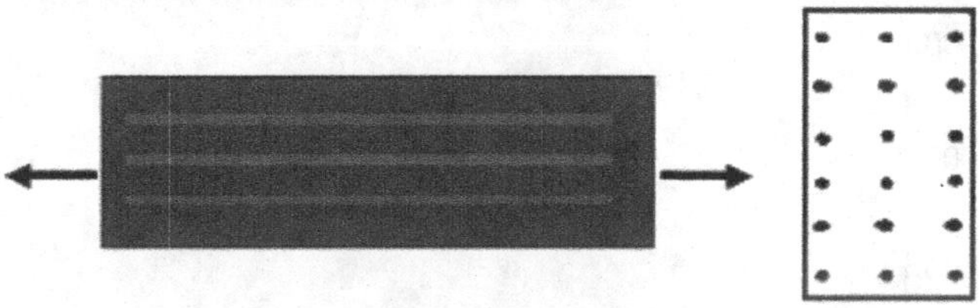

Figure 3.7 Finite element model of reinforced concrete constructed by a structured mesh, with nodes that belong to concrete and nodes that belong to steel.

Table 3.2 The Material Properties of Concrete, Steel and the Bond-Slip

Material properties		
Concrete	Steel	Bond-slip
$E_c = 45700$ MPa	$E_s = 210000$ MPa	$K_b = 25400$ MPa
$\nu_c = 0.20$	$\nu_s = 0.30$	$G_b = 19000$ MPa
$G_f = 193$ N/m		$\tau_y = 6$ MPa
$\bar{\sigma}_f = 2.77$ MPa		

Figure 3.8 RC specimen used in simple tension validation test: Chosen finite element mesh.

bond-slip field along the steel reinforcement, are indicated in a light gray color and positioned all along the three steel reinforcement layers. The bond-slip is prevented at both ends of each steel reinforcement layer, which is used in order to model the anchors and ensure the corresponding zero value of bond-slip.

The computations are carried out under imposed end displacement. The computed global response, in terms of an imposed strain–force diagram is presented in Figure 3.9 (top curve). The very good performance of the proposed model and very satisfying results of the numerical simulations with respect to those measured experimentally (see top graph in Figure 3.10), in regard to the prediction of the global force–displacement curve, are observed. It is obvious that the model is capable of capturing the initial elastic part and FPZ creation, followed by the first fully developed crack that leads to a sudden drop in resistance. The model is also able to express further stress redistribution and the resulting recovery of the specimen, before the next macro-crack develops, as well as the same scenario repeating for the final crack appearance in this specimen. It is important to note that it is not only the force values but also the imposed strain values, computed at each new crack, that match the experimental results very well.

Validation of the local results pertains to the cracks' width. Comparing the bottom curves in Figure 3.9 against those in Figure 3.10, it is clear that the computed cracks' width is capable of matching quite well with the experimental results. All the salient features of local response are captured. Namely, a first sudden jump is seen when opening the first macro-crack due to the brittle localized failure of the concrete, at an imposed strain value

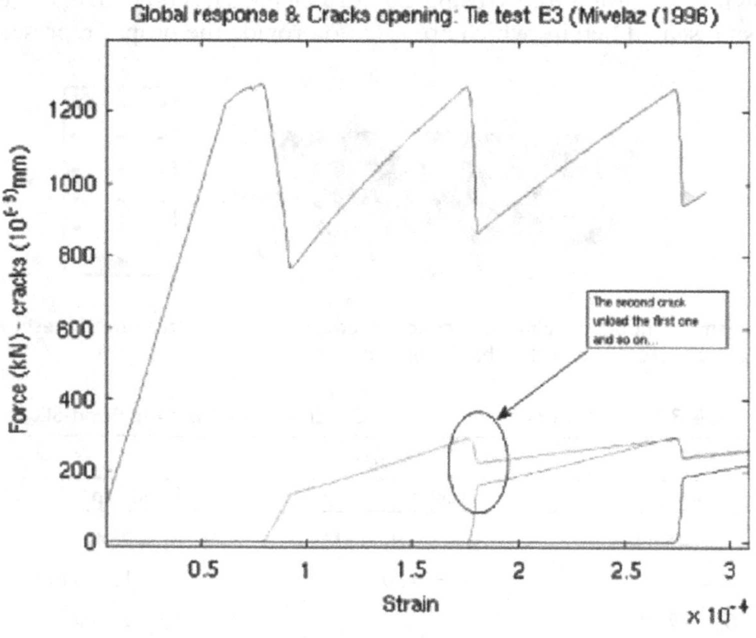

Figure 3.9 Computed force–displacement response and crack width.

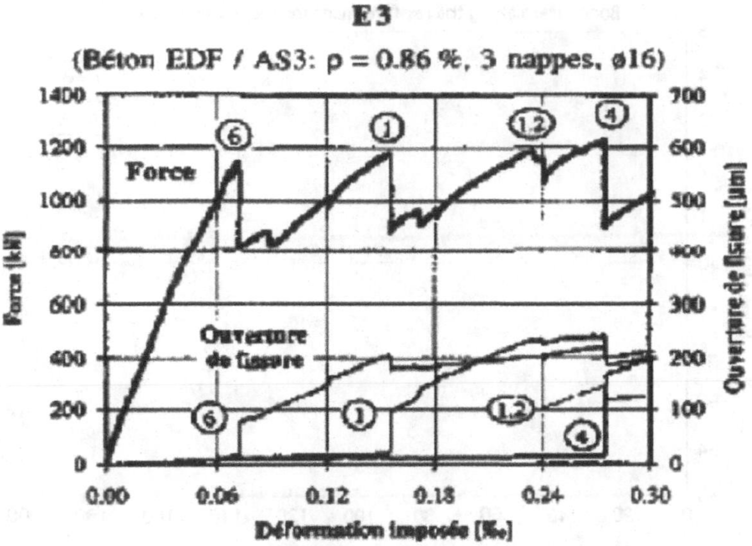

Figure 3.10 Experimentally obtained force–displacement response and crack width.

of 0.75×10^{-4}. This crack-opening keeps growing (almost linearly) until an imposed strain value of 1.75×10^{-4} where the second crack appears. The sudden appearance of the second crack will lead to a partial closing of the first crack. From then on, both existing cracks will keep growing, until the sudden appearance of the third crack at imposed strain value of 2.75×10^{-4}, when they both repeat the same scenario of partial closing again while the new crack takes off. The final picture for crack openings is thus reached, with the final values of roughly 0.2 to 0.25 mm.

The computational results presented in Figure 3.11 illustrate quite well the fundamental role of the bond-slip in internal force transfer and resulting crack-spacing. Figure 3.11 presents the bond-stresses along the particular reinforcement layer, recorded at different stages of the simulation for one, two and three activated macro-cracks, in two Gauss numerical integration points (GNP) in each bond-slip element. It is seen that the concrete and the steel share roughly the same strain value before the first crack appears and that the bond-slip has pretty much no effect. The bond-slip gets activated once the first crack appears. The bond-stress keeps growing up to the yield value quickly spreading in the crack neighborhood (in accordance with the perfect plasticity model representation), leading to plastic reinforcement sliding relative to concrete. The direction of plastic sliding is not the same on either side of the crack, on one side it is positive, and on the other side negative sliding is observed.

For example, on the left side of the crack, the concrete moves sharply to the left as the activation of the element embedded discontinuity, whereas steel remains in place. Therefore, the positive relative sliding is observed resulting in the positive value of bond-stress. The opposite signs and the same scenario apply to the right side, resulting in the negative value of bond-stress.

The bond-stress subsequent to activation of each of three cracks is illustrated in Figure 3.11. The first one opens near GNP 140, leading to different directions of relating sliding on both sides of the cracks as described above. The second crack appears near GNP 20, clearly leading to unloading at the first crack. And finally, the last crack appearing near GNP 100 unloads each of the first two cracks. Figure 3.12 shows the relative displacement field together with strong discontinuity field indicating the macro-cracks' locations.

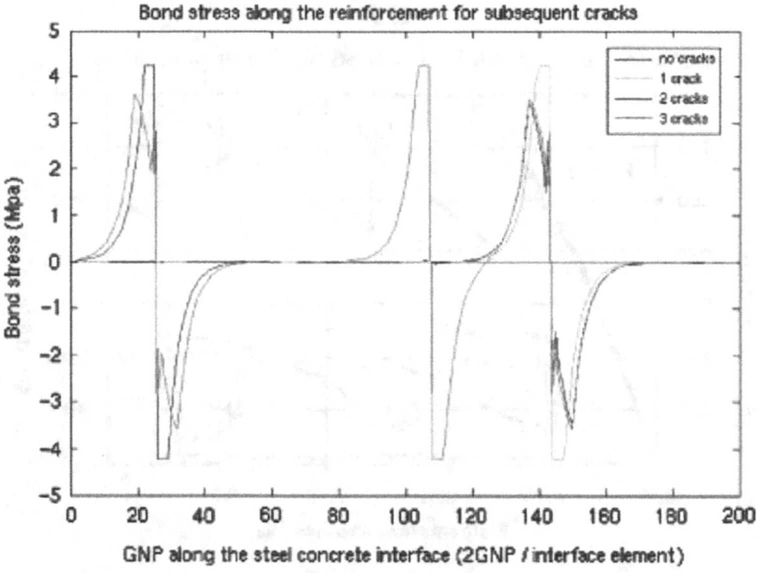

Figure 3.11 Bond-stress distribution along reinforcement layer at different Gauss numerical integration points (GNP).

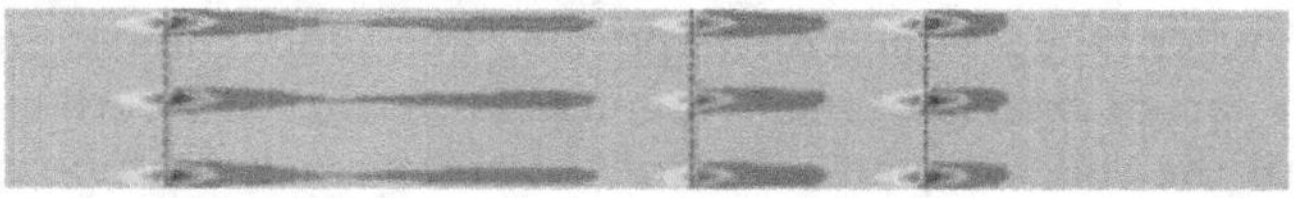

Figure 3.12 Relative displacement field contours and final spacing of macro-cracks (indicated by solid lines).

The efficiency in the iterative procedure is shown in Figure 3.13 (relative vs. absolute bond-slip displacement approach). Namely, in Figure 3.13, we give the total number of iterations needed to carry out the computations by using the relative bond-slip displacements (solid-line) vs. the absolute bond-slip displacements (dashed-line). We recall that the former always start with the frozen value of the bond-slip when predicting the RC displacement in the subsequent step, as opposed to the latter when the initial displacements of concrete and reinforcement are advanced independently.

Once the first crack appears and the bond-slip gets activated, the relative bond-slip displacement approach remains fairly stable with a nearly constant number of iterations in each time step. Whereas the absolute bond-slip displacement approach faces more and more pronounced difficulties in converging with each new crack that is activated, ending with an ever increasing number of iterations because initial displacements of concrete and reinforcement are advanced independently.

3.4.3.2 Four-Point Bending Test and Crack-Spacing Computations

Here, a for four-point bending test (very frequently used) for quasi-brittle materials is elaborated. Here a qualitative illustration pertaining to crack patterns for different reinforced cases, with either heavy or weak reinforcement case is provided.

a) Crack-Spacing Computations for Heavy Reinforcement

The selected specimen is presented in Figure 3.14. For this RC specimen, only one steel layer is modeled (2×16 mm located at 30 mm from the bottom line).

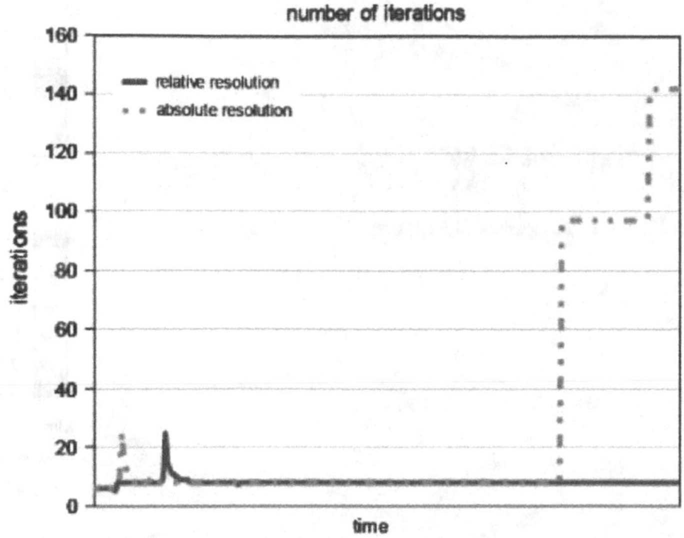

Figure 3.13 Total number of iterations in each time step for relative (solid-line) vs. absolute (dashed-line) bond-slip displacement approach.

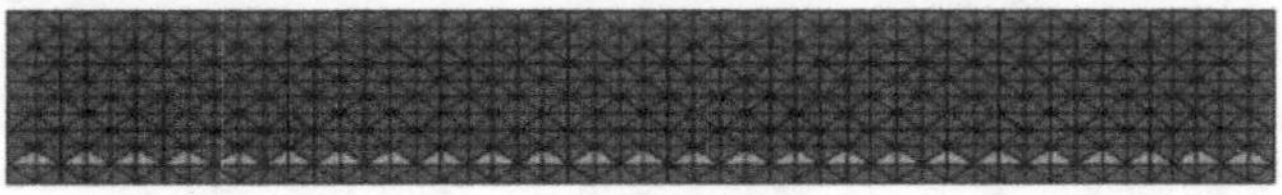

Figure 3.14 Four-point bending test on RC specimen: geometry, loading and finite element mesh.

In the same way, as in the previous example, we can observe different mechanisms at different stages of the loading process. Figure 3.15 clearly illustrates the appearance of the first macro-cracks in-between the section where the loading is applied.

The crack pattern keeps growing along with the bond-slip increase, with some cracks developing up to the core of the structure. Figure 3.16 illustrates the crack pattern at an advanced post critical scenario.

Finally, the global response (Figure 3.17) clearly shows a loss of rigidity due to the damaged concrete structure. After an elastic part, subsequent loading/unloading sessions are observed when cracks appear and keep growing. As the structure is heavily reinforced, large discontinuity in the response is not observed. Namely, the steel plays for this case a dominant role in ensuring the structural resistance.

b) Crack-Spacing Computations for Weak Reinforcement

Here the steel reinforcement is only half of the one from the previous example, while all other properties still remain the same.

The crack pattern represented in Figure 3.18 shows the cracks that are developed more regularly than in the previous case with high reinforcement percentage. This phenomenon is due to the weaker role that steel plays in the structural process. Figure 3.19 shows the role played by the bond-slip. On both parts of a crack, positive and negative sliding is observed. On the left side, concrete moves sharply to the left due to the discontinuity, whereas steel remains at the same place. As a consequence, positive relative sliding is observed. The opposite signs and the same scenario apply to the right side.

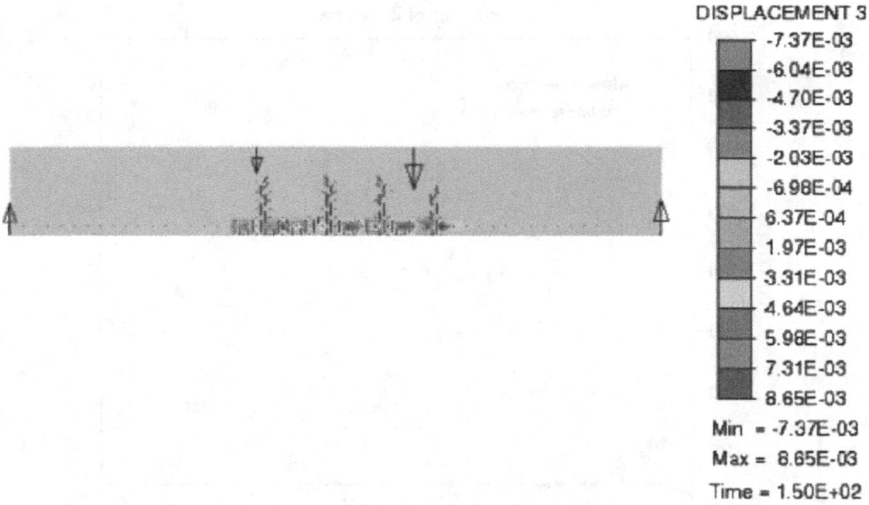

Figure 3.15 Bond-slip along the bar and crack pattern for the first macro-cracks.

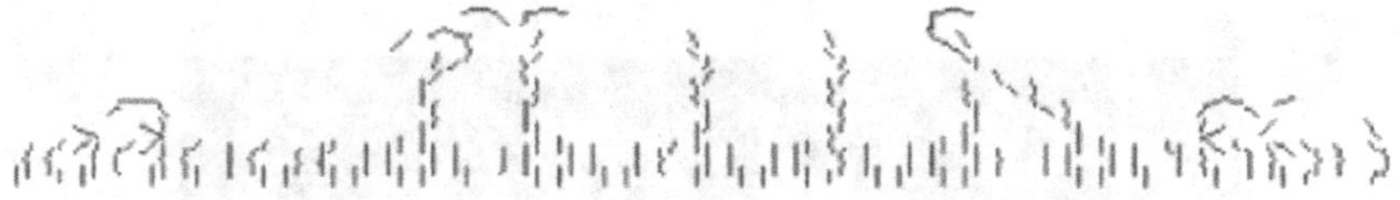

Figure 3.16 Crack pattern at an advanced loading stage.

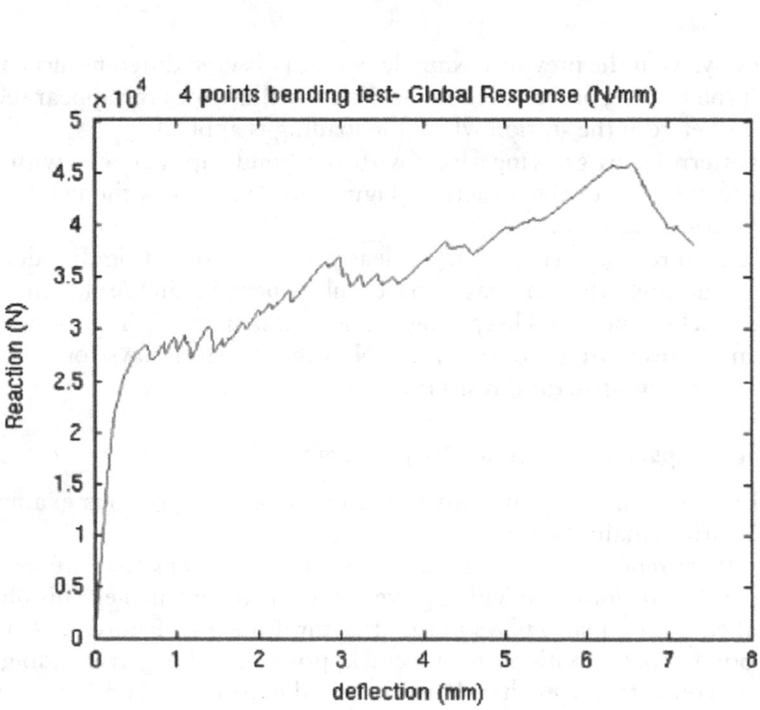

Figure 3.17 Global response.

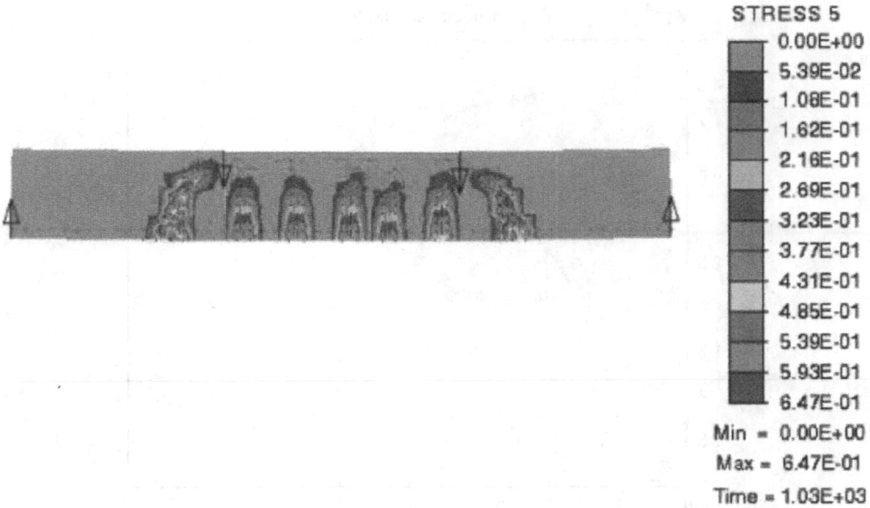

Figure 3.18 FPZ and cracks' opening.

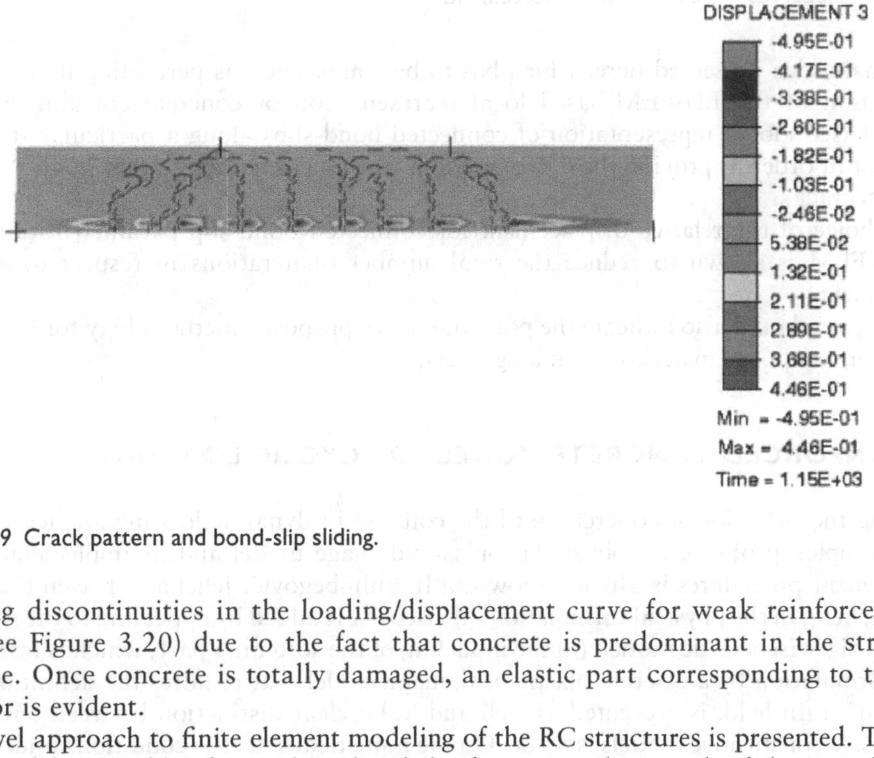

Figure 3.19 Crack pattern and bond-slip sliding.

Strong discontinuities in the loading/displacement curve for weak reinforcement is seen (see Figure 3.20) due to the fact that concrete is predominant in the structural response. Once concrete is totally damaged, an elastic part corresponding to the steel behavior is evident.

A novel approach to finite element modeling of the RC structures is presented. The proposed model is capable of providing detailed information about each of the ingredients of RC; the resulting information on crack-spacing and opening, which is of great importance for durability problems.

The problem of interest here concerns the case where the bond-slip is not excessive (no steel reinforcement pull-out), where even the simple Drucker–Prager perfect plasticity model can be used for representing the bond-slip constitutive behavior. Even a simple model for bond-slip is shown to ensure the proper description of internal force redistribution upon concrete cracking and determines to a large extent the crack-spacing.

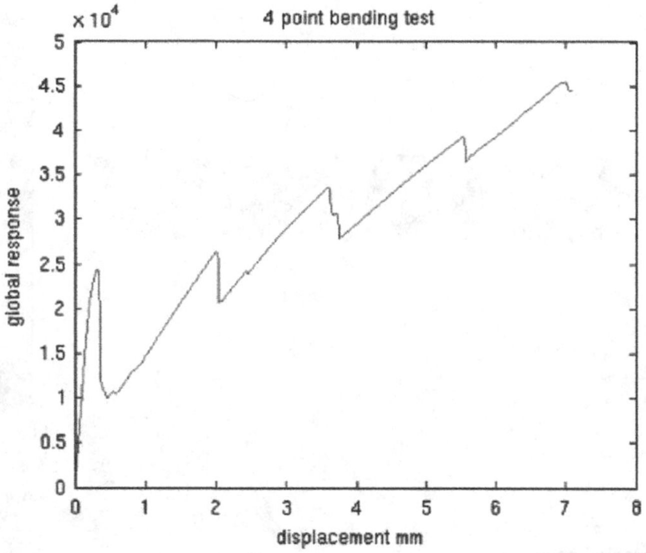

Figure 3.20 Global response for weak reinforcement.

The main idea presented here, which has to be emphasized, is pertaining to a judicious combination of the ED-FEM-based local representation of concrete cracking with the X-FEM-based global representation of connected bond-slips along a particular steel reinforcement in order to provide the corresponding strain field representation in an RC structure with cracks.

The choice of the relative displacement for connected bond-slip parametrization based upon X-FEM is shown to reduce the total number of iterations in respect to absolute displacement.

Finally, we should also indicate the potential of the proposed methodology for RC to deal with other composite materials with long fibers.

3.5 REINFORCED CONCRETE MODEL FOR CYCLIC LOADING

Modeling the behavior of concrete until the collapse in dynamic loading applications is a rather complex problem. A robust elastoplastic-damage model and its implementation in finite element procedures is already shown in Ibrahimbegović, Jehel and Daven (2007). In addition, here, in the physical formulation: (i) viscosity is added by appealing to the so-called Kelvin–Voigt viscoelastic model; (ii) the modeling of the softening part, which is introduced in the formulation by a discrete plastic or damage model that requires the definition of an enhanced strain field, is presented as well and (iii) a clear distinction between tensile and compressive constitutive models is made. On the numerical side, the conditions under which the model can enter the generalized standard materials class (Halphen and Nguyen 1975), whose elements benefit from good robustness properties, are identified.

Here, it is interesting to note the motivation of viscosity introduction into the material model. Viscosity in concrete can be associated to its loading-rate-dependent macroscopic behavior that results from several nano and microscopic phenomena such as the complex interaction between moisture and the micro-structural solid skeleton, micro-cracking process, Stefan effect, micro-inertia of the material surrounding the crack tip (Pedersen, Simone and Sluys 2008). Second, viscosity can also be associated to creep phenomena in highly dissipative materials such as asphalt concrete (Panoskaltsis et al. 2008). Thirdly, its introduction leads to the explicit appearance of a time step in the governing equations of the discretized problem. The possibility of determining a critical time step then provides a tool to regularize numerical problems that can be ill-posed in the presence of materials that exhibit strain softening behaviors (Oliver and Huespe 2004). And finally, viscosity can also be used to introduce a source of energy dissipation at the material level: it can lump the energy dissipation that physically comes from unknown or not (accurately) modeled nonlinear physical phenomena in the materials.

3.5.1 Constitutive Concrete Model

3.5.1.1 Thermodynamics with Internal Variables (TIV)

To represent the salient phenomena observed in the experimental cyclic response of a concrete representative elementary volume (REV) (Figure 3.21), a set α of internal variables is chosen. The phenomena that are modeled: (i) loading-rate-dependent behavior, (ii) strain hardening, (iii) strain softening, (iv) appearance of residual deformation, (v) stiffness and strength degradations, (vi) hysteresis loops. It is considered that the phenomena (ii), (iv) and (vi) occur only in compression, that strain softening in compression is due to the localization of permanent deformation whereas strain softening in tension is due to the localization of deformation that completely disappears after unloading. Internal variables are associated with each of these phenomenologically identified mechanisms.

The set of internal variables is defined in Table 3.3. Continuum quantities are denoted as $\bar{\bullet}$ whereas discrete quantities being localized, are marked with $\bar{\bar{\bullet}}$.

A noteworthy advantage of this material model is that all its parameters have a clear meaning and can, therefore, be easily identified according to experimental curves. The viscous parameter η can be seen as a material property and identified from free-vibration tests; the tensile softening curve coefficient *a* a is related to the fracture energy G_f, a fracture mechanics concept (Hillerborg, Modéer and Petersson 1976; Bažant, Yu and Zi 2002) that quantifies the total amount of energy that has to be furnished in tension to a concrete section between the time when displacement begins to localize and the time when the section is completely broken; all the other parameters can be identified from stress–strain curves of concrete specimen in quasi-static cyclic loading.

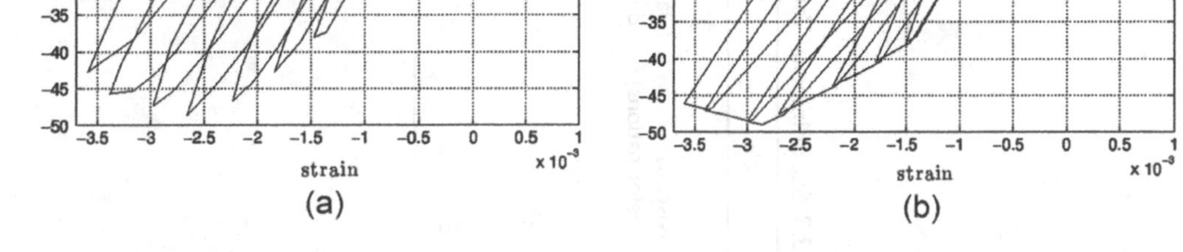

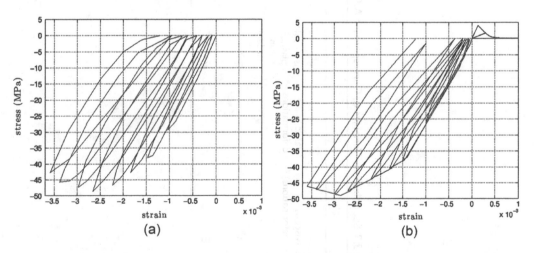

Figure 3.21 (a) Experimental (adapted from Ramtani 1990) and (b) numerical (by using the proposed constitutive concrete model) cyclic behavior of concrete in quasi-static loading.

Table 3.3 Set of Internal Variables α, Corresponding Affinities $\mathcal{A}$ and Phenomenological Coefficients PCs (Parameters of the Model)

	t/c				*Compression*				*Tension*	
α	$\bar{\varepsilon}^{\upsilon}$ (viscous deformations)	$\bar{\varepsilon}^{p}$ (plastic deformations)	$\bar{\xi}^{p}$ (isotropic plastic strain hardening)	$\bar{\lambda}^{p}$ (kinematic plastic strain hardening)	$\bar{D}$ (stiffness degradation)	$\bar{\xi}^{p}$ (isotropic damage strain hardening)	$\bar{\bar{u}}^{p}$ (plastic deformations)	$\bar{\bar{\xi}}^{p}$ (plastic strain softening)	$\bar{\bar{D}}$ (stiffness degradation)	$\bar{\bar{\xi}}^{d}$ (damage strain softening)
$\mathcal{A}$	σ^{υ}	σ	$\bar{q}^{p}$	$\bar{\kappa}^{p}$	$\frac{\sigma^2}{2}$	$\bar{q}^{d}$	t	$\bar{\bar{q}}^{p}$	$\frac{t^2}{2}$	$\bar{\bar{q}}^{d}$
PC	E, η	$\bar{E}$	$\bar{K}^{p}$	$\bar{H}^{p}$	$\bar{E}^{-1}$	$\bar{K}^{d}$	$\bar{\bar{E}}$	$\bar{\bar{K}}^{p}$	$\bar{\bar{E}}^{-1}$	σ_{∞}, a

3.5.1.2 Governing Equations of the Constitutive Model

3.5.1.2.1 Basic Ingredients

The three basic ingredients for developing a constitutive model in the framework of the TIV are defined as:

1. The split of the total strain into a viscoelastic, a plastic, a damage and another discrete part due to the likely localization of the displacement. $\mathcal{A}$-space is divided into two parts: the terms involved in tension $\sigma \geq 0$ are mentioned by the sign $\bullet^{+}$and those involved in compression $\sigma < 0$ by the sign $\bullet^{-}$

$$u(x,t) = \bar{u}(x,t) + \bar{\bar{u}}(t)\mathcal{H}_{\bar{x}}(x)$$
$$\varepsilon(x,t) = \frac{\partial u(x,t)}{\partial x} = \bar{\varepsilon}^{v} + \bar{\varepsilon}^{p} + \left(\bar{\bar{u}}^{\Diamond} + \bar{\bar{u}}^{p}\right)\delta_{\bar{x}} \tag{3.50}$$

where $\bar{\varepsilon}(x,t) = \dfrac{\partial \bar{u}(x,t)}{\partial x} = \bar{\varepsilon}^{v} + \bar{\varepsilon}^{p}$ is the continuum part of the strain and $\bar{\bar{u}}\delta_{\bar{x}} = \left(\bar{\bar{u}}^{\Diamond} + \bar{\bar{u}}^{p}\right)\delta_{\bar{x}}$ – with $\bar{\bar{u}}^{\Diamond} = \bar{\bar{u}}^{d}$ in tension and $\bar{\bar{u}}^{\Diamond} = \bar{\bar{u}}^{e}$ in compression—its discrete part introduced by the Dirac function $\delta_{\bar{x}}$ that is the derivative of the Heaviside's function $\mathcal{H}_{\bar{x}}$ that introduces the displacement jump $\bar{\bar{u}}$ in the description of the displacement. $\bar{\varepsilon}^{v}$ corresponds to the deformation in a Kelvin–Voigt rheological model: a spring–elastic in tension ($\bar{\varepsilon}^{v} = \bar{\varepsilon}^{p}$) and damaging in compression ($\bar{\varepsilon}^{v} = \bar{\varepsilon}^{d}$)– and a dashpot in parallel. For simplicity, it is assumed that the displacement jumps only take place in one section; extension to several localization sections is straightforward.

2. Helmholtz free energy functional

$$\psi^{+}(u,\alpha) = \bar{\psi}^{e}\left(\bar{u}^{+}\right) + \left(\bar{\bar{\psi}}^{d}\left(\bar{\bar{u}}^{+},\bar{\bar{D}}\right) + \bar{\bar{\Xi}}^{d}\left(\bar{\bar{\xi}}^{d}\right)\right)\delta_{\bar{x}}$$
$$\psi^{-}(u,\alpha) = \bar{\psi}^{d}\left(\bar{u}^{-},\bar{\varepsilon}^{p},\bar{D}\right) + \bar{\Xi}^{p}\left(\bar{\xi}^{p}\right) + \bar{\Lambda}^{p}\left(\bar{\lambda}^{p}\right) + \bar{\Xi}^{d}\left(\bar{\xi}^{d}\right) + \left(\bar{\bar{\psi}}^{e}\left(\bar{\bar{u}}^{-},\bar{\bar{u}}^{p}\right) + \bar{\bar{\Xi}}^{p}\left(\bar{\bar{\xi}}^{p}\right)\right)\delta_{\bar{x}} \tag{3.51}$$

3. The nonlinear plasticity and damage mechanisms are activated for a positive value of the corresponding plasticity and damage criteria
   - For $\sigma \geq 0$ (tension), the dissipative mechanisms that can be involved in the evolution process are activated according to the following criteria:

$$\bar{\phi}^{v,+}(\sigma) = \sigma^{v} = \sigma - \sigma^{e} \leq 0$$
$$\bar{\bar{\phi}}^{d,+}\left(t,\bar{\bar{q}}^{d}\right) = t - \left(\sigma_{u}^{t} - \bar{\bar{q}}^{d}\right) \leq 0 \tag{3.52}$$

   where t is the stress at the discontinuity and σ_{u}^{t} is the ultimate stress in tension

   - For $\sigma < 0$ (compression), the criteria functions are:

$$\bar{\phi}^{v,-}(\sigma) = -\sigma^{v} = -\left(\sigma - \sigma^{d}\right) \leq 0$$
$$\bar{\phi}^{p,-}\left(\sigma,\bar{q}^{p},\bar{\kappa}^{p}\right) = \left|\sigma + \bar{\kappa}^{p}\right| - \left(\sigma_{y} - \bar{q}^{p}\right) \leq 0$$
$$\bar{\phi}^{d,-}\left(\sigma,\bar{q}^{d}\right) = -\sigma - \left(\sigma_{f} - \bar{q}^{d}\right) \leq 0 \tag{3.53}$$
$$\bar{\bar{\phi}}^{p,-}\left(t,\bar{\bar{q}}^{p}\right) = -t - \left(\sigma_{u}^{c} - \bar{\bar{q}}^{p}\right) \leq 0$$

   where σ_{y}, σ_{f}, and σ_{u}^{c} are yield, fracture and ultimate stresses in compression.

State Equations of the System

The expression of the energy dissipated by this viscoelastic-plastic-damage model with different hardenings/softenings within a time unit can now be written. This can be computed as the difference between the total amount of energy imparted to the system and the amount of energy stored by the system during this time unit.

In tension:

$$\dot{\mathcal{D}}^{+} = \left(\sigma\dot{\bar{\varepsilon}}^{+} + t\dot{\bar{\bar{u}}}^{+}\right) - \dot{\psi}^{+}\left(\bar{u}^{+}, \bar{\bar{u}}^{+}, \bar{\bar{D}}, \bar{\bar{\xi}}^{d}\right) =$$

$$\left(\sigma^{e} + \sigma^{v}\right)\dot{\bar{\varepsilon}}^{+} + t\dot{\bar{\bar{u}}}^{+}\delta_{\bar{x}} - \frac{\partial\bar{\Psi}^{e}}{\partial\bar{\varepsilon}^{+}}\dot{\bar{\varepsilon}}^{+} - \left(\frac{\partial\bar{\bar{\Psi}}^{d}}{\partial\bar{\bar{u}}^{+}}\dot{\bar{\bar{u}}}^{+} + \frac{\partial\bar{\bar{\Psi}}^{d}}{\partial\bar{\bar{D}}}\dot{\bar{\bar{D}}} + \frac{\partial\bar{\bar{\Xi}}^{d}}{\partial\bar{\bar{\xi}}^{d}}\partial\dot{\bar{\bar{\xi}}}^{d}\right)\delta_{\bar{x}}$$

It is clear that $\dot{\bar{\varepsilon}}^{+} = \dot{\bar{\varepsilon}}^{e} = \dot{\bar{\varepsilon}}^{v}$ and $\dot{\bar{\bar{u}}}^{+} = \dot{\bar{\bar{u}}}^{d}$ because $\bar{\bar{u}}^{p}$ cannot evolve in tension according to equations (3.52) and (3.53). Thus:

$$\dot{\mathcal{D}}^{+} = \left(\sigma^{e} - \frac{\partial\bar{\Psi}^{e}}{\partial\bar{\varepsilon}^{e}}\right)\dot{\bar{\varepsilon}}^{e} + \sigma^{v,+}\dot{\bar{\varepsilon}}^{v,+} + \left(t - \frac{\partial\bar{\bar{\Psi}}^{d}}{\partial\bar{\bar{u}}^{d}}\right)\dot{\bar{\bar{u}}}^{d}\delta_{\bar{x}} - \frac{\partial\bar{\bar{\Psi}}^{d}}{\partial\bar{\bar{D}}}\dot{\bar{\bar{D}}}\delta_{\bar{x}} - \frac{\partial\bar{\bar{\Xi}}^{d}}{\partial\bar{\bar{\xi}}^{d}}\dot{\bar{\bar{\xi}}}^{d}\delta_{\bar{x}} \tag{3.54}$$

In compression:

$$\dot{\mathcal{D}}^{-} = \left(\sigma\dot{\bar{\varepsilon}}^{-} + t\dot{\bar{\bar{u}}}^{-}\right) - \dot{\psi}^{-}\left(\bar{u}^{-}, \bar{\bar{u}}^{-}, \bar{\varepsilon}^{p}, \bar{\xi}^{p}, \bar{\lambda}^{d}\bar{D}, \bar{\xi}^{d}, \bar{\bar{u}}^{p}, \bar{\bar{\xi}}^{p}\right) =$$

$$\left(\sigma^{d} + \sigma^{v}\right)\dot{\bar{\varepsilon}}^{-} + t\dot{\bar{\bar{u}}}^{-}\delta_{\bar{x}} - \left(\frac{\partial\bar{\Psi}^{d}}{\partial\left(\bar{\varepsilon}^{-} - \bar{\varepsilon}^{p}\right)}\left(\dot{\bar{\varepsilon}}^{-} - \dot{\bar{\varepsilon}}^{p}\right) + \frac{\partial\bar{\Psi}^{d}}{\partial\bar{D}}\dot{\bar{D}} + \frac{\partial\bar{\Xi}^{p}}{\partial\bar{\xi}^{p}}\partial\dot{\bar{\xi}}^{p} + \frac{\partial\bar{\Lambda}^{p}}{\partial\bar{\lambda}^{p}}\dot{\bar{\lambda}}^{p} + \frac{\partial\bar{\Xi}^{d}}{\partial\bar{\xi}^{d}}\partial\dot{\bar{\xi}}^{d}\right)$$

$$-\left(\frac{\partial\bar{\bar{\Psi}}^{e}}{\partial\left(\dot{\bar{\bar{u}}} - \bar{\bar{u}}^{p}\right)}\left(\dot{\bar{\bar{u}}}^{-} - \dot{\bar{\bar{u}}}^{p}\right) + \frac{\partial\bar{\bar{\Xi}}^{p}}{\partial\bar{\bar{\xi}}^{p}}\partial\dot{\bar{\bar{\xi}}}^{p}\right)\delta_{\bar{x}}$$

It is clear that $\dot{\bar{\varepsilon}}^{-} = \dot{\bar{\varepsilon}}^{v} + \dot{\bar{\varepsilon}}^{p}$ with $\dot{\bar{\varepsilon}}^{v} = \dot{\bar{\varepsilon}}^{d}$ and $\dot{\bar{\bar{u}}}^{-} = \dot{\bar{\bar{u}}}^{e} + \dot{\bar{\bar{u}}}^{p}$. Thus:

$$\begin{aligned}\dot{\mathcal{D}}^{-} = &\left(\sigma^{d} - \frac{\partial\bar{\Psi}^{d}}{\partial\left(\bar{\varepsilon}^{-} - \bar{\varepsilon}^{p}\right)}\right)\dot{\bar{\varepsilon}}^{d} + \sigma^{v}\dot{\bar{\varepsilon}}^{v,-} + \sigma\dot{\bar{\varepsilon}}^{p} - \frac{\partial\bar{\Psi}^{d}}{\partial\bar{D}}\dot{\bar{D}} - \frac{\partial\bar{\Xi}^{p}}{\partial\bar{\xi}^{p}}\partial\dot{\bar{\xi}}^{p} - \frac{\partial\bar{\Lambda}^{p}}{\partial\bar{\lambda}^{p}}\dot{\bar{\lambda}}^{p}\\ &-\frac{\partial\bar{\Xi}^{d}}{\partial\bar{\xi}^{d}}\partial\dot{\bar{\xi}}^{d} + \left(t - \frac{\partial\bar{\bar{\Psi}}^{e}}{\partial\left(\bar{\bar{u}}^{-} - \bar{\bar{u}}^{p}\right)}\right)\dot{\bar{\bar{u}}}^{e}\delta_{\bar{x}} + t\dot{\bar{\bar{u}}}^{p}\delta_{\bar{x}} - \frac{\partial\bar{\bar{\Xi}}^{p}}{\partial\bar{\bar{\xi}}^{p}}\partial\dot{\bar{\bar{\xi}}}^{p}\delta_{\bar{x}}\end{aligned} \tag{3.55}$$

Note that $\bar{\varepsilon}^{e}, \bar{\bar{u}}^{d}, \bar{\varepsilon}^{d}, \bar{\bar{u}}^{e}$ are not internal variables. Moreover, when there is no evolution of any internal variable, the dissipation is null. Therefore, according to equations (3.54) and (3.55), the following state equations are valid:

$$\sigma^{e} = \frac{\partial\bar{\Psi}^{e}}{\partial\bar{\varepsilon}^{e}}, \sigma^{d} = \frac{\partial\bar{\Psi}^{d}}{\partial\bar{\varepsilon}^{d}}, t^{+} = \frac{\partial\bar{\bar{\Psi}}^{d}}{\partial\bar{\bar{u}}^{d}}, t^{-} = \frac{\partial\bar{\bar{\Psi}}^{e}}{\partial\bar{\bar{u}}^{e}} \tag{3.56}$$

For the thermodynamic forces, expressions that will then lead to an accurate reproduction of the experimentally observed behavior of a concrete REV, the different terms in the expression of the Helmholtz free energy are defined as follows:

$$\begin{aligned}
&\bar{\psi}^e = \frac{1}{2}\bar{\varepsilon}^e \bar{E} \bar{\varepsilon}^e && \bar{\psi}^d = \frac{1}{2}\left(\bar{\varepsilon}^- - \bar{\varepsilon}^p\right)\bar{D}^{-1}\left(\bar{\varepsilon}^- - \bar{\varepsilon}^p\right) \\
&\bar{\bar{\psi}}^d = \frac{1}{2}\bar{\bar{u}}^d \bar{\bar{D}}^{-1} \bar{\bar{u}}^d && \bar{\Xi}^p = \frac{1}{2}\bar{\xi}^p \bar{K}^p \bar{\xi}^p \\
&\bar{\bar{\Xi}}^d = -\left(\sigma_u^t - \sigma_\infty\right)\left(\bar{\bar{\xi}}^d + \frac{1}{a}e^{-a\bar{\bar{\xi}}^d}\right) && \bar{\Lambda}^p = \frac{1}{2}\bar{\lambda}^p \bar{H}^p \bar{\lambda}^p \\
& && \bar{\Xi}^d = \frac{1}{2}\bar{\xi}^d \bar{K}^d \bar{\xi}^d \\
& && \bar{\bar{\psi}}^e = \left(\bar{\bar{u}}^- - \bar{\bar{u}}^p\right)\bar{\bar{E}}\left(\bar{\bar{u}}^- - \bar{\bar{u}}^p\right) \\
& && \bar{\bar{\Xi}}^p = \frac{1}{2}\bar{\bar{\xi}}^p \bar{\bar{K}}^p \bar{\bar{\xi}}^p
\end{aligned} \tag{3.57}$$

The state equations of the internal variables are defined from the Helmholtz free energy functional that plays the role of a thermodynamic potential:

$$\mathcal{A}_i = -\frac{\partial \psi}{\partial \alpha_i} \tag{3.58}$$

Finally, the state equations of the system, that is the equations that characterize the state of a concrete material point at a given instant in $\mathcal{A}$-space, are given in Table 3.4.

3.5.1.2.2 Equations of Evolution of the Internal Variables

Here, the equations that govern the evolution of the internal variables of this concrete phenomenological constitutive law are determined. The same as in the models for steel, it is assumed that the evolution is driven by the principle of maximum dissipation. Table 3.4 presents the state equations.

Taking this into account the dissipation equations (3.54) and (3.55) can be rewritten as:

$$\begin{aligned}
\dot{\mathcal{D}} = {} & \underbrace{\sigma^v \dot{\bar{\varepsilon}} + \sigma \dot{\bar{\varepsilon}}^p}_{\dot{\bar{D}}^v} + \underbrace{\bar{q}^p \dot{\bar{\xi}}^p + \bar{\kappa}^p \dot{\bar{\lambda}}^p}_{\dot{\bar{D}}^p} + \underbrace{\frac{1}{2}\sigma \dot{\bar{D}} \sigma + \bar{q}^d \dot{\bar{\xi}}^d}_{\dot{\bar{D}}^d} \\
& + \underbrace{\left(t \dot{\bar{\bar{u}}}^p + \bar{\bar{q}}^p \dot{\bar{\bar{\xi}}}^p\right)\delta_{\bar{x}}}_{\dot{\bar{\bar{D}}}^p} + \underbrace{\left(\frac{1}{2} t \dot{\bar{\bar{D}}} t + \bar{\bar{q}}^d \dot{\bar{\bar{\xi}}}^d\right)\delta_{\bar{x}}}_{\dot{\bar{\bar{D}}}^d}
\end{aligned} \tag{3.59}$$

To characterize the evolution of the system, except the viscous one, the principle of maximum dissipation is to this model and thus the flow of the internal variables is computed by assuming that they maximize the dissipation. A problem of constrained maximization that involves inequalities has to be solved, recalling that the criteria functions equations (3.52) and (3.53) must be satisfied. Denoting $f = \left(\sigma, t, q^p, \kappa^p, q^d, \bar{\bar{q}}^p, \bar{\bar{q}}^d\right)$, the following problem has to be solved:

$$\max_f \dot{\mathcal{D}} \text{ with constraints } \phi^p(f) \le 0, \phi^d(f) \le 0, \bar{\bar{\phi}}^p(f) \le 0, \bar{\bar{\phi}}^d(f) \le 0 \tag{3.60}$$

This problem is solved by utilizing the Lagrange multipliers method and the Lagrangian is defined as:

$$\phi^p(f) \le 0, \phi^d(f) \le 0, \bar{\bar{\phi}}^p(f) \le 0, \bar{\bar{\phi}}^d(f) \le 0 \tag{3.61}$$

where $\dot{\gamma} = \left(\dot{\gamma}^p, \dot{\gamma}^d, \dot{\bar{\bar{\gamma}}}^p, \dot{\bar{\bar{\gamma}}}^d\right)$ is the set of Lagrange's multipliers associated to each constraint. According to the Kuhn–Tucker defined in Chapter 2 equation (1.11), and here repeated due to consistency. If the set $f^* = \left\{f_i^*\right\}$ is a solution of the problem, then there exists a unique set of Lagrange multipliers such that the following relations are verified for all i and for all j where the last three conditions are referred to as the loading/unloading conditions.

$$\frac{\partial \mathcal{L}\left(f^*, \dot{\gamma}^*\right)}{\partial f_i} = 0, \frac{\partial \mathcal{L}\left(f^*, \dot{\gamma}^*\right)}{\dot{\gamma}_j} \geq 0, \dot{\gamma}_j^* \geq 0, \dot{\gamma}_j^* \phi_j\left(f^*\right) = 0 \tag{3.62}$$

Finally, the equations of evolution of the internal variables are given in Table 3.5.

For the viscous internal variable $\bar{\varepsilon}^v$, the evolution is expressed, in an associative form to (Perzyna 1966):

$$\dot{\bar{\varepsilon}}^v = \frac{\bar{\phi}^v}{\eta} \frac{\partial \bar{\phi}^v}{\partial \sigma} \tag{3.63}$$

Table 3.4 State Equations of the System

Tension	*Compression*
$\sigma^e = \bar{E}\bar{\varepsilon}^e$	$\sigma^d = \bar{D}\bar{\varepsilon}^d$
$t = \bar{\bar{D}}^{-1}\bar{\bar{u}}^d$	$t = \bar{\bar{E}}\bar{\bar{u}}^e$
$\bar{q}^d = -\frac{\bar{\Xi}^d}{\bar{\bar{\xi}}^d} = \left(\sigma_u^t - \sigma_\infty\right)\left(1 - e^{-a\bar{\xi}^d}\right)$	$\bar{q}^p = -\frac{\bar{\Xi}^p}{\bar{\xi}^p} = -\bar{K}^p\bar{\xi}^p$
$\frac{1}{2}t^2 = -\frac{\partial \bar{\bar{\psi}}^d}{\partial \bar{\bar{D}}}$	$\bar{\kappa}^p = -\frac{\bar{\Lambda}^p}{\bar{\lambda}^p} = -\bar{H}^p\bar{\lambda}^p$
	$\bar{q}^d = -\frac{\bar{\Xi}^d}{\bar{\xi}^d} = -\bar{K}^d\bar{\xi}^d$
	$\frac{1}{2}\sigma^2 = -\frac{\bar{\bar{\Xi}}^p}{\bar{\bar{\xi}}^p} = -\bar{\bar{K}}^p\bar{\bar{\xi}}^p$

Table 3.5 Equations of the Evolution of the Internal Variables

Continuous plasticity	*Continuous damage*
$\dot{\varepsilon}^p = \dot{\gamma}^p \frac{\partial \phi^p}{\partial \sigma}$	$\dot{D}\sigma = \dot{\gamma}^d \frac{\partial \phi^d}{\partial \sigma}$
$\dot{\xi}^p = \dot{\gamma}^p \frac{\partial \phi^p}{\partial q^p}$	$\dot{\xi}^d = \dot{\gamma}^d \frac{\partial \phi^d}{\partial q^d}$
$\dot{\lambda}^p = \dot{\gamma}^p \frac{\partial \phi^p}{\partial \kappa^p}$	
$\dot{\gamma}^p \geq 0, \phi^p \leq 0, \dot{\gamma}^p\phi^p = 0$	$\dot{\gamma}^d \geq 0, \phi^d \leq 0, \dot{\gamma}^d\phi^d = 0$
Discrete plasticity	Discrete damage
$\dot{\bar{\bar{u}}}^p = \dot{\bar{\bar{\gamma}}}^p \frac{\partial \bar{\bar{\phi}}^p}{\partial t}$	$\dot{\bar{\bar{D}}}t = \dot{\bar{\bar{\gamma}}}^d \frac{\partial \bar{\bar{\phi}}^d}{\partial t}$
$\dot{\bar{\bar{\xi}}}^p = \dot{\bar{\bar{\gamma}}}^p \frac{\partial \bar{\bar{\phi}}^p}{\partial \bar{\bar{q}}^p}$	$\dot{\bar{\bar{\xi}}}^d = \dot{\bar{\bar{\gamma}}}^d \frac{\partial \bar{\bar{\phi}}^d}{\partial \bar{\bar{q}}^d}$
$\dot{\bar{\bar{\gamma}}}^p \geq 0, \bar{\bar{\phi}}^p \leq 0, \dot{\bar{\bar{\gamma}}}^p\bar{\bar{\phi}}^p = 0$	$\dot{\bar{\bar{\gamma}}}^d \geq 0, \bar{\bar{\phi}}^d \leq 0, \dot{\bar{\bar{\gamma}}}^d\bar{\bar{\phi}}^d = 0$

where η is the viscous parameter assumed here to be rate-insensitive although this might not be the case (Pedersen, Simone and Sluys 2008). It is interesting to see that (3.63) is in accordance with the rheological relation in a dashpot $\sigma^v = \dot{\varepsilon}^v \eta$.

Note also that concrete is in general not an associative material. However, in the 1D context of this paper, we nevertheless assume that it is the case. All the equations of evolution of the internal variables are thus expressed in an associative form. The normality of the flow rule to the loading surface is in particular implied by the appeal to the maximum dissipation principle.

3.5.1.3 Euler–Lagrange Equations of a Concrete 1D Structure in Dynamic Loading

The local constitutive model is used as the basis for the formulation of the structural, mechanical problem. This latter, illustrated here in the 1D case, has to be adapted to the local constitutive model and requires the definition of an enhanced displacement field.

3.5.1.3.1 Enhanced Displacement Field Kinematics

Figure 3.22 illustrates that the displacement field is written as the sum of a smooth linear part $\tilde{u}(x,t)$ enhanced by an additional part $\bar{\bar{u}}(x,t)$ to take into account the possible appearance of a displacement jump $\bar{\bar{u}}(x,t)$ in the 1D structure (Garikipati and Hughes 1998; Ibrahimbegović and Brancherie 2003)

$$u(x,t) = \tilde{u}(x,t) + \underbrace{\left(\mathcal{H}_{\bar{x}}(x) - \varphi(x)\right)\bar{\bar{u}}(t)}_{\bar{\bar{u}}(x,t)} \tag{3.64}$$

which can also be written, in the form adapted for identification with (3.50), as:

$$u(x,t) = \underbrace{\tilde{u}(x,t) - \varphi(x)\bar{\bar{u}}(t)}_{\bar{u}(x,t)} + \bar{\bar{u}}(t)\mathcal{H}_{\bar{x}}(x) \tag{3.65}$$

Then with $\varepsilon(x,t) = \bar{\varepsilon}(x,t) + \bar{\bar{u}}(t)\delta_{\bar{x}}$

$$\bar{\varepsilon}(x,t) = \frac{\partial \bar{u}(x,t)}{\partial x} = \frac{\partial \tilde{u}(x,t)}{\partial x} - \frac{d\varphi(x)}{dx}\bar{\bar{u}}(t) \tag{3.66}$$

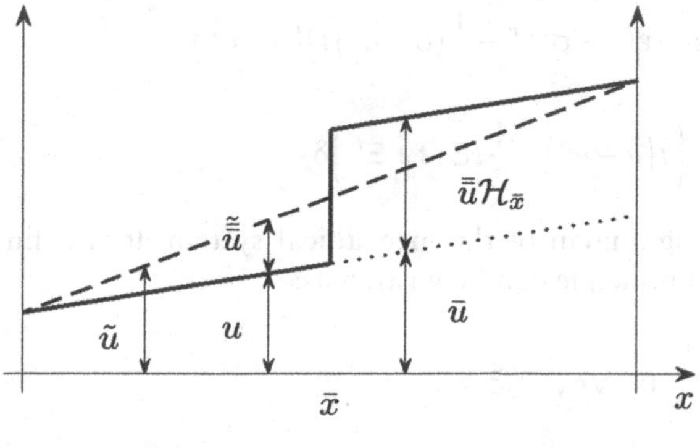

Figure 3.22 Enhanced displacement field in a 1D structure. The solid, dashed and dotted lines represent the enriched displacement field, the smooth displacement field and the continuum displacement field.

In this 1D context, only failure mode I is considered; we refer to Dujc et al. (2010) for a presentation of the kinematics with an embedded crack of quadrilateral 2D finite elements where several failure modes are considered.

3.5.1.3.2 Complementary Lagrangian Variational Formulation

The kinetic energy, assumed to depend only on the smooth part of the displacement, is:

$$T(\tilde{u},t)=\frac{1}{2}\int_{\Omega}\left(\frac{\partial\tilde{u}(x,t)}{\partial t}\right)d\Omega \tag{3.67}$$

where the volume mass ρ in the volume Ω is considered as constant.

The total potential energy can be written:

$$U\left(\tilde{u},\bar{\bar{u}},\alpha,t\right)=\int_{\Omega}\psi\left(\tilde{u},\bar{\bar{u}},\alpha,t\right)d\Omega-U^{\text{ext}}\left(\tilde{u},t\right) \tag{3.68}$$

To introduce the stress fields σ and t as other independent variables of the problem and thus give a more general setting that can be helpful to derive the governing equations of the problem (Markovic and Ibrahimbegović 2006; Simo, Kennedy and Taylor 1989), it is necessary to appeal to the partial Legendre transformation of the Helmholtz free energy:

$$\begin{aligned}\psi^{+}&=\sigma^{e}\bar{\varepsilon}^{e}-\frac{1}{2}\sigma^{e}\bar{E}\sigma^{e}+\left(t\bar{\bar{u}}^{d}-\frac{1}{2}t\bar{\bar{D}}t+\bar{\bar{\Xi}}^{d}\right)\delta_{\bar{x}}\\ \psi^{-}&=\sigma^{d}\bar{\varepsilon}^{d}-\frac{1}{2}\sigma^{d}\bar{D}\sigma^{d}+\bar{\Xi}^{p}+\bar{\Lambda}^{p}+\bar{\Xi}^{d}+\left(t\bar{\bar{u}}^{e}-\frac{1}{2}t\bar{\bar{E}}^{-1}t+\bar{\bar{\Xi}}^{p}\right)\delta_{\bar{x}}\end{aligned} \tag{3.69}$$

Recalling that $\bar{\varepsilon}^{e}=\bar{\varepsilon}^{v}$ in tension, $\bar{\varepsilon}^{d}=\bar{\varepsilon}^{v}$ in compression, $\sigma=\sigma^{v}+\sigma^{e/d}$ and according to equations (3.50) and (3.66), these equations can be rewritten with respect to $\tilde{u}(x,t),\bar{\bar{u}}(t),\sigma(x,t),t(t)$ and α:

$$\begin{aligned}\psi^{+}&=\sigma\left(\frac{\partial\tilde{u}}{\partial x}-\frac{d\varphi}{dx}\bar{\bar{u}}-\bar{\varepsilon}^{p}\right)-\sigma^{v}\bar{\varepsilon}^{v}-\frac{1}{2}\left(\sigma-\sigma^{v}\right)\bar{E}^{-1}\left(\sigma-\sigma^{v}\right)+\\&+\left(t\left(\bar{\bar{u}}-\bar{\bar{u}}^{p}\right)-\frac{1}{2}t\bar{\bar{D}}t+\bar{\bar{\Xi}}^{d}\right)\delta_{\bar{x}}\\ \psi^{-}&=\sigma\left(\frac{\partial\tilde{u}}{\partial x}-\frac{d\varphi}{dx}\bar{\bar{u}}-\bar{\varepsilon}^{p}\right)-\sigma^{v}\bar{\varepsilon}^{v}-\frac{1}{2}\left(\sigma-\sigma^{v}\right)\bar{D}\left(\sigma-\sigma^{v}\right)\\&++\bar{\Xi}^{p}+\bar{\Lambda}^{p}+\bar{\Xi}^{d}+\left(t\left(\bar{\bar{u}}-\bar{\bar{u}}^{p}\right)-\frac{1}{2}t\bar{\bar{E}}^{-1}t+\bar{\bar{\Xi}}^{p}\right)\delta_{\bar{x}}\end{aligned} \tag{3.70}$$

L is denoted as the Lagrangian of the mechanical system. It is defined as $L=T-U$ and Lagrange's variational principle can be written as:

$$\int_{t_1}^{t_2}\delta L\left(\tilde{u},\bar{\bar{u}},\sigma,\mathrm{t},t\right)dt=0\quad\forall t_1,\forall t_2\dot{\xi} \tag{3.71}$$

with $\delta\tilde{u}(x,t),\delta\bar{\bar{u}}(t),\delta\sigma(x,t),\delta t(t)=0$ when $t=t_1$ and $t=t_2$.

The following set of governing equations available for tension and compression are obtained after some mathematical operations:

$$\begin{aligned}
&\int_{\Omega} \rho \frac{\partial^2 \tilde{u}}{\partial t^2} \delta\tilde{u} d\Omega + \int_{\Omega} \frac{\partial \psi}{\partial \tilde{u}} \delta\tilde{u} d\Omega - \delta U^{ext} = 0 \\
&\int_{\Omega} \frac{\partial \psi}{\partial \bar{\bar{u}}} d\Omega \delta\bar{\bar{u}} = 0 \\
&\int_{\Omega} \frac{\partial \psi}{\partial \sigma} \delta\sigma d\Omega = 0 \\
&\int_{\Omega} \frac{\partial \psi}{\partial t} d\Omega \delta t = 0
\end{aligned} \tag{3.72}$$

which leads, with Γ denoting the section in which the displacement is likely to occur and by remembering that neither $\bar{\bar{u}}$ nor t depends on the position x (they are only defined at the position x_Γ of the section Γ), to the set of Euler–Lagrange equations of a concrete 1D structure written in Table 3.6.

The first equation in Table 3.6 enforces a satisfying level of global force equilibrium (either for tension or compression); the second equation gives the condition of compatibility between the continuum σ and discrete t stresses; the third and fourth equations correspond to the weak form of the local continuum and discrete constitutive models.

One can take advantage of the dependency between the continuum and the discrete stresses that appear in Table 3.6: $\left(\int_{\Omega} \frac{d\varphi}{dx} \sigma - t\delta_{\bar{x}} \right) d\Omega = 0 \Rightarrow t = t(\sigma)$ and give expressions of the Lagrangian in the form:

$$\begin{aligned}
&\int_{\Omega} \psi^{+} d\Omega = \int_{\Omega} \left(\sigma \left(\frac{\partial \tilde{u}}{\partial x} - \varepsilon^{v} - \varepsilon^{p} \right) + \left(-t\bar{u}^{p} - \frac{1}{2} t\bar{\bar{D}} t + \bar{\bar{\Xi}}^{d} \right) \delta_{\bar{x}} \right) d\Omega \\
&\int_{\Omega} \psi^{-} d\Omega = \int_{\Omega} \left(\sigma \left(\frac{\partial \tilde{u}}{\partial x} - \varepsilon^{v} - \varepsilon^{p} \right) + \Xi^{p} + \Lambda^{p} + \Xi^{d} + \left(-t\bar{u}^{p} - \frac{1}{2} t\bar{\bar{E}}^{-1} t + \bar{\bar{\Xi}}^{p} \right) \delta_{\bar{x}} \right) d\Omega
\end{aligned} \tag{3.73}$$

Table 3.6 Euler–Lagrange Equations of a Concrete 1D Structure

Tension	*Compression*
$\int_{\Omega} \left(\rho \frac{\partial^2 \tilde{u}}{\partial t^2} \delta\tilde{u} + \frac{\partial \delta\tilde{u}}{\partial x} \sigma \right) d\Omega - \delta U^{ext} = 0$	$\int_{\Omega} \left(\rho \frac{\partial^2 \tilde{u}}{\partial t^2} \delta\tilde{u} + \frac{\partial \delta\tilde{u}}{\partial x} \sigma \right) d\Omega - \delta U^{ext} = 0$
$\left(\int_{\Omega} \frac{d\varphi}{dx} \sigma - t\delta_{\bar{x}} \right) d\Omega \delta\bar{\bar{u}} = 0$	$\left(\int_{\Omega} \frac{d\varphi}{dx} \sigma - t\delta_{\bar{x}} \right) d\Omega \delta\bar{\bar{u}} = 0$
$\int_{\Omega} \delta\sigma \left(\frac{\partial \tilde{u}}{\partial x} - \frac{d\varphi}{dx} \bar{\bar{u}} - \varepsilon^{v} - \varepsilon^{p} \right) d\Omega = 0$	$\int_{\Omega} \delta\sigma \left(\frac{\partial \tilde{u}}{\partial x} - \frac{d\varphi}{dx} \bar{\bar{u}} - \varepsilon^{v} - \varepsilon^{p} \right) d\Omega = 0$
$\int_{\Gamma} \left(\bar{\bar{u}} - \bar{\bar{u}}^{p} - \bar{\bar{D}} t \right) d\Gamma \delta t = 0$	$\int_{\Gamma} \left(\bar{\bar{u}} - \bar{\bar{u}}^{p} - \bar{\bar{E}}^{-1} t \right) d\Gamma \delta t = 0$

Moreover, the response of the localization section is supposed to be infinitely rigid before the stress reaches an ultimate value and therefore $\bar{\bar{E}}^{-1} \to 0$. Which, finally, denotes A_Γ as the area of the section Γ where displacement localization can take place, the simplified form of these equations is written in Table 3.7. Only two unknown fields remain: the smooth displacement $\tilde{u}(x,t)$ and the continuum stress $\sigma(x,t)$.

3.5.2 Numerical Examples

As in the previous examples, this numerical implementation was performed using the general purpose finite element computer program FEAP. Due to the fact any external distributed loading is involved (self-weight is, in particular, neglected) in these applications, the concrete structure is always modeled with constant stress bar elements. Figure 3.23 shows the 1D structure that was tested.

Neglecting the concrete mass $m = \rho AL$ as compared to the added mass M, the stiffness k, the fundamental pulsation w and period T of this structure is computed as:

$$k = \frac{ES}{L} = 1.4\,\text{GPa}; \omega = \sqrt{\frac{k}{m}} = 44.7\,\text{s}^{-1}; T = \frac{2\pi}{\omega} = 0.14\,\text{s} \tag{3.74}$$

3.5.2.1 *Identification of a Concrete Law*

All the parameters of the 1D cyclic concrete model presented here have a clear physical meaning. It will be shown that (i) the viscous parameter η is related to the material critical damping ratio ξ, (ii) the tensile softening law parameter a is related to the fracture energy G_f and (iii) all the other parameters can be directly identified from the experimental stress–strain response of a concrete specimen in quasi-static cyclic loading.

Table 3.7 Final Euler–Lagrange Equations of a Concrete 1D Structure

Tension	*Compression*
$\int_\Omega \left(\rho \frac{\partial^2 \tilde{u}}{\partial t^2} \delta\tilde{u} + \frac{\partial \delta\tilde{u}}{\partial x} \sigma \right) d\Omega - \delta U^{ext} = 0$	$\int_\Omega \left(\rho \frac{\partial^2 \tilde{u}}{\partial t^2} \delta\tilde{u} + \frac{\partial \delta\tilde{u}}{\partial x} \sigma \right) d\Omega - \delta U^{ext} = 0$
$\int_\Omega \delta\sigma \left(\frac{\partial \tilde{u}}{\partial x} - \bar{\varepsilon}^v - \bar{\varepsilon}^p - \left(\bar{\bar{u}}^p + \bar{\bar{D}}t \right) \frac{dt}{d\sigma} \delta_{\bar{x}} \right) d\Omega = 0$	$\int_\Omega \delta\sigma \left(\frac{\partial \tilde{u}}{\partial x} - \bar{\varepsilon}^v - \bar{\varepsilon}^p - \bar{\bar{u}}^p \frac{dt}{d\sigma} \delta_{\bar{x}} \right) d\Omega = 0$
with $t = t(\sigma) = \frac{1}{A} \int_\Omega \frac{d\varphi}{dx} \sigma d\Omega$	with $t = t(\sigma) = \frac{1}{A} \int_\Omega \frac{d\varphi}{dx} \sigma d\Omega$

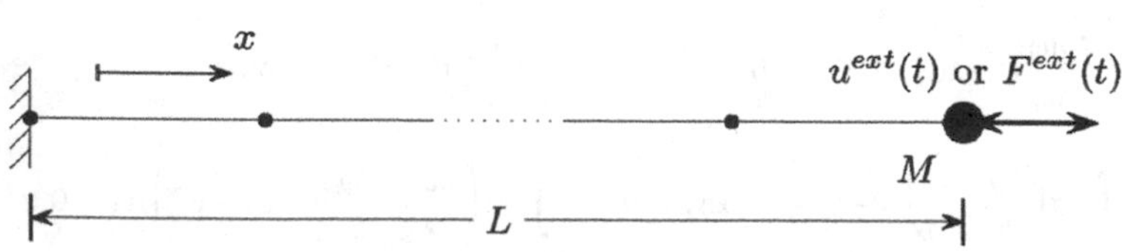

Figure 3.23 1D concrete structure model used for the numerical applications. The cross-section area $S = 0.04\ \text{m}^2$, the length $L = 1$ m, the added mass $M = 7.0\ \text{E}5$ kg, the volume mass $\rho = 2400\ \text{kg/m}^3$ and the concrete elastic modulus $E = 35$ GPa.

a) Identification of the Viscous Parameter η

The viscous parameter η can be interpreted as a material property that can be identified according to, among others, the experimental results of free low-amplitude—so that the structure remains elastic—vibration tests (Figure 3.24):

1. First, denoting ξ as the critical damping ratio and c as the structural viscous parameter, if one assumes that $\xi << 1$, one can write (Chopra 2001):

$$c = 2\xi m\omega \tag{3.75}$$

2. Then, the material viscous parameter η can be connected to structural parameter c starting from the 1D local form of the equilibrium, and the Kelvin–Voigt model constitutive equations:

$$\rho\ddot{u} + \frac{\partial\sigma}{\partial x} = 0; \sigma = E\frac{\partial u}{\partial x} + \eta\frac{d}{dt}\left(\frac{\partial u}{\partial x}\right) \tag{3.76}$$

For a tested structure with constant cross-section area S and strain field $(\partial u / \partial x = u / L)$, the introduction of $(3.76)_1$ into $(3.76)_2$ and then the integration of the resulting local equation over the whole structure gives:

$$\underbrace{\rho SL}_{m}\ddot{u} + \underbrace{\frac{S}{L}\eta}_{c}\dot{u} + \underbrace{\frac{ES}{L}}_{k}u = 0 \tag{3.77}$$

3. The Kelvin–Voigt viscoelastic model exhibits a hysteretic—energy dissipative—behavior in cyclic loading as illustrated in Figure 3.25 for excitation of the form with, except for other indicated values, $A = 0.3$ mm, $\Omega = 157.08$ s–1 and $\tau = 0.001$ s. The amount of energy dissipated per cycle D^{cyc}—the area of the elliptical loop—can be analytically written as (Wang 2009):

$$D^{cyc} = \pi c\Omega A^2 \tag{3.78}$$

The main drawback of the Kelvin–Voigt model is that D^{cyc} is dependent on the loading frequency Ω. It is indeed not realistic in particular for seismic excitations

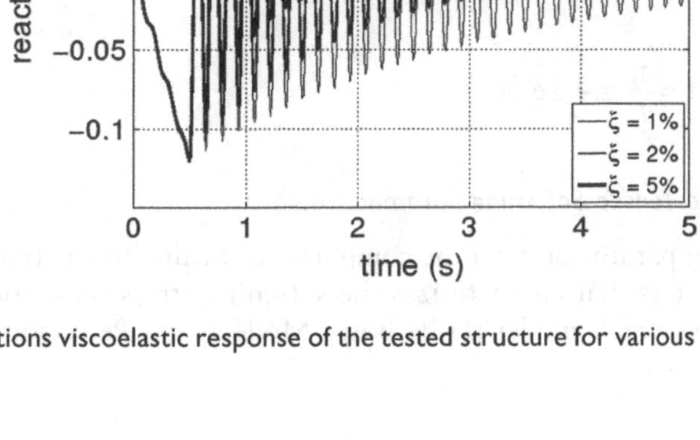

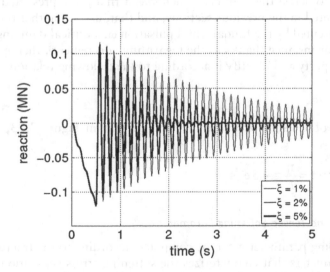

Figure 3.24 Free-vibrations viscoelastic response of the tested structure for various critical damping ratios.

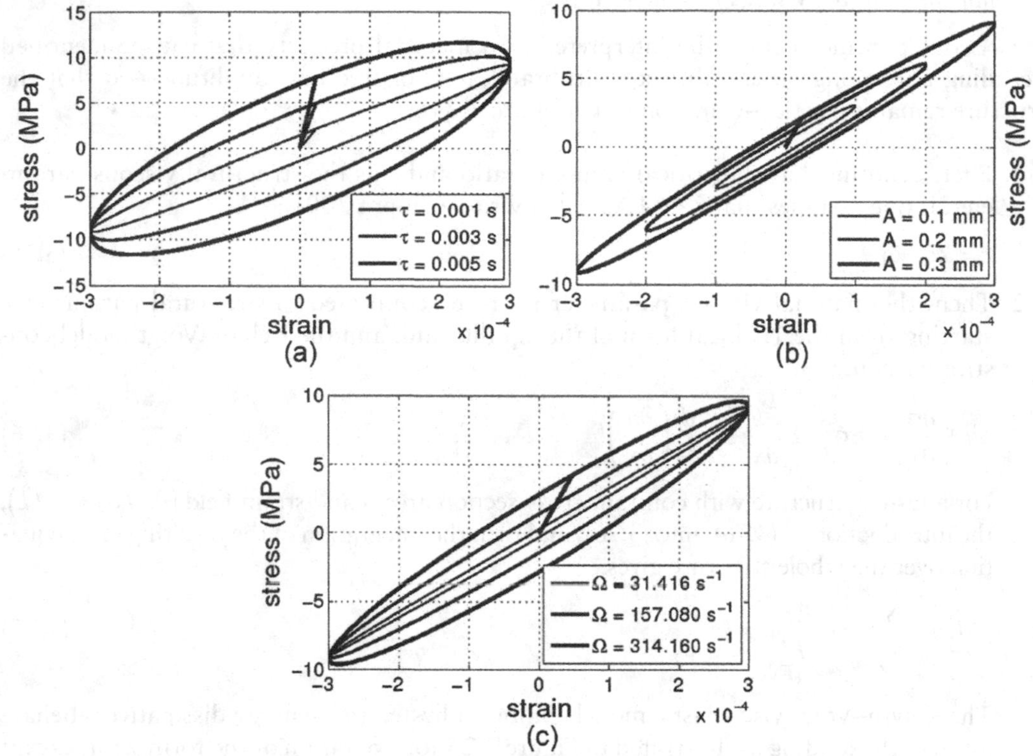

Figure 3.25 Hysteretic response of the viscoelastic model in quasi-static sine loading for several (a) viscous parameters, (b) loading amplitudes and (c) forcing pulsations.

(Wang 2009). This dependency is only negligible in the cases where the response is primarily represented by the resonant pulsation, that is $\Omega \simeq \omega$.

4. If it is supposed that a concrete structure is excited by an impulsion that is weak enough not to active the nonlinear phenomena that are represented by the concrete model presented in the previous sections, and that, assuming that the response is primarily represented by the fundamental pulsation ω, a critical damping ratio $\xi = 1\%$ is measured (for instance thanks to the logarithmic decrement), then η can be seen as a material property and identify it according to the following relation:

$$\eta = \frac{2\xi m\omega C}{S} \tag{3.79}$$

In the case of a tested structure identical to that in Figure 3.23, the following is computed:

$$\eta = 15.7\,\text{MPa}; \tau = \frac{\eta}{E} \cong 4.5\,\text{e}^{-4}\text{s} \tag{3.80}$$

b) Identification of the Tensile Softening Parameter a

The tensile softening parameter a can be computed according to the fracture energy G_f. G_f is one of the parameters that characterizes the softening stress-separation (crack-opening) curve of the cohesive crack model (Hillerborg, Modéer and Petersson 1976; Bažant, Yu and Zi 2002) and represents the total amount of energy that has to be furnished in tension to a section between the time when displacement begins to localize and the time when the local softening problem becomes (numerically) ill-posed, because $\bar{\bar{K}}^d < 0$; physically, t_{cri} has to coincide with the time when the section is completely broken: when all the energy G_f is consumed. In the case of this concrete model, one can demonstrate that t_{cri} is always defined and that:

$$G_f = \int_{t_{\text{loc}}}^{t_{\text{cri}}} t\dot{\bar{\bar{u}}}dt = \frac{\sigma_u^t}{2a} \Rightarrow a = \frac{\sigma_u^t}{2G_f} \tag{3.81}$$

In addition, there is an empirical relation (Bažant and Becq-Giraudon 2002) that allows for computing G_f according to material characteristics:

$$G_f = 2.5\alpha_0\left(\frac{\sigma_u^c}{0.051}\right)^{0.46}\left(1+\frac{d_a}{11.27}\right)^{0.22}\left(\frac{w}{c}\right)^{-0.30} \tag{3.82}$$

with σ_u^c in MPa and where $\alpha_0 = 1$ for rounded aggregates, $\alpha_0 = 1.44$ for crushed or angular aggregates, d_a is the maximum aggregate size in mm and w/c is the water-cement ratio. Finally, for concrete with: $\alpha_0 = 1.44, d_a = 25\,\text{mm}, w/c = 0.5$, computed fracture energy is $G_f = 136\,\text{N/m}$ that leads, from (3.80) with $\sigma_u^t = 3\,\text{MPa}$, to:

$$a = 22060\ \text{m}^{-1} \tag{3.83}$$

c) Identification of the Other Parameters

The remaining parameters of the model can be identified from the stress–strain experimental response in Figure 3.21 (a).

1. Identify Young's modulus for tensile and compressive parts. It is considered that they have the same value: $E = \bar{E} = 35$ GPa.
2. Identify the set of stresses so as to characterize the changes in the slope of the backbone curve. It is chosen that: $\sigma_y = 3$ MPa, $\sigma_f = 37$ MPa, $\sigma_u^c = 49$ MPa, $\sigma_u^t = 0.6 \cdot \sigma_u^c = 3$ MPa σ_y=3 MPa, σ_f=37 MPa, =49 MPa and =0.6 · =3 MPa.
3. Simultaneously identify $\bar{K}^p$ and $\bar{H}^p$ to describe both the strain hardening phase of the backbone curve in the range $\sigma_y < \sigma < \sigma_f$ and the shape of the local hysteresis loops.
4. Identify $\bar{K}^d$ so as to describe the remaining strain hardening phase of the backbone curve in the range of $\sigma_f < \sigma < \sigma_u^c$.
5. Identify $\bar{\bar{K}}^p$ so as to describe the softening part of the backbone curve.

Figure 3.21 (b) is plotted with the parameters indicated above. Note that although all the parameters have a clear physical meaning, they are also all connected and the results of the first identification process often thus need to be refined. Indeed, the slope of the hardening part of the backbone curve in the range of $\sigma_y < \sigma < \sigma_f$ is $C_1 = (\bar{E}\bar{K}^p)/(\bar{E}+\bar{K}^p)$, in the range of $\sigma_f < \sigma < \sigma_u^c$ is $C_2 = (C_1\bar{K}^d)/(C_1+\bar{K}^d)$and the slope of the softening part of the backbone curve is :$C_3 = (C_2\bar{\bar{K}}^p)/(C_2+\bar{\bar{K}}^p)$. Note also that although we observed that activating plasticity before damage leads to better identification, it is possible to invert the roles of σ_y and σ_f.

d) Mesh Objectivity

In the formulation of this model, we introduced a strong discontinuity—a displacement jump—and thus concentrated the localization in a zero-length zone (Ibrahimbegović and Brancherie 2003; Simo, Oliver and Armero 1993). This method leads to a formulation that does not need any characteristic length and thus satisfies the mesh objectivity requirement, that is the unicity of the solution.

e) Loading-Rate-Sensitive Response

The four main effects of the loading strain rate on the response of concrete are: the increases of (i) the compressive strength (Bischoff and Perry 1995; Lu and Xu 2004), (ii) the tensile strength (Weerheijm and Van Doormaal 2007), (iii) Young's modulus (Weerheijm and Van Doormaal 2007) and (iv) the brittle behavior (Dilger, Koch and Kowalczyk 1984). For seismic excitations, the strain rates are comprised between 10^{-5} 1/s and 10^{-2} 1/s, which respectively corresponds to an increase of (i) the compression strength between 0 and 30% (Bischoff and Perry 1995; Lu and Xu 2004), (ii) the tensile strength between 0 and 60% (Weerheijm and Van Doormaal 2007) and (iii) Young's modulus between 0 and 10% (Weerheijm and Van Doormaal 2007).

The numerical applications presented in Figure 3.26(a) show that Young's modulus computed for a loading strain rate of 10^{-2} 1/s is about 5.7% larger than the one computed for a loading strain rate of 10^{-5} 1/s. The increase of Young's modulus is thus represented by the model but a little underestimated here. However, this increase is related to the viscous parameter τ computed in these numerical applications from the damping ratio ξ=1% whose value has been arbitrarily chosen: one can thus hope that, in reality, the right value for the damping ratio would lead to correct modeling of the increase of elastic stiffness. The compressive response is presented in Figure 3.26(b). The proposed concrete model has not yet been developed to represent the rate-dependent tensile and compressive strengths by itself. This can be done in a simplified manner with the commonly accepted practice in engineering where the concrete strength parameters are majorated for seismic analysis.

3.5.3 Numerical Examples Seismic Application

a) Global Seismic Response

It is only the compressive part of the concrete law that is in focus. Thus, a static compressive normal force in imposed to the bar before the seismic excitation. In the absence of

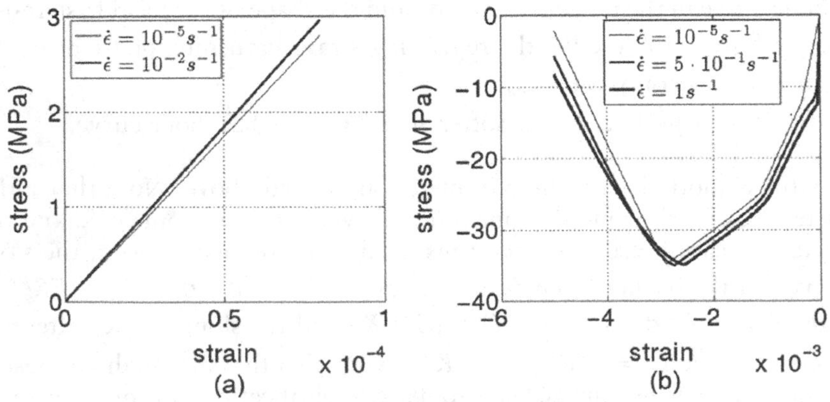

Figure 3.26 Tensile (a) and (b) compressive concrete response for several loading rates.

steel, once the strength degradation process has begun the structure is no longer capable of resisting the seismic force demand and global equilibrium can no longer be satisfied; therefore, it is necessary to couple the concrete bar to an elastic steel bar with cross-section area $S_s = 0.02 \cdot S_c = 8e^{-4} m^2$ and Young's modulus $E_s = 210$ GPa. The behavior law for concrete is that presented in Figure 3.20(b). The loading pattern is shown in Figure 3.27 and the global structural response in Figure 3.28.

b) Local Response and Intrinsic Dissipated Energies

TIV provides a useful framework for quantifying the material intrinsic dissipation. Indeed, the amount of continuum viscous, plastic and damage energies dissipated throughout the structure and the discrete plastic and damage energies dissipated in the section where displacement localizes can be computed according to (3.59)

$$
\begin{aligned}
\bar{E}^v &= \int_0^\tau \int_\Omega \dot{\bar{\mathcal{D}}}^v d\Omega dt = \int_0^\tau \int_\Omega \sigma^v \dot{\bar{\varepsilon}}^v d\Omega dt \\
\bar{E}^p &= \int_0^\tau \int_\Omega \dot{\bar{\mathcal{D}}}^p d\Omega dt = \int_0^\tau \int_\Omega \left(\sigma \dot{\bar{\varepsilon}}^p + \bar{q}^p \dot{\bar{\xi}}^p + \bar{\kappa}^p \dot{\bar{\lambda}}^p \right) d\Omega dt \\
\bar{E}^d &= \int_0^\tau \int_\Omega \dot{\bar{\mathcal{D}}}^d d\Omega dt = \int_0^\tau \int_\Omega \left(\frac{1}{2} \sigma \dot{\bar{D}} \sigma + \bar{q}^p \dot{\bar{\xi}}^p \right) d\Omega dt \\
\bar{\bar{E}}^p &= \int_0^\tau \int_\Omega \dot{\bar{\bar{\mathcal{D}}}}^p d\Omega dt = \int_0^\tau \int_\Gamma \left(t \dot{\bar{\bar{u}}}^p + \dot{\bar{\bar{q}}}^p \dot{\bar{\bar{\xi}}}^p \right) d\Gamma dt \\
\bar{\bar{E}}^d &= \int_0^\tau \int_\Omega \dot{\bar{\bar{\mathcal{D}}}}^d d\Omega dt = \int_0^\tau \int_\Gamma \left(\frac{1}{2} t \dot{\bar{\bar{D}}} t + \bar{\bar{q}}^d \dot{\bar{\bar{\xi}}}^d \right) d\Gamma dt
\end{aligned}
\tag{3.84}
$$

Note that the total intrinsic dissipation is composed of a volume and a surface part in the section Γ where displacement will localize.

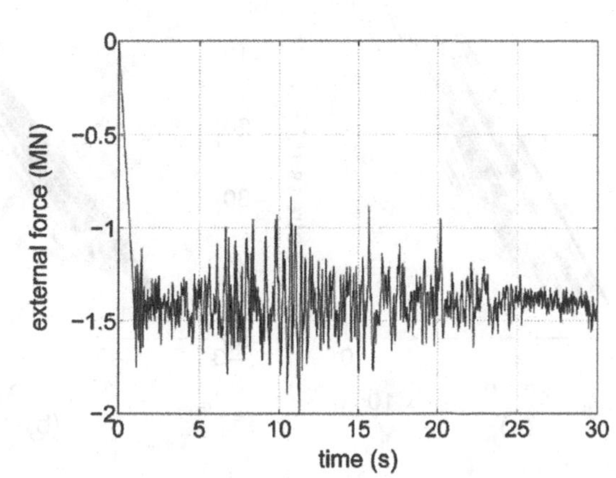

Figure 3.27 Loading pattern: static loading + seismic loading time-histories.

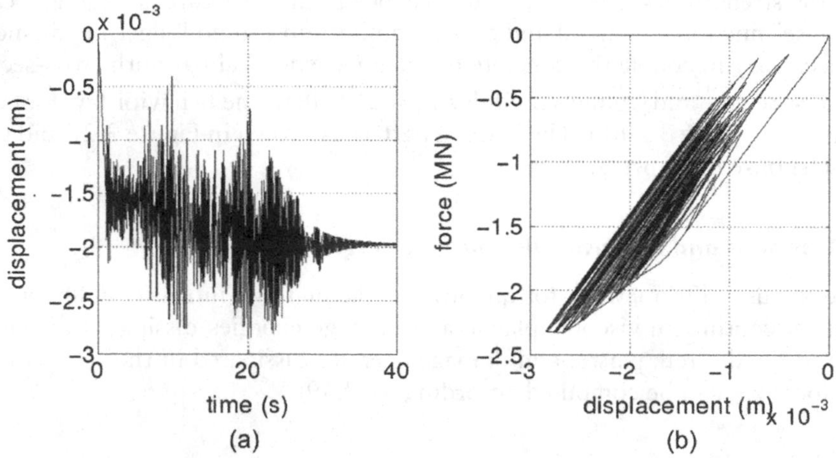

Figure 3.28 (a) Right-node displacement and (b) reaction-displacement time-histories.

Figure 3.29 shows the local response of material points located outside and inside Γ: the localization of the displacement in Γ leads to a larger amount of dissipated energy in this section. The amount and source of the intrinsic energy dissipation within the seismic excitation are detailed in Figure 3.29. Around the time $t=11$ s, a big amount of seismic energy is imparted to the structure, so as this latter does not collapse, it has to dissipate this input energy. The input energy is converted both into stored energy in the material and dissipated energy (irreversible mechanisms). Figure 3.30 illustrates the amount and sources of intrinsic energy dissipation within the seismic excitation.

Taking into account the material identification test for concrete (Ramtani 1990, see Figure 3.12, UPMC thesis on experimental identification of dynamic response on concrete), allowed us to suggest the most appropriate model combining viscosity, plasticity, damage and softening. As seen the model, it can capture both the change of elastic response (due to damage) and residual deformation (due to plasticity) in both the hardening and softening phase. More importantly, the model can capture the three salient phases of dynamic response in an earthquake, with small-amplitude vibrations for minor earthquakes, strong

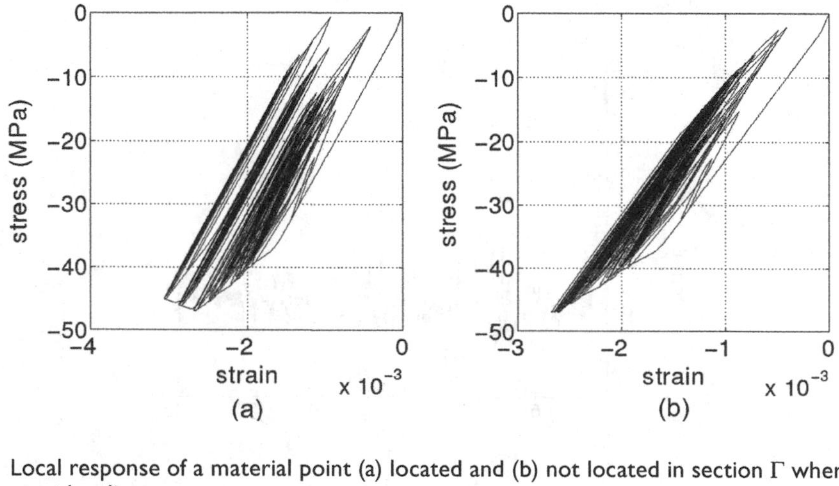

Figure 3.29 Local response of a material point (a) located and (b) not located in section Γ where displacement localizes.

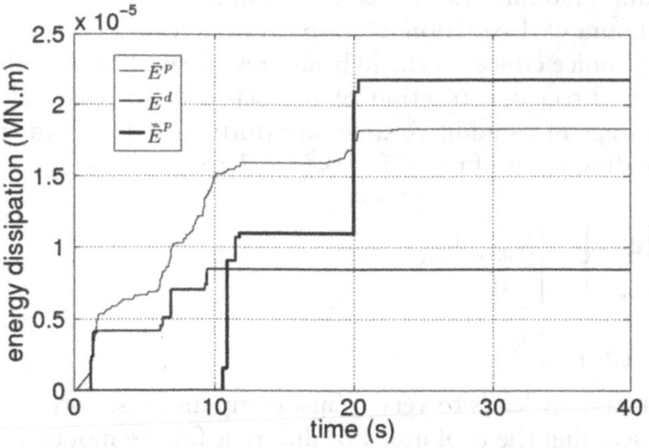

Figure 3.30 Continuum plastic and damage, and discrete plastic intrinsic dissipated energies time-histories.

shaking for a major earthquake, followed by free-vibrations of a damaged structure in a post-earthquake regime (Figure 3.31).

It is important that the proposed inelastic structural model and additional damping model is able to represent the salient phenomena corresponding to each of these three phases. Phase 1 represents an elastic phase with no or only a few incursions passing in the inelastic domain. This means that the energy dissipation comes from the friction in the cracks that appeared when applying a dead load and from other mechanisms always present in mechanical systems. The intensified ground motion is seen in phase 2 where a large amount of seismic energy is imparted to the structure and some parts of the structure then exhibit inelastic behavior. The part of the energy that is not accounted for in the inelastic structural model is introduced with the additional damping model. In the third and final phase, the structure suffered irreversible degradations that modified its dynamic properties. Due to the formation of cracks and friction in the cracks that were formed during the second phase the energy dissipative mechanisms are different.

Our strategy consists of providing a comprehensive coupled plasticity-damage model with both hardening and softening, where the evolution of all the different mechanisms is defined through the corresponding failure (plasticity or damage) criterion. For simplicity,

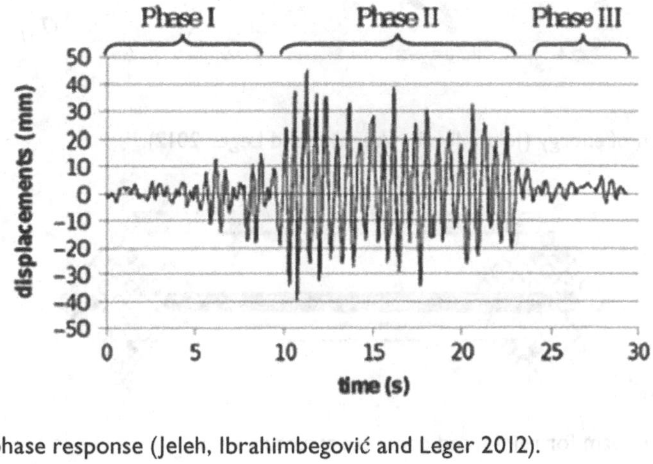

Figure 3.31 Three-phase response (Jeleh, Ibrahimbegović and Leger 2012).

the hardening/softening moduli are kept constant, and the identification problem seeks to address the same amount of dissipation in a typical hysteresis loop (Figure 3.32).

Another judicious choice concerns the hybrid-stress formulation of the proposed model, which allows for the direct construction of the elasto-visco-plastic-damage compliance, with the main advantage in its additive structure (format), which can easily be inverted to obtain the corresponding value of the stiffness (e.g. 3.85).

$$\begin{pmatrix} c_t\mathbf{M}^e & \mathbf{G}^e \\ \mathbf{G}^{e,T} & -\hat{H} \end{pmatrix} \begin{Bmatrix} \mathbf{d}_{n+1}^{(i)} \\ \beta_{n+1} \end{Bmatrix} = \begin{Bmatrix} \mathbf{R}_{\mathbf{d},n+1}^{(i)} \\ 0 \end{Bmatrix} \Rightarrow \qquad (3.85)$$

$$K^{r,\tan(i)} = c_t\mathbf{M}^e + \mathbf{G}^e\hat{H}^{-1}\mathbf{G}^{e,T}$$

Such computational strategy leads to very robust computations, with local computations of internal variables governing the evolutions of different failure modes would need only one iteration for each of them. Clearly, such computational strategy leads to very robust local computations.

The proposed model was implemented in a multi-fiber beam for reinforced concrete structures (see Figure 3.33), leading to typical results. The fiber element and the constitutive behavior law were implemented in the finite element computer program FEAP (2002). Fibers were defined as a regular mesh of rectangles.

It is interesting to note that the proposed coupled damage-plasticity model can also be brought in correspondence with a dynamic response at the structural level typical of viscous

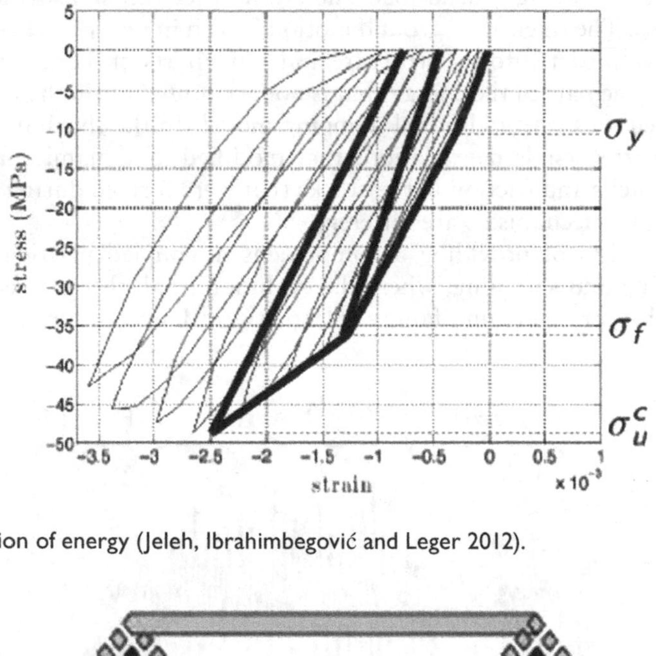

Figure 3.32 Dissipation of energy (Jeleh, Ibrahimbegović and Leger 2012).

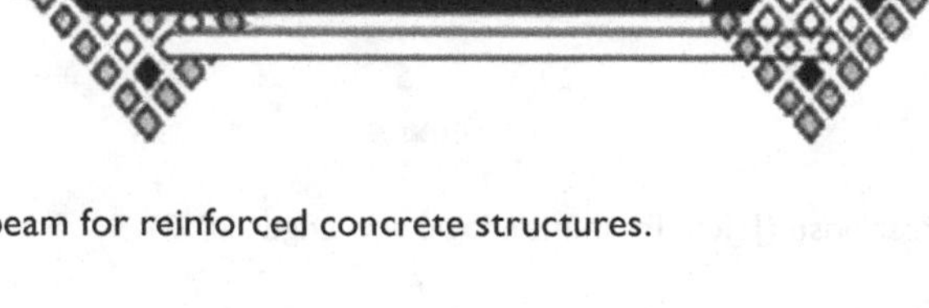

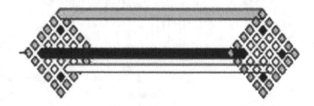

Figure 3.33 Multi-fiber beam for reinforced concrete structures.

damping (with an exponential-type of decay of the vibration amplitude), if we account for heterogeneities common to concrete structures (Figure 3.34), which is defined as a lognormal random variable leading to spatial variability in the material behavior law. Although the yield stress is considered heterogeneous, the loading and unloading slopes remain constant throughout (Jehel and Cottereau 2012).

Figure 3.35 shows the displacement time-history of the free-end of the cantilever beam for three simulations of the same beam model but with a different variance parameter of the lognormal random field describing the spatial variability of $\sigma_y(x)$ and $\sigma_u(x)$. It can be seen from Figure 3.35 that almost all the imparted energy is dissipated after 14 s when $\tau = 1.06$. This model is able to well represent the material energy dissipated during seismic loading. This kind of analysis leads to very robust results and a short computational time.

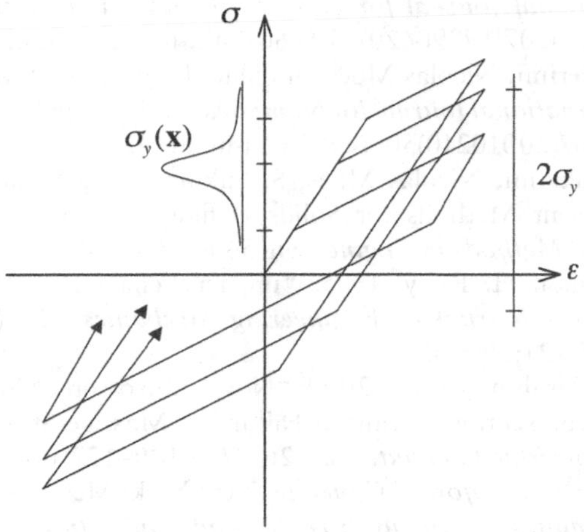

Figure 3.34 **Local elastoplastic behavior law with linear kinematic hardening. The yield stress varies according to the position in the beam.**

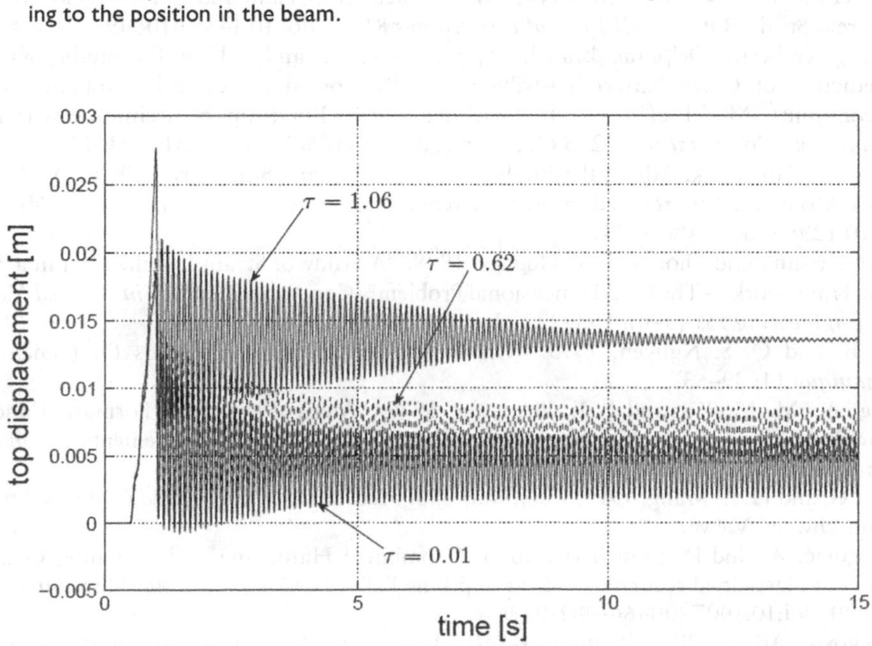

Figure 3.35 **Dynamic response in free-vibration: free-end displacement time-history (Jehel and Cottereau 2012).**

REFERENCES

Armero, F. 1997. "Localized Anisotropic Damage of Brittle Materials". In: Owen D. R. J., Hinton E. (eds). Onate E. *COMPLAS — Computational Plasticity: Fundamentals and Applications.* Proceedings of the 5th International Conference, Spain Barcelona. 635–640.

Bažant, Z. P., Q. Yu and G. Zi. 2002. "Choice of Standard Fracture Test for Concrete and Its Statistical Evaluation". *International Journal of Fracture* 118: 303–337.

Bažant, Zdeněk P. 1986. "Mechanics of Distributed Cracking". *Applied Mechanics Reviews* 30 (5): 675–705. doi:10.1115/1.3143724.

Bažant, Zdeněk P and Emilie Becq-Giraudon. 2002. "Statistical Prediction of Fracture Parameters of Concrete and Implications for Choice of Testing Standard". *Cement and Concrete Research* 32 (4): 529–56. doi:10.1016/s0008-8846(01)00723-2.

Belytschko, T. and T. Black. 1999. "Elastic Crack Growth in Finite Elements with Minimal Remeshing". *International Journal for Numerical Methods in Engineering* 45 (5): 601–20. doi:10.1002/(sici)1097-0207(19990620)45:5<601::aid-nme598>3.0.co;2.

Belytschko, Ted, Chandu Perimi, Nicolas Moës and Shuji Usui. 2001. "Arbitrary Discontinuities in Finite Elements". *International Journal for Numerical Methods in Engineering* 50 (4): 993–1013. doi:10.1002/1097-0207(20010210)50:4<993::aid-nme164>3.0.co;2-m.

Belytschko, Ted, Chandu Perimi, Nicolas Moës, S. Sukumar and Shuji Usui. 2002. "Structured Extended Finite Element Methods for Solids Defined by Implicit Surfaces". *International Journal for Numerical Methods in Engineering* 56 (4): 609–3.

Bischoff, Peter H. and Simon H. Perry. 1995. "Impact Behavior of Plain Concrete Loaded in Uniaxial Compression". *Journal of Engineering Mechanics* 121 (6): 685–93. doi:10.1061/(asce)0733-9399(1995)121:6(685).

Brancherie, D. and Adnan Ibrahimbegović. 2009. "Novel Anisotropic Continuum-discrete Damage Model Capable of Representing Localized Failure of Massive Structures". Edited by Adnan Ibrahimbegović. *Engineering Computations* 26 (1/2): 100–127. doi:0.1108/02644400910924.

Chen, W. F. 1982. *Plasticity in Reinforced Concrete.* New York: McGraw-Hill.

Chopra, Anil K. 2001. *Dynamics of Structures: Theory and Applications to Earthquake Engineering.* Upper Saddle River: Prentice-Hall.

Dilger, W. H., R. Koch and R. Kowalczyk. 1984. "Ductility of Plain and Confined Concrete Under Different Strain Rates". *ACI Journal Proceedings* 81 (1). doi:10.14359/10649.

Dominguez, Norberto, Delphine Brancherie, Luc Davenne and Adnan Ibrahimbegović. 2005. "Prediction of Crack Pattern Distribution in Reinforced Concrete by Coupling a Strong Discontinuity Model of Concrete Cracking and a Bond-slip of Reinforcement Model". *Engineering Computations* 22 (5/6): 558–82. doi:10.1108/02644400510603014.

Dujc, Jaka, Bostjan Brank, Adnan Ibrahimbegović and Delphine Brancherie. 2010. "An Embedded Crack Model for Failure Analysis of Concrete Solids". *Computers & concrete* 7 (4): 331–46. doi:10.12989/cac.2010.7.4.331.ž.

Garikipati, Krishna and Thomas J. R. Hughes. 1998. "A Study of Strain Localization in a Multiple Scale Framework — The One-Dimensional Problem". *Computer Methods in Applied Mechanics and Engineering* 159 (3–4): 193–222. doi:10.1016/s0045-7825(97)002.

Halphen, B. and Q. S. Nguyen. 1975. "On generalized standard materials (in French)". *J. de Mecanique* 14: 39–63.

Hillerborg, A., M. Modéer and P. E. Petersson. 1976. "Analysis of Crack Formation and Crack Growth in Concrete by Means of Fracture Mechanics and Finite Elements". *Cement and Concrete Research* 6 (6): 773–81. doi:10.1016/0008-8846(76)90007-7.

Hofstetter, G. and H. A. Mang. 1995. *Computational Mechanics of Reinforced Concrete Structures.* Braunschweig: Vieweg.

Ibrahimbegović, A. and D. Brancherie. 2003. "Combined Hardening and Softening Constitutive Model of Plasticity: Precursor to Shear Slip Line Failure". *Computational Mechanics* 31 (1–2): 88–100. doi:10.1007/s00466-002-0396-x.

Ibrahimbegović, Adnan. 2009. *Nonlinear Solid Mechanics: Theoretical Formulations and Finite Element Solution Methods.* Berlin: Springer.

Ibrahimbegović, Adnan and Edward L. Wilson. 1992. "Unified Computational Model for Static and Dynamic Frictional Contact Analysis". *International Journal for Numerical Methods in Engineering* 34 (1): 233–47. doi:10.1002/nme.1620340115.

Ibrahimbegović, Adnan and E. L. Wilson. 1991. "A Modified Method of Incompatible Modes". *Communications in Applied Numerical Methods* 3 (7): 187–94. doi:10.1002/cnm.1630070303.

Ibrahimbegović, Adnan and François Frey. 1993a. "Stress Resultant Finite Element Analysis of Reinforced Concrete Plates". *Engineering Computations* 10 (1): 15–30. doi:10.1108/eb023892.

Ibrahimbegović, Adnan and François Frey. 1993b. "An Efficient Implementation of Stress Resultant Plasticity in Analysis of Reissner–Mindlin Plates". *International Journal for Numerical Methods in Engineering* 36 (2): 303–20. doi:10.1002/nme.

Ibrahimbegović, Adnan and Sergiy Melnyk. 2007. "Embedded Discontinuity Finite Element Method for Modeling of Localized Failure in Heterogeneous Materials with Structured Mesh: An Alternative to Extended Finite Element Method". *Computational Mechanics* 40 (1): 149–55. doi:10.1007/s00466-006-0091-4.

Ibrahimbegović, Adnan, Pierre Jehel and Luc Daven. 2007. "Coupled Damage-Plasticity Constitutive Model and Direct Stress Interpolation". *Computational Mechanics* 42 (1): 1–11. doi:10.1007/s00466-007-0230-6.

Jehel, P. and R. Cottereau. 2012. "On damping Created by the Heterogeneity of the Mechanical Properties in RC Frame Seismic Analysis". *15th World Conference on Earthquake Engineering.* Lisbon.

Jeleh, P, A Ibrahimbegović and P Leger. 2012. "Efficient Inelastic Fiber Frame Element and Consistent Damping Model for Seismic Time-History Analyses". *15th World Conference on Earthquake Engineering.* Lisbon. 1–9.

Jirásek M. and T. Belytschko 2002. "Computational Resolution of Strong Discontinuities". *WCCM — World Congress on Computational Mechanics.* Vienna, Austria. 1–7.

Jirásek, Milan and Thomas Zimmermann. 2001. "Embedded Crack Model. Part II: Combination with Smeared Cracks". *International Journal for Numerical Methods in Engineering* 50 (6): 1291–305. doi:10.1002/1097-0207(20010228)50:6<1291::aid-nme1>.

Kucerova, A., D. Brancherie, J. Zeman and Z. Bitt. 2009. "Novel Anisotropic Continuum-Discrete Damage Model Capable of Representing Localized Failure of Massive Structures". Edited by Adnan Ibrahimbegović. *Engineering Computations* 26 (1/2): 128–44. doi:10.1108/02644400910924834.

Lu, Yong, and Kai Xu. 2004. "Modelling of Dynamic Behaviour of Concrete Materials Under Blast Loading". *International Journal of Solids and Structures* 41 (1): 131–43. doi:10.1016/j.ijsolstr.2003.09.019.

Markovic, Damijan and Adnan Ibrahimbegović. 2006. "Complementary Energy Based FE Modelling of Coupled Elasto-Plastic and Damage Behavior for Continuum Microstructure Computations". *Computer Methods in Applied Mechanics and Engineering* 195 (37–50): 5077–93. doi:10.1016/j.cma.2005.05.058.

Melenk, J. M. and I. Babuška. 1996. "The Partition of Unity Finite Element Method: Basic Theory and Applications". *Computer Methods in Applied Mechanics and Engineering* 139 (1–7): 289–314. doi:10.1016/s0045-7825(96)01087-0.

Mivelaz, P. 1996. Waterproofness of Reinforced Concrete Structures: Leaking Through the Cracks. *Ph.D. Thesis (in French).* Lausanne, Switzerland: EPFL.

Oliver, J. 2000. "On the Discrete Constitutive Models Induced by Strong Discontinuity Kinematics and Continuum Constitutive Equations". *International Journal of Solids and Structures* 37 (48–50): 7207–29. doi:10.1016/s0020-7683(00)00196-7.

Oliver, J., A. E. Huespe and P. J. Sánchez. 2006. "A Comparative Study on Finite Elements for Capturing Strong Discontinuities: E-FEM vs. X-FEM". *Computer Methods in Applied Mechanics and Engineering* 195 (37–40): 4732–52. doi:10.1016/j.cma.2005.0.

Oliver, J. and A. E. Huespe. 2004. "Theoretical and Computational Issues in Modelling Material Failure in Strong Discontinuity Scenarios". *Computer Methods in Applied Mechanics and Engineering* 193 (27–29): 2987–3014. doi:10.1016/j.cma.2003.08.007.

Panoskaltsis, V. P., K. D. Papoulia, S. Bahuguna and I. Korovajchuk. 2008. "The Generalized Kuhn Model of Linear Viscoelasticity". *Mechanics of Time-Dependent Materials* 11 (3–4): 217–30. doi:10.1007/s11043-007-9044-3.

Pedersen, R. R., A. Simone and L. J. Sluys. 2008. "An Analysis of Dynamic Fracture in Concrete with a Continuum Visco-Elastic Visco-Plastic Damage Model". *Engineering Fracture Mechanics* 75 (13): 3782–805. doi:10.1016/j.engfracmech.2008.02.

Perzyna, P. 1966. "Fundamental Problems in Viscoplasticity". *Advances in Applied Mechanics (Academic Press)* 9: 243–377.

Ramtani, S. 1990. *Contribution to the modeling of the multi-axial behavior of damaged concrete with description of the unilateral characteristics*. Ph.D. thesis, Paris 6: University.

Simo, J. C., J. G. Kennedy and R. L. Taylor. 1989. "Complementary Mixed Finite Element Formulations for Elastoplasticity". *Computer Methods in Applied Mechanics and Engineering* 74 (2): 177–206. doi:10.1016/0045-7825(89)90102-3.

Simo, J. C., J. Oliver and F. Armero. 1993. "An Analysis of Strong Discontinuities Induced by Strain-Softening in Rate-Independent Inelastic Solids". *Computational Mechanics* 12 (5): 277–96. doi:10.1007/bf00372173.

Stein, E. and D. Tihomirov. 1999. "Anisotropic damage-plasticity modelling of reinforced concrete". *ECCM'99 — European Conference on Computational Mechanics*. Munchen, Germany. 1–19.

Wang, Judith. 2009. "Intrinsic Damping: Modeling Techniques for Engineering Systems". *Journal of Structural Engineering* 135 (3): 282–91. doi:10.1061/(asce)0733-9445(2009)135:3(282).

Weerheijm, J. and J. C. A. M. Van Doormaal. 2007. "Tensile Failure of Concrete at High Loading Rates: New Test Data on Strength and Fracture Energy from Instrumented Spalling Tests". *International Journal of Impact Engineering* 34 (3): 609–26.

Zienkiewics, O. C. and R. L. Taylor. 2005. *The Finite Element Method*. 6th ed. New York: Elsevier/Butterworth-Heinemann.

Chapter 4

Masonry structures

Masonry structures are defined as structures made from individual units of different shapes, sizes and layouts (bricks, stones, marble, granite, travertine, limestone, cast stone, concrete block etc.), with or without mortar as a bonding material. The masonry structures are well known for being easy to construct and less costly compared to steel and reinforced concrete structures, as well as their benefits in respect to thermal insulation, fire protection, low maintenance, widespread geographic availability and durability. This method of construction is the oldest (Abrams 2001) and simplest; however, one which is still the least understood in terms of strength and deformation characteristics even under static loading.

The development of design rules for masonry structures have not kept pace with the advance of other civil engineering materials, such as reinforced concrete and steel as they became more competitive, and masonry has been put aside. There were several reasons for this. First, new buildings in developed countries are mainly made of reinforced concrete and steel; second, there is a lack of insight into and models for the complex behavior of masonry units, mortar, joints and masonry as a complex material. However, the situation is completely different in developing countries where structural masonry is still widely used. Additionally, in the last decades, the need for preservation of old historic buildings has increased in the sense that old buildings have intrinsic value; they are reminders of a city's culture and complexity and attract people and business. This all has had an influence on the research on masonry structures. It is clear that research has to be two-sided, experimental and numerical which is the basis for validation and further improvement of existing models. Besides the checking of the ULS (ultimate limit state) and SLS (serviceability limit state) in the new masonry structures assessment, strengthening of existing masonry structures, especially historical monuments, have opened a new phase in the research of masonry structures and their modeling.

The complexity of these structures is even more emphasized in conditions of seismic loading. Characteristics of masonry structures depend on their individual components (units and mortars) as well as the interface between the units and mortar joints. Additionally, the behavior of head and bed joints are different. This all results in modeling difficulties and still remains an open challenge for engineers due to the nature of masonry structures, which are characterized by complex and particularly nonlinear behavior. The nonlinear behavior of masonry walls is highly complicated. Nonlinearity is mainly dependent on the cracking and crushing of masonry, which is rather heterogeneous and anisotropic. Softening and dilatancy play a curial rule in the nonlinear processes (Lourenço 1996). For one to define the nonlinear behavior of a masonry wall, it is necessary to first define the failure modes. It is well known that there are four primary in-plane failure modes: rocking, bed joint sliding, diagonal tension failure along masonry units or along the head and bed joints in a stair-stepped fashion and toe-crushing. These four primary in-plane failure modes are not sufficient to define the inelastic behavior of masonry wall, and additional coupled failure modes are needed, as defined in FEMA 306. Through experimental studies, it has been seen that different failure modes exhibit different hysteretic energy dissipation, displacement capacity and strength degradation. From the experimental results, it was observed that the aspect ratio (height/length) and vertical stress are the most important factors that have a direct impact on the type of the failure mode of the masonry walls.

4.1 COMPUTATIONAL MODELING OF MASONRY STRUCTURES

It is only in recent years that numerical modeling techniques based on the finite element method have emerged for the analysis of masonry structures. The complexity lies in the strong nonlinear behavior of masonry; modeling needs to be done by two- (2D) or three-dimensional (3D) elements which makes it more complex in regard to steel and concrete structures. The main problem is the development of robust and reliable models capable of predicting the nonlinear behavior of such structures.

Masonry represents a composite material made of units and mortar joints. Modeling, depending on the required accuracy and desired simplicity, can have different modeling strategies (Lourenço, Computer Strategies for Masonry Structures 1996):

- Detailed micro-modeling—units and mortar in the joints are represented by continuum elements whereas the unit–mortar interface is represented by discontinuous elements;
- Simplified micro-modeling—expanded units are represented by continuum elements whereas the behavior of the mortar joints and unit–mortar interface is lumped in discontinuous elements;
- Macro-modeling—units, mortar and unit–mortar interface are smeared out in the continuum.

More details regarding macro-modeling can be found in Rots (1991) and for micro-modeling, either detailed or simplified, see Lourenço (1996). Preferring one model over the other is not possible as they are to be used on different scales. Micro-modeling studies are necessary for a better understanding of the local behavior of masonry structures, mainly if one is to be interested in the strong heterogeneous states of stress and strain. On the other hand, macro-models are applicable when the structure is composed of solid walls with sufficiently large dimensions, so that the stresses across or along a macro-length will be essentially uniform. This type of modeling is most valuable when a compromise between accuracy and efficiency is needed (Lourenço 1996).

Accuracy is proportional to model complexity and inversely proportional to computer memory and time. In this respect, detailed micro-modeling is formulated in such detail that its applicability can be seen only for small samples. In order to make this type of modeling more usable, a simplified model Figure 4.1(c)) was developed where the mortar joints and interface are lumped into discontinuous elements, whereas the units remain as continuum elements as in Figure 4.1(b) but are further expanded. In this respect, masonry is viewed as composed of elastic blocks that are bonded by potential fracture/slip lines at the joints. Anthoine (1992), in his work, gave a very detailed review of different micro-modeling attempts by different authors.

It should be mentioned that some authors have studied only the joint (Rosson and Suelter 2001; Nwofor 2011), and others investigated and modeled the blocks (Gambarotta and Lagomarsino 1997; Guinea et al. 2000).

Finally, global structural behavior modeling at such a level of detail is impractical but above all the interface modeling can be disregarded as it has been shown that its effect are

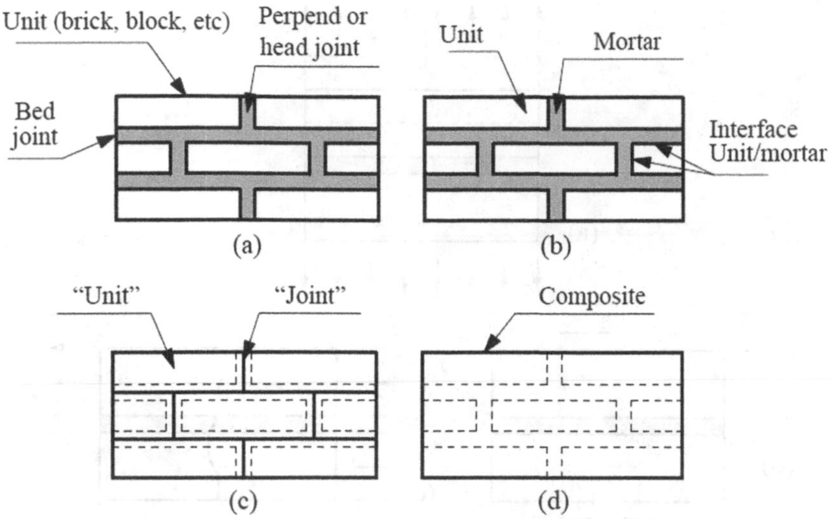

Figure 4.1 Modeling strategies for masonry structures: (a) masonry sample; (b) detailed micro-modeling; (c) simplified micro-modeling; (d) macro-modeling. (Lourenço 1996).

negligible. In this respect, many authors have turned to macro-modeling. Here, material is referred to as an anisotropic composite, and in forming constitutive relations one deals with average stresses and strains (Lourenço, Rots and Blaauwendraad 1998; Papa and Nappi 1997; Yan, Thambiratnam and Corderoy 1998). In order to describe the material behavior to a full degree, orthotropic material needs to be implemented, which is very complex due to the fact that only a few material models of this kind exist (Lourenço 1996; Anthoine 1992). If the structures are made in such a way that there is a rather regular pattern of units and joints, then homogeneous models based either on plasticity (Lopez, Oller and Lubliner 1999; Ma, Hao and Lu 2001) or damage (Raimondo and Sacco 1997; Pegon and Anthoine 1997; Marfia and Secco 2001) have been seen as applicable.

In order to make an accurate model regardless of its scale material parameters are required. It is quite obvious that the diversity of these characteristics is large; from material properties of individual components, the arrangement of bed and head joints, their existence or non-existence, interface between mortar and units, the anisotropy of units, quality of workmanship, the degree of curing, atmospheric effects, age etc.

Due to all these reasons, the analysis and modeling of masonry structures currently represents one of the most important fields in civil engineering.

4.2 MICRO-MODELING OF MASONRY

Here, a micro-model for masonry will be presented. The starting point is the identification of various local failure modes, as clearly performed by Lourenço and Rots (1997). As illustrated in Figure 4.2a and b, which represent joint mechanisms, formerly associated with joint tension cracking and later with sliding along the bed joint due to low values of normal stress. Brick mechanisms are illustrated in Figure 4.2c represented by unit direct tension cracking. Figures 4.2d and 4.2e represent the combined brick and joint mechanics. The former is associated with the diagonal tension cracking of masonry units at values of normal stress sufficient to cause friction in the joints, and the latter is connected to masonry being crushed as a result of mortar dilatancy at high values of normal stress.

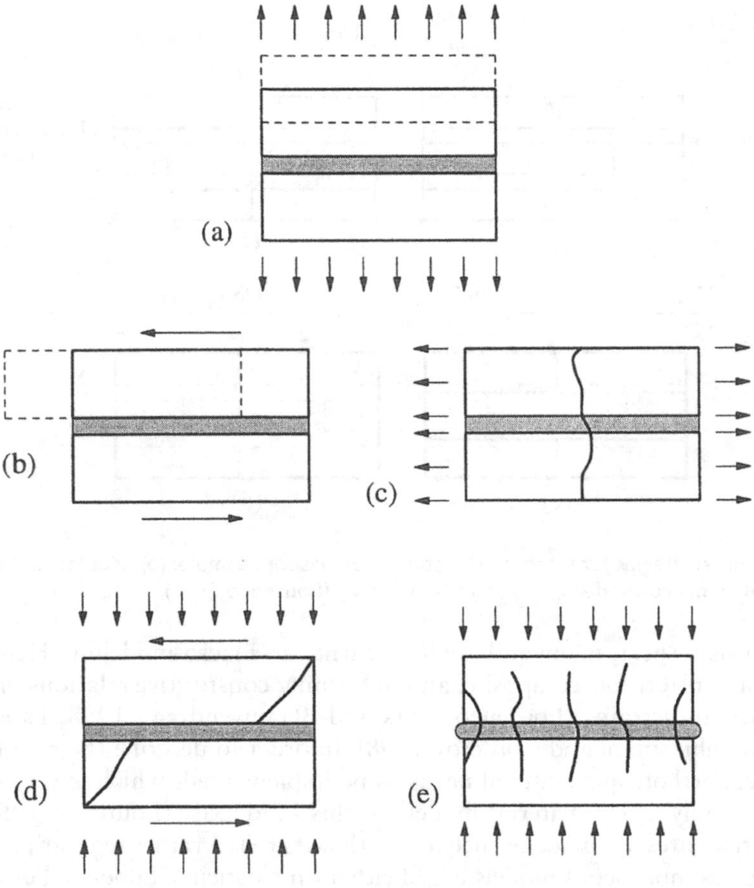

Figure 4.2 Local modes of failure: (a) joint tension cracking, (b) joint slip, (c) unit direct tension crack, (d) unit diagonal tension crack, and (e) masonry crack. (Lourenço and Rots 1997).

Based on the work of Alfano, Rosati and Valoroso (2000), it seems that these patterns are of great importance in the search for a predictive model. However, the softening behavior of materials led to some difficulties, such as the lack of digital objectivity. In Lourenço and Rots (1997), softening in compression is located in the mortar joints and in the brick elements; pure tension cracks in the units is placed vertically in the middle of each unit and using interface elements to model the joints this has given good results in comparison to the tests that were carried out. The approach that will be presented here models the two materials using continuous elements, while reintegrating the incompatible modes in the blocks; this is done using strong displacement discontinuities and thus solving the problem of digital objectivity. In this respect, the model is explained below.

4.3 STRAIN LOCALIZATION PHENOMENA

Softening and localization have a clear physical origin in microstructure evolution even though they are observed in a macroscopic material response. Materials that exhibit softening (concrete, masonry, soli-cohesive materials) are usually described with a stress–strain relation in numerical analysis. Until now, it has been well known (de Borst 1999) that localized deformations, closely related to material softening, is one of the primary difficulties for classical continuum models when representing inelastic deformation and assuring mesh independency. Here it leads to an ill-posed boundary value problem, since equilibrium or motion equations do not remain elliptic or hyperbolic once the descending branch is entered. The only way to define the localization is to set the measure to zero. This leads to mesh-dependency. There are two ways to resolve this issue. One possibility is to concentrate the deformation on interfaces or discrete cracks. Here, constitutive relations are a function of the tractions on relative displacements of two interface sides. This is done using interface finite elements (Rots 1988). Here, it is necessary to predict the location of the crack. The meshless paradigm has provided new insights into the finite element method (Babuška and Melenk 1996; Duarte and Oden 1996; Fries and Belytschko 2010) by using so-called extended finite elements, placed within the partition of unity framework. This partition of unity framework (Babuška and Melenk 1996) is a powerful technique for modeling discontinuities and singularities. This is done through element domains owing to an enrichment of the approximation basis with Heaviside-type functions.

Here, a model which takes into account both micro and macrocracking is presented, as this is physically closer to reality. Additionally, it interprets, in a correct manner, the nature of the associated inelastic energy dissipation.

4.3.1 Modeling of In-Plane and Out-Of-Plane Loading on a Brick Wall

4.3.1.1 Modeling of Failure Mechanisms in Bricks

The novelty of this model is that the nonlinear softening behavior of these elements is modeled by introducing strong displacement discontinuities via jump functions (Oliver et al. 2002; Simo, Oliver and Armero 1993; Colliat, Davenne and Ibrahimbegovic, 2002). This means that potential cracks in the units must always be included in micro-modeling in order for the correct behavior of the masonry structures to be captured. In this respect, the displacement field is written in such a way that it is allowed to have a discontinuity (crack or opening), and it can be written as:

$$\underline{u}(\underline{x},t)=\underline{\bar{u}}(\underline{x},t)+[\![\underline{\bar{\bar{u}}}]\!](t)H_{\bar{\underline{x}}}(\underline{x}) \tag{4.1}$$

where $\underline{\bar{u}}(\underline{x},t)$ is the smooth part, $H_{\bar{\underline{x}}}(\underline{x})$is the Heaviside function and $[\![\underline{\bar{\bar{u}}}]\!]$ is the displacement jump corresponding to the crack-opening.

The strain field obtained from (4.1) and the well-known relation from the kinematics equation:

$$\underline{\underline{\varepsilon}}(\underline{x},t)=\frac{\partial\underline{u}(\underline{x},t)}{\partial\underline{x}} \tag{4.2}$$

is therefore expressed by:

$$\underline{\underline{\varepsilon}}(\underline{x},t)=\frac{\partial\underline{u}(\underline{x},t)}{\partial\underline{x}}=\frac{\partial\underline{\bar{u}}(\underline{x},t)}{\partial\underline{x}}+[\![\underline{\bar{\bar{u}}}]\!](t)\delta_{\bar{\underline{x}}}(\underline{x}) \tag{4.3}$$

where, $\delta_{\bar{\underline{x}}}(\underline{x})$ is the Dirac-delta function.

The elastoplastic softening behavior is modeled using the theory of plasticity on the dissipation surface; this defines an elastic field in the well-known form of:

$$\phi(\underline{\sigma},q)=|t|-(\sigma_y-q) \tag{4.4}$$

where t is the normal stress at the displacement discontinuity, q is the hardening variable and σ_y is the yield stress, separating the elastic and plastic domains. If (4.4) is less or equal to zero, this assumes that there is an elastic domain where no change of the inelastic deformation will occur. Additive decomposition of the total deformations into elastic and plastic parts is acceptable for small deformations in plasticity models (Lubliner 1990). It follows that the stress depends on the elastic deformation part, while the plastic flow can be written as:

$$\dot{\underline{\underline{\varepsilon}}}(\underline{x},t)=\frac{\dot{\underline{\underline{\sigma}}}}{E}+\dot{\underline{\gamma}}(\underline{x},t)\operatorname{sign}(\underline{\underline{\sigma}}) \tag{4.5}$$

where:
E is Young's modulus and
γ is the plastic multiplier.

Once the results obtained in (4.3) are introduced in (4.5), and taking into account the smoothness of the stress field, it is obvious that the following must be true:

$$\underline{\gamma}(\underline{x},t)=\underline{\bar{\gamma}}(\underline{x},t)+\underline{\bar{\bar{\gamma}}}(\underline{x},t)\delta_{\underline{\bar{x}}}(\underline{x}) \tag{4.6}$$

Going back to (4.6) and introducing (4.3) it follows that:

$$\dot{\underline{\underline{\sigma}}}(\underline{x},t)=E\left[\frac{\partial\dot{\underline{\bar{u}}}(\underline{x},t)}{\partial\underline{x}}-\underline{\bar{\gamma}}(\underline{x},t)\operatorname{sign}(\underline{\underline{\sigma}})\right] \tag{4.7}$$

Leading to:

$$\underline{\bar{\bar{\gamma}}}(t)=[\![\dot{\underline{\bar{u}}}]\!] \tag{4.8}$$

The consistency conditions follow from (4.4) and read:

$$\begin{aligned}0=\dot{\phi}(\underline{\underline{\sigma}},q)&=\operatorname{sign}(\underline{\underline{\sigma}})E\frac{\partial\dot{\underline{\bar{u}}}(\underline{x},t)}{\partial\underline{x}}-\underline{\bar{\gamma}}(\underline{x},t)E\left[\operatorname{sign}(\underline{\underline{\sigma}})\right]^2\\&+\dot{q}+\underbrace{E\left([\![\underline{\bar{\bar{u}}}]\!]\operatorname{sign}(\underline{\underline{\sigma}})_{\underline{\bar{x}}}-\underline{\bar{\bar{\gamma}}}\right)}_{0}\delta_{\underline{\bar{x}}}(\underline{x})\end{aligned} \tag{4.9}$$

It is assumed that the micro and macro-cracks will appear subsequently; the first phase can be defined as:

$$\underline{\bar{\gamma}}(\underline{x},t)=\frac{\operatorname{sign}(\underline{\underline{\sigma}})E\dfrac{\partial\dot{\underline{\bar{u}}}(\underline{x},t)}{\partial\underline{x}}}{E\left[\operatorname{sign}(\underline{\underline{\sigma}})\right]^2+K} \tag{4.10}$$

From this it implies that the non-local form of the consistency condition leads to:

$$\dot{q}=-\operatorname{sign}(\underline{\underline{\sigma}})\dot{\underline{\underline{\sigma}}} \tag{4.11}$$

indicating that the hardening stress-line variable has to be bonded. Taking into account that $\dot{q}=-K\dot{\xi},\dot{\xi}=\gamma$, where K is the hardening modulus and ξ is the hardening variable. Considering the equality set in (4.8) and the above mentioned qualities, it follows that the second part in (4.6) becomes:

$$\overline{\overline{\gamma}}(\underline{x},t)\delta_{\overline{x}}(\underline{x}) = [\![\overline{\overline{u}}]\!]\delta_{\overline{x}}(\underline{x}) = -K^{-1}\dot{q} \tag{4.12}$$

This additional implies that:

$$K^{-1} = \bar{K}^{-1}\delta_{\overline{x}}(\underline{x}) \tag{4.13}$$

This plasticity model is incorporated in the following manner: four node elements are considered and one discontinuity, meaning the potential crack is placed in the center of the element and the interpolation is constructed as shown in Figure 4.3. In addition to the usual interpolation of the displacement field, a form of the function called the "Incompatible mode" M(x) is superimposed.

The potential crack is placed in the center of the real strain field and is constructed as follows:

$$\underline{\underline{\varepsilon}}(\underline{x},t) = \sum_{a=1}^{4} \nabla^s N_a(\underline{x}) d_a + G(\underline{x})\alpha \tag{4.14}$$

where α is the parameter for incompatible mode, and G is produced from the chosen displacement interpolation by applying the same differential operator to the smooth and discontinuous part.

The interpolation of the virtual strain is then constructed with:

$$\delta\underline{\underline{\varepsilon}}(\underline{x},t) = \sum_{a=1}^{4} \nabla^s N_a(\underline{x}) w_a + G(\underline{x})\beta \tag{4.15}$$

where $\hat{G}$ is a modified incompatible mode strain–displacement matrix with a zero-mean, which should guarantee satisfaction of patch test conditions (Ibrahimbegović and Wilson 1991). As the same shape for the virtual strain field is chosen, the discrete formulation of the problem is of the standard form of the equations of the incompatible mode method (Ibrahimbegović and Wilson 1991), where the set of global equilibrium equations is replaced by element based local equilibrium equations. The advantage of this approach is that that the fracture energy G_f of the material corresponds to the given mode; meaning that the total amount of dissipated energy is independent of the foreseen element sizes, and in this way the problem of numerical objectivity is answered.

4.3.1.2 Modeling of Mortar Joints in a Brick Wall

The modeling of the interfaces between the unit elements (in this specific case bricks) in the structural assemblies is of utmost importance. Its contribution is crucial to the strength of

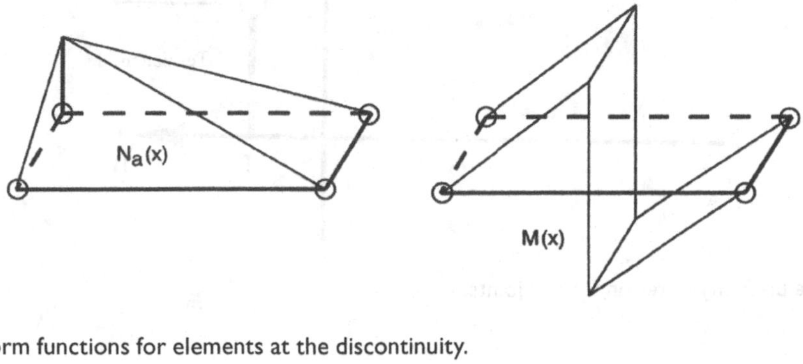

Figure 4.3 Form functions for elements at the discontinuity.

masonry structures. Two issues have to be taken into account here: the size of the interface elements in respect to the size of the structure and the weak mechanical properties of mortar in existing structures. The complexities of modeling masonry structures are numerous: from the irregularity of the structure, the constitutive equations for different materials (stone for example is a rigid block), the presence or absence of mortar as well as the thickness of the mortar etc. Until now either phenomenological modeling or deductive modeling was used. In the former, the thickness of the interface is equated as zero (Lourenço 1996). This means that the mortar and the adhesion surface between the brick and the mortar are modeled as a unique interface. Whereas, in the later, interphase elements are modeled at the micromechanical level, mortar is specifically modeled as a continuum material and the joint (adhesion between the mortar and the unit) by a specific interface (Sacco and Toti 2010).

In order to model the mortar between the two blocks, it is clear that the nonlinear behavior of this zone is associated with two phenomena: one linked to the opening in tension, and the other to shear. So, the important feature of the composite yield surface is the coupling of tension and shear softening due to the fact that both phenomena are related to the bond or adhesion between the brick units and the mortar. For the latter, one should also note the importance of the dilatancy of the mortar joints as they have a low dilatancy so it is necessary to formulate the model with non-associated plasticity. The dilatancy angle is considered as a function of the plastic-relative shear displacement and the normal confining pressure. Under the increasing values of these quantities, the dilatancy angle tends to zero. This is physically realistic, given the microstructure of the unit–mortar interface and it being confirmed by experiments. Also, softening behavior should be included for all the modes of the composite yield surface.

The theory of plasticity is used as the basis for the development of this model (Colliat, Davenne and Ibrahimbegovic 2002); the elastic range is defined by a multisurface (see Figure 4.4). The same basis was used by Giuseppe and Di Gati (1997) where a simple, cohesive interface

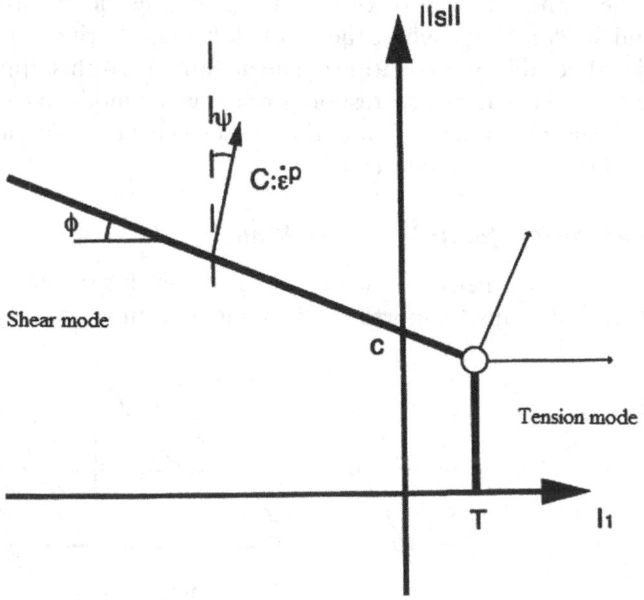

Figure 4.4 The plasticity criterion for the joints.

model adopted a yield surface using a classical bi-linear Coulomb condition with a tension cut-off and with a non-associative flow law.

A perfect plasticity model for joints has been chosen, and the constant dilatancy angle is part of this approach. This treatment has been developed through the use of the strong discontinuities of displacement under the incompatible modes method.

For the shear mode a Drucker–Prager criterion is implemented in the form:

$$\phi_1 = \left\| \underline{\underline{s}} \right\| + \beta I_1 - c\psi \tag{4.16}$$

While a tensile limit for the opening of the joints is defined as:

$$\phi_2 = I_1 - T\psi \tag{4.17}$$

where I_1 and $\underline{\underline{s}}$ are the hydrostatic and deviatoric part of the stress tensor and β is the material parameter which can characterize the internal friction.

All elements for the multisurface criteria are described in *Non-Smooth Multisurface Plasticity and Viscoplasticity. Loading/Unloading Conditions and Numerical Algorithms* (Simo, Kennedy and Govindjee 1988). The equal degradation of strength can be assumed as both phenomena occur due to the breakage of the same bridge that exists at the micro-level between the unit and mortar.

The law of plastic flow has the usual form:

$$\underline{\underline{\dot{\varepsilon}}}^p = \sum_{\alpha=1}^{m} \dot{\gamma}^\alpha \frac{\partial \phi_\alpha^*}{\partial \underline{\underline{\sigma}}} \tag{4.18}$$

where $\dot{\gamma}^\alpha$ are the different plasticity multiplications.

Indeed, it is shown in Simo, Kennedy and Govindjee (1988) that $\phi_\alpha^{\text{trial}} > 0$ does not imply $\gamma^\alpha > 0$ or which may have $\phi_\alpha^{\text{trial}} > 0$, but $\phi_{\alpha,n+1} > 0$.

In fact, γ^α can be seen as the contravariant components of $\underline{\underline{C}} : \Delta\underline{\underline{\varepsilon}}_{n+1}^p$ with the base $\left\{ \underline{\underline{C}}, \frac{\partial \phi_\alpha^*}{\partial \underline{\underline{\sigma}}} \right\}$. The condition $\gamma^\alpha > 0°$ for $\alpha = 1,2$ defines a mode in the corner.

a) Drucker Mode

One has:

$$\frac{\partial \phi_1}{\partial \underline{\underline{\sigma}}} = \underline{\underline{n}} + \frac{\beta}{3}\underline{\underline{1}} \text{ and } \frac{\partial \phi_1^*}{\partial \underline{\underline{\sigma}}} = \underline{\underline{n}} + \frac{\psi}{3}\underline{\underline{1}} \text{ with } \underline{\underline{n}} = \frac{\underline{\underline{s}}}{\left\| \underline{\underline{s}} \right\|} \tag{4.19}$$

The consistency condition gives the value of the plastic multiplier in this mode:

$$\gamma_1 = \frac{\phi_1^{\text{trial}}}{2G + \beta K \psi} \tag{4.20}$$

which now gives the first part of the new state of constraints:

$$I_{1,n+1} = I_1^{\text{trial}} - \gamma_1 K \psi \tag{4.21}$$

$$\underline{\underline{s}}_{n+1} = \underline{\underline{s}}_{n+1}^{\text{trial}} - 2G\gamma_1 \underline{\underline{n}} \tag{4.22}$$

and, the other part, the elastoplastic part being consistent (Simo and Taylor 1985) with:

$$\alpha = \frac{2G\left\|\underline{\underline{s}}_{n+1}\right\|}{\left\|\underline{\underline{s}}_{n+1}\right\| + 2G\gamma_1} \text{ and } \bar{\alpha} = K - \frac{\alpha}{3} \tag{4.23}$$

$$\underline{\underline{\underline{C}}}^{ep} = \alpha I + \bar{\alpha}\underline{\underline{1}} \otimes \underline{\underline{1}} - \frac{1}{\alpha + \beta K\psi}\left[\alpha^2 \underline{\underline{n}} \otimes \underline{\underline{n}} + \alpha\beta K \underline{\underline{n}} \otimes \underline{\underline{1}} + \alpha\psi K \underline{\underline{1}} \otimes \underline{\underline{n}} + \beta K^2 \psi \underline{\underline{1}} \otimes \underline{\underline{1}}\right] \tag{4.24}$$

b) Mixed Mode (Corresponding to Corner)

In addition to the above equations, the following expression is added:

$$\frac{\partial \phi_2}{\partial \underline{\underline{\sigma}}} = \frac{1}{3}\underline{\underline{1}} \tag{4.25}$$

The consistency condition can express both plastic multipliers:

$$\gamma_1 = \frac{\left\|\underline{\underline{s}}^{\text{trial}}\right\| - c + \beta T}{2G} \text{ and } \gamma_2 = \frac{I_1^{\text{trial}} - T}{K} - \psi\gamma_1 \tag{4.26}$$

from which the new state of stress is obtained:

$$I_{1,n+1} = I_1^{\text{trial}} - K\psi\gamma_1 - K\gamma_2 \tag{4.27}$$

$$\underline{\underline{s}}_{n+1} = \underline{\underline{s}}_{n+1}^{\text{trial}} - 2G\gamma_1 \underline{\underline{n}} \tag{4.28}$$

and the elastoplastic module reads:

$$\underline{\underline{\underline{C}}}^{ep} = \alpha \underline{\underline{\underline{I}}} - \frac{\alpha}{3}\bar{\alpha}\underline{\underline{1}} \otimes \underline{\underline{1}} - \alpha \underline{\underline{n}} \otimes \underline{\underline{n}} \tag{4.29}$$

So, the plastic flow rule used is:

$$\phi_1^* = \left\|\underline{\underline{s}}\right\| + \psi I_1 \tag{4.30}$$

and

$$\dot{\underline{\underline{\varepsilon}}}^p = \frac{\partial \phi_1^*}{\partial \underline{\underline{\sigma}}} = \underline{\underline{n}} + \frac{1}{3}\psi \underline{\underline{1}} \tag{4.31}$$

Most of the research was focused on the in-plane failure of masonry structures (Figures 4.5 and 4.6) due to earthquake actions.

It is necessary for the wall to have a sufficient capacity to withstand out-of-plane failure as it has been seen in past earthquakes that it can trigger partial or complete collapse of the structure. It has been observed in past earthquakes from Banja Luka, the Bosnia and Herzegovina earthquake in 1969 (Figure 4.7) to the Christchurch earthquake (Figure 4.8) in 2011 (Ingham and Griffith 2011). The lack of sufficient connections between wall to wall and wall to floor results in out-of-plane damage and the failure of masonry members. This is something that is characteristic of the old masonry structures being more vulnerable than the new structures with rigid diaphragms. This means that even in cases where connections are shown to be effective, this type of failure can be triggered due to the excessive vertical

(b)

(a)

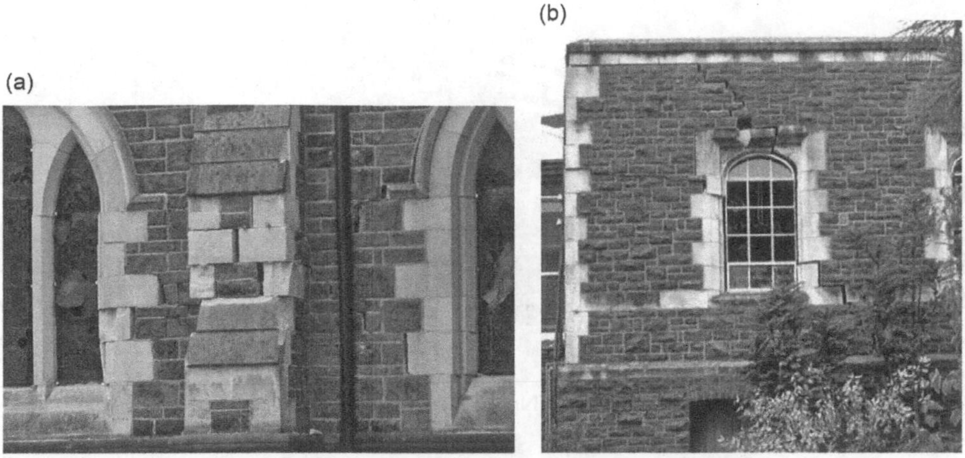

Figure 4.5 In-plane failure—Christchurch (New Zealand) (Ingham and Griffith 2011).

Figure 4.6 In-plane failure—Structural damage to masonry building resulting from the 1994 Northridge, California, earthquake (NAS 2003).

(a) (b)

Figure 4.7 Out-of-plane failure—Banja Luka earthquake (Bosnia and Herzegovina).

(a) (b)

Figure 4.8 Out-of-plane failure—Christchurch (New Zealand) (Ingham and Griffith 2011).

and/or horizontal slenderness of the wall. In this respect, the out-of-plane failure of masonry is mainly related to the geometric stability rather than to the strength of the materials. In these cases, the out-of-plane failure mechanisms are activated by lower values of ground motion than the in-plane ones. An additional element that needs to be taken into account is the question of the type of floors, with flexible or rigid diaphragms. Diaphragm flexibility significantly increases the out-of-plane displacement response. Considerable experimental work is also reported for the strength and response of brick walls under out-of-plane loads (Griffith et al. 2007; Derakhshan, Ingham and Griffith 2009; Meisle, Elwood and Ventura 2007; Maheri, Najafgholipour and Rajabi 2011).

Even though there have been many attempts and recommendations in modern codes for avoiding out-of-plane failures, it is still an existing problem and an important field of research. Masonry is still an open area for further research and especially out-of-plane failure which even now is not fully understood. As Paulay and Priestley (1992) state, out-of-plane failure is "one of the most complex and ill-understood areas of seismic analysis".

4.3.2 Examples

Traditionally, experiments on shear walls have been adopted by the masonry structures research community as full scale in-plane tests. Results are compared with the ones obtained in Lourenço and Rots (1997) and CUR (1994).

4.3.2.1 Example 1

The shear walls tested have a width/height ratio of 1 with dimensions of 990×1000 [mm^2], and are built up with 18 courses, having 16 courses active while 2 courses are clamped in steel beams. The walls are made of wire-cut solid clay blocks with dimensions of 210×52×100 [mm^2] and the thickness of the mortar is 10 [mm].

First of all vertical pre-compression is applied and then the horizontal load is monolithically increased under top displacement as indicated in Figure 4.9. The boundary conditions are formulated in such a way that both the bottom and top boundaries are horizontal, preventing vertical movement and rotations.

The material properties are summarized in Table 4.1.

It should be noted that the choice of zero tensile strength is due to the use of a perfect plasticity model; open joints will thus transmit the force. The dilatancy angle ψ is a measure of the uplift of one unit over the other upon shearing. The average value of $\tan\psi$ falls into the

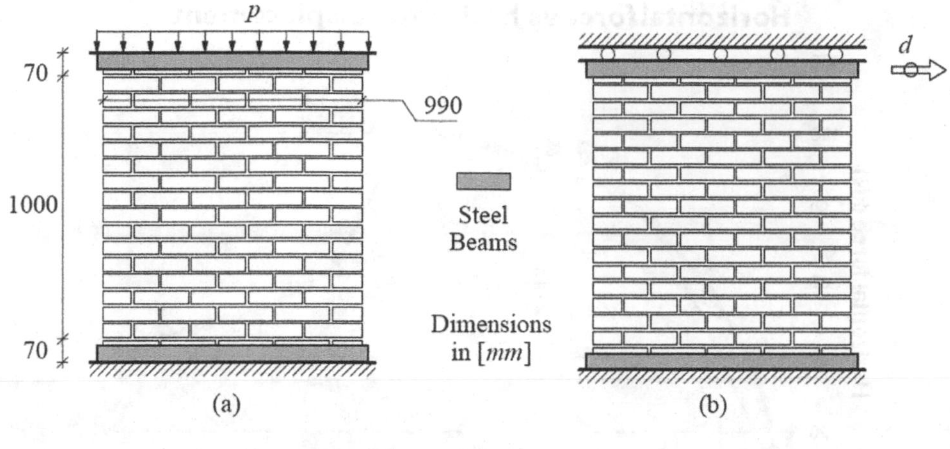

Figure 4.9 Shear walls tested (a) vertical loading—phase 1, (b) horizontal loading under displacement control-phase 2 (Lourenço 1996).

Table 4.1 Material Properties

Bricks	*Joints*	
E(MPa) = 16700	E(MPa) = 20000	
υ-Poisson ratio = 0.15	υ-Poisson ratio = 0.20	
Compression (vertical)	c = 0.35(MPa) =	$\tan\varphi = 0.75$
$\sigma_y = 8$(MPa)		
$G_c = 220$(J/m^2)		
Tension (horizontal)	$\tan\psi$=0.35	T = 0(MPa)
$\sigma_y = 1$(MPa)		
$G_f = 80$(J/m^2)		

range of 0.2 to 0.7 for low confining pressure, depending on the roughness of the unit surface. With increasing slip, $\tan\psi$ also decreases to zero due to the smoothening of the sheared surface (Lourenço and Rots 1997). As for the dilatancy angle, sensitivity tests have shown that regarding normal stresses it has little influence, which agrees with Lourenço and Rots (1997), where it has been seen that after the horizontal displacement of d = 7.5 mm the solution with $\tan\psi = 0$ and $\tan\psi = 0.4$ are similar because the residual compressive strength controls the response.

The comparison between the experimental results and the model is given in Figure 4.10 shows the total horizontal total vs. displacement of the top of the wall. The proposed model fits well with the experimental tests. It may be noted that the initial stiffness of the wall is essentially correct without any need for the modification of the elastic parameters; this is an advantage in view of a predictive model. Regarding the peak value of the force this is satisfactory; however a greater displacement gives the wall a more ductile behavior by modeling than it is observed in the experiment.

The failure is caused by crushing. In the reinforced concrete shear walls investigated in Feenstra and De Borst (1996), it was seen that higher initial vertical loads lead to an increase of strength but decrease of ductility, meaning an increase of brittleness. In shear walls, the increase of confinement does not lead to an increase of ductility in the case where the softening regime is generally governed by the failure of the compressed toes of the wall.

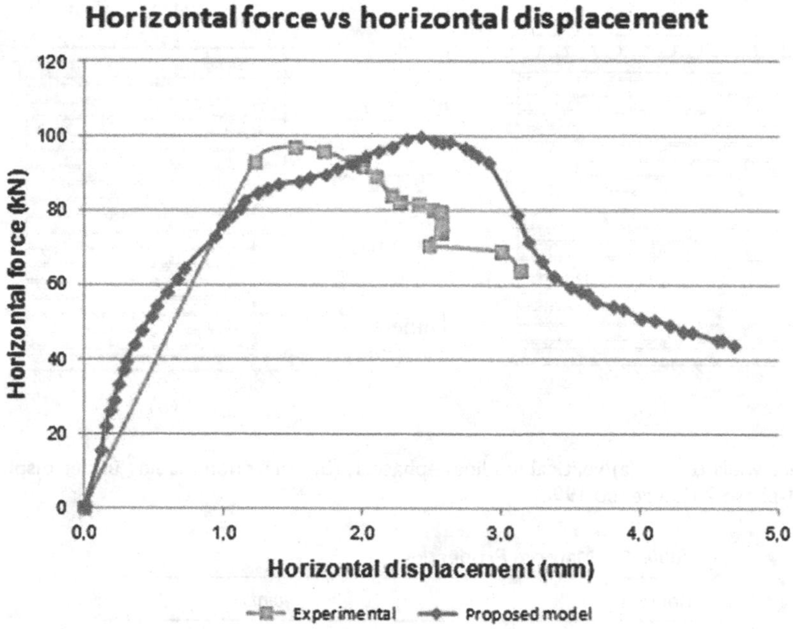

Figure 4.10 Comparison based on the CUR program testing; pre-compression of vertical loading 2 MPa.

As the material can always fall into the out-of-plane failure this can happen. However, in masonry, shear wall experiments showed that higher initial loads lead to the increasing strength of the shear walls, but higher failure loads did not correspond to increasing brittleness. This can be explained by the internal force distribution at failure (for more information see Lourenço (1996)).

4.3.2.2 *Example 2*

This test program was conducted by the Centre Scientifique et Technique du Bâtiment; the masonry was made out of different types of burned brick elements and was unreinforced. Here, only the case where the walls were loaded horizontally in their plane will be discussed.

The considered wall was made of brick elements having dimensions of 500×200×200 [mm^3], the mechanical characteristics of vertical compression are summarized in Table 4.2. It was composed of 12 horizontal rows with seven bricks in each row. Details on the tests can be found in Capra et al. (2001).

The calculation was carried out by imposing horizontal displacement on the upper part of the wall and without any possible rotation thereof. The results are plotted in Figure 4.11. The obtained numerical results were reasonably consistent with the measurements obtained from experimental tests.

Table 4.2 Material Properties

Bricks
$E = 2300$(MPa)
$\upsilon = 0.12$-Poisson ratio
$\sigma_c = 2.70$(MPa)

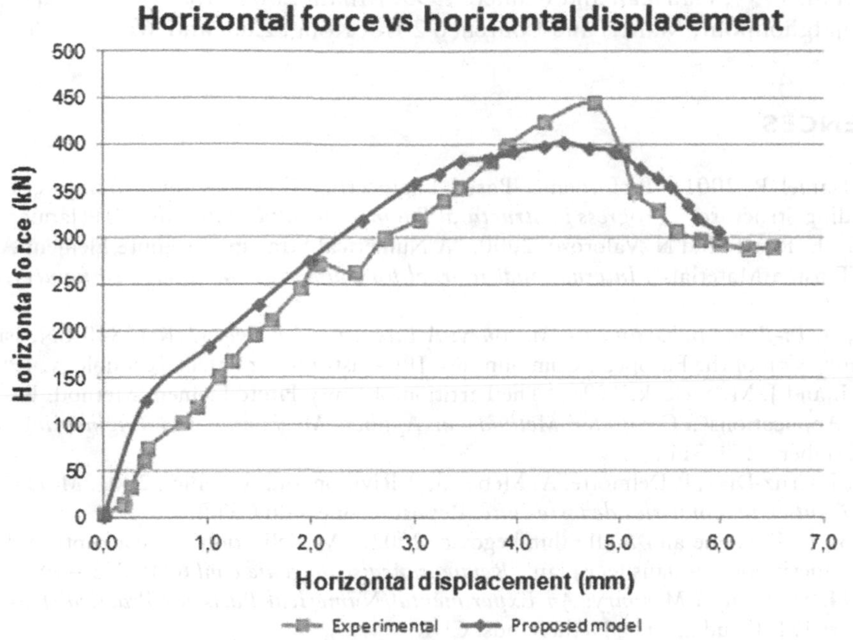

Figure 4.11 Comparison based on the CSTB program testing and numerical model.

The first pseudo-linear phase coincided with the nonlinear behavior of the joints (opening and shear); it was observed that the initial stiffness of the wall was correct, and had an absence of timing parameters (elastic and inelastic for the brick joints).

The failure of the wall occurred with the crushing of the bricks, mainly in the two corners, which was seen in the experiments. The experimental behavior was satisfactorily reproduced and the collapse load could be estimated within a 10% range of the experimental values, which was very good. From the diagram, it is evident that the displacement of the top of the wall was quite consistent with the values obtained in the model. The good match between the proposed model and measurements is explained by the choice of the brick pattern.

The importance of local failure modes and particularly those related to the compression resistant blocks was strongly emphasized.

The implementation of this model in the code of the finite element FEAP allowed one to obtain very satisfactory results in the case of hollow brick walls. This is consistent with the assumption made previously about the importance of compression failure modes. It is quite difficult to solve larger problems using micro-modeling and in this respect it is necessary to move to macro-modeling. It should be noted that the micro-modeling has shown applicability for small walls (Rekik and Lebon 2010; Rekik and Lebon 2012).

4.4 FURTHER RESEARCH

Most of the available models for masonry structures provide analyses in either the in-plane or out-of-plane direction. However, in reality, most loading occurs as a combination of in-plane and out-of-plane motion. Very few studies were carried out on the numerical response under simultaneous in-plane and out-of-plane loading. Therefore, a comprehensive modeling strategy is needed to simulate the behavior of masonry structures for 3D cyclic loading taking into account the in-plane and out-of-plane interaction (Dolatshahi and Aref 2011;

Shaprio et al. 1994; Flanagan and Bennett 1999; Milani 2008; Maheri and Najafgholipour 2012; Najafgholipour, Maheri and Lourenço 2013; Kong, Zhai and Wang 2016).

REFERENCES

Abrams, Daniel P. 2001. "Performance-Based Engineering Concepts for Unreinforced Masonry Building Structures". *Progress in Structural Engineering and Materials* 3 (1) (January): 48–56.

Alfano, G., L. Rosati and N. Valoroso. 2000. "A Numerical Strategy for Finite Element Analysis of No-Tension Materials". *International Journal for Numerical Methods in Engineering* 48 (3): 317–50.

Anthoine, A. *In-Plane Behaviour Of Masonry: A Literature Review.* EUR 13840 EN, Ispra, Italy: Commission of the European Communities, JRC-Institute for Safety Technology, 1992.

Babuška, I. and J. M. Melenk. 1996. "The Partition of Unity Finite Element Method: Basic Theory and Applications". *Computer Methods in Applied Mechanics and Engineering* 139 (1–7) (December): 289–314.

Capra, B., J. Cruz-Diaz, P. Delmotte, A. Mebarki, P. Rivillon and A. Sellier. 2001. *Murs de contreventement en maconneries de terre cuite.* Report, Cahiers du CSTB.

Colliat J.B., L. Davenne and A. Ibrahimbegovic. 2002. "Modelisation jusqu'a rupture de murs en maconnerie charges dans leur plan". *Revue francaise de genie civil* 6 (4): 395–408.

CUR. 1994. *Structrural Masonry: An Experimental/Numerical Basis for Practical Design Rules.* Report 171, Goudam, The Netherlands: CUR.

de Borst, R. 1999. "Computational Methods for Localisation and Failure". *European Conference on Computational Mechanics.* Munchen, 150.

Derakhshan, H., J. M. Ingham and M. C. Griffith. 2009. "Tri-Linear Force-Displacement Models Representative of Out-Of-Plane Unreinforced Masonry Wall Behaviour". *11th Canadian Masonry Symposium.* Toronto.

Dolatshahi, Kiarash M. and Amjad J. Aref. 2011. "Three Dimensional Modeling of Masonry Structures and Interaction of In-Plane and Out-of-Plane Deformation of Masonry Walls". *Engineering Mechanics Institute Conference* 1–6.

Duarte, Armando C. and Tinsley J. Oden. 1996. "An H-P Adaptive Method Using Clouds". *Computer Methods in Applied Mechanics and Engineering* 139 (1–4) (December): 237–62.

Feenstra, Peter H. and René De Borst. 1996. "A Composite Plasticity Model for Concrete". *International Journal of Solids and Structures* 33 (5) (February): 707–30.

Flanagan, R. D. and R. M. Bennett. 1999. "Bidirectional Behaviour of Structural Clay Tile Infilled Frames". *Journal of Structural Engineering, ASCE,* 125 (3): 236–44.

Fries, Thomas-Peter and Ted Belytschko. 2010. "The Extended/generalized Finite Element Method: An Overview of the Method and Its Applications". *International Journal for Numerical Methods in Engineering* 84 (3) (August): 253–304.

Gambarotta, L. and S. Lagomarsino. 1997. "Damage Models for the Seismic Response of Brick Masonry Shear Walls. Part I: The Mortar Joint Model and Its Applications". *Earthquake Engineering & Structural Dynamics* E26 (4) (April): 423–39.

Giuseppe, Giambanco and Leonardo Di Gati. 1997. "A Cohesive Interface Model for the Structural Mechanics of Block Masonry". *Mechanics Research Communications* 24 (5) (September): 503–12.

Griffith, M. C., J. Vaculik, N. T. K. Lam, J. Wilson and E. Lumantarna. 2007. "Cyclic Testing of Unreinforced Masonry Walls in Two-Way Bending". *Earthquake Engineering and Structural Dynamics* 36: 801–21.

Guinea, G. V., G. Hussein, M. Elices and J. Planas. 2000. "Micromechanical Modeling of Brick-Masonry Fracture". *Cement and Concrete Research* 30 (5) (May): 731–37.

Ibrahimbegović, Adnan and E. L. Wilson. 1991. "A Modified Method of Incompatible Modes". *Communications in Applied Numerical Methods* 7 (3) (April): 187–94.

Ingham, Jason M. and Michael C. Griffith. 2011. *The Performance of Unreinforced Masonry Buildings in the 2010/2011 Canterbury Earthquake Swarm*. Technical report, Canterbury: Canterbury Earthquake Royal Commission of Inquiry.

Kong, Jingchang, Changhai Zhai and Xiaomin Wang. 2016. "In-Plane Behavior of Masonry Infill Wall Considering Out-of-Plane Loading". *Periodica Polytechnica Civil Engineering* 60 (2): 217–21.

Lopez, L., S. Oller and J. Lubliner. 1999. "A Homogeneous Consitutive Model for Masonry". *International Journal for Numerical Methods in Engineering* 46 (10).

Lourenço, Paulo B. and Jan G. Rots. 1997. "Multisurface Interface Model for Analysis of Masonry Structures". *Journal of Engineering Mechanics* 123 (7) (July): 660–68.

Lourenço, Paulo B., Jan B. Rots and Johan Blaauwendraad. 1998. "Continuum Model for Masonry: Parameter Estimation and Validation". *Journal of Structural Engineering* 124 (6) (June): 642–52.

Lourenço, Paulo Jose Brandao Barbosa. 1996. *Computational strategies for Masonry Structures*. PhD Thesis, Delft: Delft University Press, Stevinweg 1, 2628 CN Delft, The Netherlands.

Lubliner, Jacob. 2005. *Plasticity Theory*. Berkeley: University of California at Berkeley, 1990, revised 2005.

Ma, Guowei, Hong Hao and Yong Lu. 2001. "Homogenization of Masonry Using Numerical Simulations". *Journal of Engineering Mechanics* 127 (5) (May).

Maheri, M. R., M. A. Najafgholipour and A. R. Rajabi. 2011. "The Influence of Mortar Head Joints on the In-Plane and Out of Plane Seismic Strength of Brick Masonry Walls". *Iranian Journal of Science and Technology* 35: 63–79.

Maheri, Mahmoud R. and M. A. Najafgholipour. 2012. "In-Plane Shear and Out-Of-Plane Bending Capacity Interaction In Brick Masonry Wall". *15WCEE*. Lisbon, 1–10.

Marfia, A. and E. Secco. 2001. "Modeling of Reinforced Masonry Elements". *International Journal of Solids and Structures* 38 (24–25): 4177–98.

Meisle, C. S., K. J. Elwood and C. E. Ventura. 2007. "Shake Table Tests on the Out-Of-Plane Response of Unreinforced Masonry Walls". *Canadian Journal of Civil Engineering* 34: 1381–92.

Milani, G. 2008. "3D Upper Bound Limit Analysis of Multi-Leaf Masonry Walls". *International Journal of Mechanical Sciences* 50 (4): 817–36.

Najafgholipour, M. A., Mahmoud R. Maheri and P. B. Lourenço. 2013. "Capacity Interaction in Brick Masonry Under Simultaneous In-Plane and Out-of-Plane Loads". *Construction and Building Materials* 38: 619–26.

NAS. 2003. *Preventing Earthquake Disasters*. National Academy of Sciences.

Nwofor, T. C. 2011. "Finite Element Stress Analysis of Brick-Mortar Masonry Under Compression". *Journal of Applied Science and Technology* 16 (1–2) (March): 48–67.

Oliver, J., A. E. Huespe, M. D. G. Pulido and E. Chaves. 2002. "From Continuum Mechanics to Fracture Mechanics: The Strong Discontinuity Approach". *Engineering Fracture Mechanics* 69 (2) (January).

Papa, Enrico and Alfonso Nappi. 1997. "Numerical Modelling of Masonry: A Material Model Accounting for Damage Effects and Plastic Strains". *Applied Mathematical Modelling* 21 (6) (June): 319–35.

Paulay, T. and M. J. N. Priestley. 1992. *Seismic Design of Reinforced Concrete and Masonry Buildings*. New York: Wiley, John & Sons.

Pegon, P. and A. Anthoine. 1997. "Numerical Strategies for Solving Continuum Damage Problems with Softening: Application to the Homogenization of Masonry". *Computers & Structures* 64 (1–4) (July): 623–42.

Raimondo, Luciano and Elio Sacco. 1997. "Homogenization Technique and Damage Model for Old Masonry Material". *International Journal of Solids and Structures* 34, (24) (August): 3191–208.

Rekik, A. and F. Lebon. 2012. "Homogenization Methods for Interface Modeling in Damaged Masonry". *Advances in Engineering Software* 46 (1) (April): 35–42.

Rekik, Amna and Frédéric Lebon. 2010. "Identification of the Representative Crack Length Evolution in a Multi-Level Interface Model for Quasi-Brittle Masonry". *International Journal of Solids and Structures* 47 (22–23) (November): 3011–21.

Rosson, Barry T. and Jason L. Suelter. 2001. "Closed-Form Equations for Hardening of Sand-Lime Mortar Joints". *Journal of Engineering Mechanics* 127 (6) (June): 574–81.

Rots, J. G. 1988. *Computational modeling of concrete fracture*. PhD Thesis, Delft: Delft University of Technology.

Rots, J. G. 1991. *Numerical Simulation of Cracking in Structural Masonry*. Heron.

Sacco, Elio and Jessica Toti. 2010. "Interface Elements for the Analysis of Masonry Structures". *International Journal for Computational Methods in Engineering Science and Mechanics* 11 (6) (November): 354–73.

Shaprio, D., J. Uzarski, M. Webster, R. Angel and D. Abrams. 1994. *Estimating out of plane strength of cracked masonry infills*. Structural Research Series No. 588, Illinois: University of Illinois at Urbana-Champaign, Civil Engineering Studies.

Simo, J. C., J. G. Kennedy and S. Govindjee. 1988. "Non-Smooth Multisurface Plasticity and Viscoplasticity. Loading/unloading Conditions and Numerical Algorithms". *International Journal for Numerical Methods in Engineering* 26 (10) (October): 2161–85.

Simo, J. C., J. Oliver and F. Armero. 1993. "An Analysis of Strong Discontinuities Induced by Strain-Softening in Rate-Independent Inelastic Solids". *Computational Mechanics* 12 (5): 277–96.

Simo, J. and R. Taylor. 1985. "Consistent Tangent Operators for Rate-Independent Elastoplasticity". *Computer Methods in Applied Mechanics and Engineering* 48 (3): 101–18.

Yan, Zhuge, David Thambiratnam and John Corderoy Corderoy. 1998. "Nonlinear Dynamic Analysis of Unreinforced Masonry". *Journal of Structural Engineering* 124 (3) (March): 270–77.

Part II

Chapter 5

Dynamics extreme loads in earthquake engineering

Considerable research has been conducted on the dynamic analysis of structure-foundation systems. However, due to the use of frequency domain analysis (inherently implied superposition), the application is restricted to linear analysis. This approach is used to facilitate the modeling of the radiation condition in a semi-infinite foundation domain. In order to conduct nonlinear analysis of structure-foundation systems, time domain analysis is required. Time domain analysis, however, implies a local approximation to the radiation condition, i.e. the transmitting boundaries are used to prevent the reflection of outgoing waves in the finite-size foundation model. For nonlinear problems, it is reasonable to expect that the dissipation of energy due to inelastic material behavior will compensate and diminish the effect of approximating the radiation condition. The paraxial boundary (Cohen and Jennings 1983; Kausel 1988; Wolf 1986) emerged as systematic local approximation of the global boundary conditions and the generalization of the ideas of Engquist and Majda (1977) from the scalar wave equation to the elasticity equations of motion. In this sense, paraxial boundary conditions represent local boundary conditions based on paraxial approximations of one-way wave equations to finite element analysis.

The models used for the dynamic analysis of structure-foundation systems are usually very complex, and their detailed nonlinear analysis represents a prohibitive computational expense. In light of the uncertainties associated with earthquake ground motion characteristics, one is prone to choose an approach that requires a moderate computational effort so that the analysis can be placed in an appropriate framework of random vibrations theory (Lin 1976).

In light of a sound earthquake-resistant design philosophy, nonlinearities should be localized in a small predetermined part of the structure-foundation system (Clough and Wilson 1979). There are different base isolation systems that can be utilized here. It is important to indicate that the locations where nonlinear displacements may occur have to be known in advance. This means that it is convenient to eliminate the purely linear response degrees of freedom by static condensation before performing the dynamic analysis; in order to implement this reduction scheme, the structure has to be idealized as an assemblage of substructures. One, which will be presented in the further sections, is the dynamic contact, a model problem for local nonlinearity, i.e. the structure is permitted to uplift. This all indicates that the specific feature of this model is that on one hand the foundation interaction effect by utilizing the dynamic substructuring concept (Wilson and Bayo 1986) is included and on the other hand the model is being reduced the dynamic substructuring is used to reduce the size of the linear part to an easily manageable form prior to nonlinear analysis. This is in sharp contrast to the vast majority of ad-hoc simplified models used for the same purpose.

The formulation is based on the added motion approach of Clough and Penzien (1993). The added motion approach is used for the analysis of the inertial interaction. By utilizing the added motion approach, the kinematic interaction phase is not avoided, except in a special case of a constant free-field motion across structure-foundation interface.

5.1 THE FREE-FIELD VS. ADDED MOTION APPROACH IN STRUCTURE-FOUNDATION INTERACTION

A general methodology for dynamic analysis of linear structure-foundation systems is presented in this section with localized nonlinearities occurring in the structure-foundation interface only. Earthquake ground motion is usually recorded at a single point, the so-called control point (Roesset and Kim 1987). In order to use this information, the added motion approach is suggested by Clough and Penzien (1993). The added motion approach for the dynamic analysis of structure-foundation systems is based on an additive split of the total motion field into two motion fields: the free-field motion (motion of foundation with no structure on it) and the added motion field (due to addition of a structure). The evaluation of the structural component behavior within the added motion approach is straightforward, since for the structural component of a structure-foundation system the added motion is the same as the total motion. On the other hand, the evaluation of the foundation component behavior becomes more complicated, since two motion fields have to be obtained. However, the model, which includes the "source" of the earthquake excitation is now chosen of sufficient size since transmitting boundaries are supplied to model the radiation condition.

The basic concepts on which the added motion approach of Clough and Penzien (1993) is based is presented in the following paragraphs, in order to make it clear before implying the nonlinear analysis. The system presented in Figure 5.1 is utilized for the derivation of the added motion approach.

The free-field motion is determined considering only the foundation part. The semi-discrete equations of motion read:

$$\mathbf{M}_f\ddot{\mathbf{v}}(t)+\mathbf{C}_f\dot{\mathbf{v}}(t)+\mathbf{K}_f\mathbf{v}(t)=\mathbf{f}_{ff}(t) \tag{5.1}$$

where:

$\mathbf{M}_f$, $\mathbf{C}_f$, $\mathbf{K}_f$ represent mass, damping and stiffness matrices of the semi-discrete foundation model;

$\mathbf{v}(t), \dot{\mathbf{v}}(t), \ddot{\mathbf{v}}(t)$ represent free-field displacement, velocity and acceleration vectors, and

$\mathbf{f}_{\mathrm{ff}}(t)$ is the far-field boundary loading (where the earthquake ground motion originated).

Adding the structure component to the foundation one, a complete structure-foundation system is formed. Now, the equations of motion for the complete system (structure and foundation) exposed to the same far-field boundary loading $\mathbf{f}_{\mathrm{ff}}(t)$ can be expressed as:

$$\left[\mathbf{M}_s+\mathbf{M}_f\right]\left(\ddot{\mathbf{v}}(t)+\ddot{\mathbf{u}}(t)\right)+\left[\mathbf{C}_s+\mathbf{C}_f\right]\left(\dot{\mathbf{v}}(t)+\dot{\mathbf{u}}(t)\right)+\left[\mathbf{K}_s+\mathbf{K}_f\right]\left(\mathbf{v}(t)+\mathbf{u}(t)\right)=\mathbf{f}_{ff}(t) \tag{5.2a}$$

$$\left[\mathbf{M}\right]\left(\ddot{\mathbf{v}}(t)+\ddot{\mathbf{u}}(t)\right)+\left[\mathbf{C}\right]\left(\dot{\mathbf{v}}(t)+\dot{\mathbf{u}}(t)\right)+\left[\mathbf{K}\right]\left(\mathbf{v}(t)+\mathbf{u}(t)\right)=\mathbf{f}_{ff}(t) \tag{5.2b}$$

where:

$\mathbf{u}(t), \dot{\mathbf{u}}(t), \ddot{\mathbf{u}}(t)$ are added motion displacement, velocity and acceleration vectors, and

$\mathbf{M}_s$, $\mathbf{C}_s$, $\mathbf{K}_s$ are structural mass, damping and stiffness matrices.

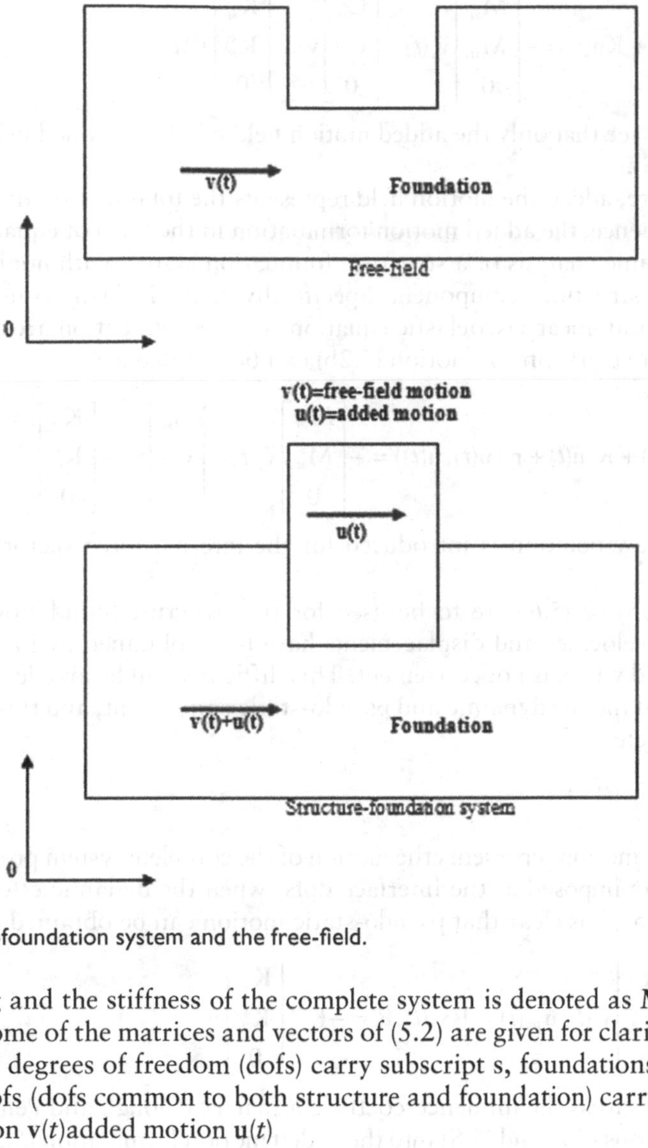

Figure 5.1 Structure-foundation system and the free-field.

Mass, damping and the stiffness of the complete system is denoted as **M**, **C**, **K**. A partitioned form for some of the matrices and vectors of (5.2) are given for clarity, with a remark that the structure degrees of freedom (dofs) carry subscript s, foundations dofs subscript *f*, while interface dofs (dofs common to both structure and foundation) carry subscript *i*.

Free-field motion $\mathbf{v}(t)$added motion $\mathbf{u}(t)$

$$\mathbf{v}(t)=\begin{pmatrix}0\\ \mathbf{v}_i(t)\\ \mathbf{v}_f(t)\end{pmatrix};\mathbf{u}(t)=\begin{pmatrix}\mathbf{u}_s(t)\\ \mathbf{u}_i(t)\\ \mathbf{u}_f(t)\end{pmatrix} \tag{5.3}$$

structure mass matrix $\mathbf{M}_s$ and foundation mass matrix $\mathbf{M}_f$

$$\mathbf{M}_s=\begin{bmatrix}\mathbf{M}_{ss} & \mathbf{M}_{si} & 0\\ \mathbf{M}_{is} & \mathbf{M}_{ii}^{s} & 0\\ 0 & 0 & 0\end{bmatrix};\mathbf{M}_f=\begin{bmatrix}0 & 0 & 0\\ 0 & \mathbf{M}_{ii}^{f} & M_{if}\\ 0 & \mathbf{M}_{fi} & \mathbf{M}_{ff}\end{bmatrix} \tag{5.4}$$

similar partitioned forms apply for damping and stiffness matrices.

The equations of motion can be written in the form of (5.5) if (5.1) and (5.2b) are combined and by taking advantage of sparsity:

$$\mathbf{M}\ddot{\mathbf{u}}(t)+\mathbf{C}\dot{\mathbf{u}}(t)+\mathbf{K}\mathbf{u}(t)=-\begin{bmatrix}\mathbf{M}_{si}\\ \mathbf{M}_{ii}^{s}\\ 0\end{bmatrix}\ddot{\mathbf{v}}_i(t)-\begin{bmatrix}\mathbf{C}_{si}\\ \mathbf{C}_{ii}^{s}\\ 0\end{bmatrix}\dot{\mathbf{v}}_i-\begin{bmatrix}\mathbf{K}_{si}\\ \mathbf{K}_{ii}^{s}\\ 0\end{bmatrix}\mathbf{v}_i(t) \tag{5.5}$$

It is interesting to see that only the added motion field will be obtained using the solution of the equations (5.5).

For the structure, added the motion field represents the total motion field (see partitioned equations (5.3)); hence, the added motion formulation in the form of equations (5.5) is amenable to the dynamic analysis of a structure-foundation system with nonlinear constitutive equations for the structural component. Specifically, if the inelastic constitutive equations for the structure and linear viscoelastic equations for the foundation are assumed, then the set of semi-discrete equations of motion (5.2b) can be restated as:

$$\mathbf{M}\ddot{\mathbf{u}}(t)+\mathbf{C}_f\dot{\mathbf{u}}(t)+\mathbf{K}_f\mathbf{u}(t)+\mathbf{r}_s\left(\mathbf{u}(t),\dot{\mathbf{u}}(t)\right)=-\begin{bmatrix}\mathbf{M}_{si}\\ \mathbf{M}_{ii}^{s}\\ 0\end{bmatrix}\ddot{\mathbf{v}}_i(t)-\begin{bmatrix}\mathbf{C}_{si}\\ \mathbf{C}_{ii}^{s}\\ 0\end{bmatrix}\dot{\mathbf{v}}_i-\begin{bmatrix}\mathbf{K}_{si}\\ \mathbf{K}_{ii}^{s}\\ 0\end{bmatrix}\mathbf{v}_i(t) \tag{5.6}$$

where the only new notation is introduced for the internal force vector for the structure dofs, $\mathbf{r}_s\left(\mathbf{u}(t),\dot{\mathbf{u}}(t)\right)$.

If equations (5.5) or (5.6) are to be used for the structure-foundation dynamic analysis, the free-field velocities and displacements have to be obtained by integrating the given acceleration record which is not convenient. This difficulty can be avoided by separating the added motion field into its dynamic and pseudo-static component, and this can be done only for the linear system:

$$\mathbf{u}(t)=\mathbf{u}_d(t)+\mathbf{u}_{ps}(t) \tag{5.7}$$

The pseudo-static motion represents the motion of the complete system produced by the free-field displacements imposed at the interface dofs, when the dynamic effects are neglected. Going back to (5.5), it is clear that pseudo-static motion can be obtained as:

$$\mathbf{K}\mathbf{u}_{ps}(t)=-\begin{bmatrix}\mathbf{K}_{si}\\ \mathbf{K}_{ii}^{s}\\ 0\end{bmatrix}\mathbf{v}_i(t);\mathbf{u}_{ps}(t)=\mathbf{R}\mathbf{v}_i(t);\mathbf{R}=-\mathbf{K}^{-1}\begin{bmatrix}\mathbf{K}_{si}\\ \mathbf{K}_{ii}^{s}\\ 0\end{bmatrix} \tag{5.8}$$

where $\mathbf{R}$ represents to as the influence-coefficient matrix (Clough and Penzien 1993).

Inserting equations (5.7) and (5.8) into the added motion formulation (5.5) (and neglect the damping proportional forcing function), the new formulation of the motion equations reads:

$$\mathbf{M}\ddot{\mathbf{u}}_d(t)+\mathbf{C}\dot{\mathbf{u}}_d(t)+\mathbf{K}\mathbf{u}_d(t)=\tilde{\mathbf{M}}\ddot{\mathbf{v}}_i(t);\tilde{\mathbf{M}}-\left\{\mathbf{M}\mathbf{R}+\begin{bmatrix}\mathbf{M}_{si}\\ \mathbf{M}_{ii}^{s}\\ 0\end{bmatrix}\right\} \tag{5.9}$$

The equations of motion (5.9) are further simplified for the embedded structures and the structures supported at the ground surface as the assumption of free-field motion of the same value at each interface nodal point can be employed. This means that the pseudo-static displacements of the structure due to unit displacement at the interface dofs, (5.8), are actually rigid body displacements. So, the equations of motion for the rigid base earthquake ground motion are now furnished:

$$\mathbf{M}\ddot{\mathbf{u}}_d(t)+\mathbf{C}\dot{\mathbf{u}}_d(t)+\mathbf{K}\mathbf{u}_d(t)=-\mathbf{M}_s\mathbf{R}_{rb}\ddot{\mathbf{v}}_i(t) \tag{5.10}$$

where $\mathbf{R}_{rb}$ is the rigid body transformation matrix (a set of rigid body displacements for unit displacement at the structure base).

This means that with the added motion approach, leading to equations of motion in (5.10), splits the total motion field into the dynamic and pseudo-static added motion for the structure, and into the dynamic, pseudo-static and free-field motion components for the foundation. However, the assumption on the constant free-field motion overall interface dofs simplifies the stress recovery in the structure, i.e. the solution of (5.10) is sufficient for that purpose.

Equation (5.6), for a linear system, can be modified in another way as well. In this assumption, the structure and the foundation are governed by linear viscoelastic constitutive equations, while only the interface constitutive equations are considered as inelastic. Figure 5.2 is given as a clear illustration of the second proposal.

In this case, the total motion field for the interface dofs on the foundation side (Figure 5.2), according to formulation (5.6), is split into the free-field motion and the added motion field. With the solution of (5.6), only the added motion field in the foundation is obtained. However, for the inelastic constitutive equations of the interface, the total motion field is normally required.

Further additive split of the structure added motion (which is equal to the total motion of the structure) is proposed between the reference motion introduced by the structure interface motion identical to the foundation interface free-field motion and the remaining part (see Figure 5.2). If the assumption of constant free-field motion over all interface dofs is introduced (equivalent to a rigid base assumption), then the reference motion caused by the free-field motion represents rigid body motion, and the corresponding stress field is zero for both the interface and the structure. Quite the opposite, if the free-field motion varies along the structure-foundation interface, the stress field will be

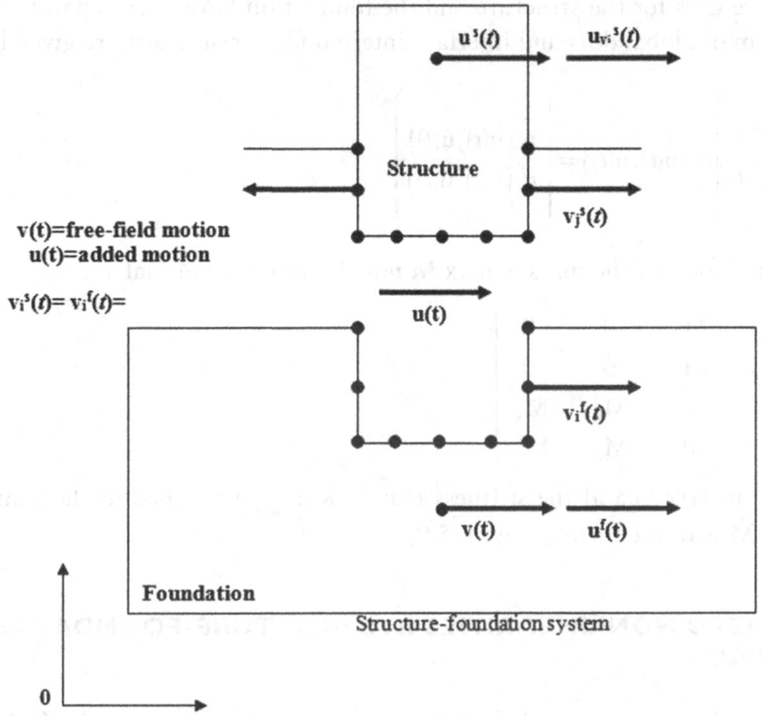

Figure 5.2 Structure-foundation system with inelastic interface-additive split of total motion field.

non-zero in both the structure and the interface. However, from the specified variation of the free-field motion, this stress field can be easily obtained (state determination), even for the inelastic interface constitutive equations, and further introduced in (5.6) as the initial stress in the interface.

This indicates that the split of the structure added motion is equivalent to the split of the total motion into the pseudo-static and the dynamic components for the linear structure-foundation system.

This requires the equality of the foundation free-field $\mathbf{v}_i^f(t)$ for the interface dofs on the foundation side (upper side) and the structure reference motion $\mathbf{v}_i^s(t)$ for the interface dofs on the structure side (lower side):

$$\mathbf{v}_i^f(t) \equiv \mathbf{v}_i^s(t) \tag{5.11}$$

Total structure motion is composed of two additive parts:

$$\mathbf{u}^t(t) = \mathbf{u}(t) + \mathbf{u}_{vi}(t) \tag{5.12}$$

where, by definition:

$$\mathbf{u}_{vi}(t) = -\mathbf{K}_{ss}^{-1}\mathbf{K}_{si}\mathbf{v}_i^s(t) = \mathbf{R}\mathbf{v}_i^f(t) \tag{5.13}$$

and additionally insertion of (5.11) into (5.13) is done.

Now the equations of motion (5.5) can be rewritten as:

$$\mathbf{M}\ddot{\mathbf{u}}(t) + \mathbf{C}\dot{\mathbf{u}}(t) + \mathbf{K}\mathbf{u}(t) + \mathbf{r}\left(\mathbf{u}(t), \dot{\mathbf{u}}(t)\right) = \tilde{\mathbf{M}}\ddot{\mathbf{v}}_i^s(t) \tag{5.14}$$

where interface dofs for the structure and the foundation have been separated. For clarity, expanded form of global dofs and interface internal force $\mathbf{r}\left(\mathbf{u}(t), \dot{\mathbf{u}}(t)\right)$are given below:

$$\mathbf{u}(t) = \begin{pmatrix} \mathbf{u}_s(t) \\ \mathbf{u}_i^s(t) \\ \mathbf{u}_i^f(t) \\ \mathbf{u}_f(t) \end{pmatrix}; \mathbf{r}\left(\mathbf{u}(t), \dot{\mathbf{u}}(t)\right) = \begin{pmatrix} 0 \\ \mathbf{r}_i^s\left(\mathbf{u}(t), \dot{\mathbf{u}}(t)\right) \\ \mathbf{r}_i^f\left(\mathbf{u}(t), \dot{\mathbf{u}}(t)\right) \\ 0 \end{pmatrix} \tag{5.15}$$

and partitioned form of the mass matrix $\mathbf{M}$ now becomes a 4x4 matrix:

$$\mathbf{M} = \begin{bmatrix} \mathbf{M}_{ss} & \mathbf{M}_{si} & 0 & 0 \\ \mathbf{M}_{is} & \mathbf{M}_{ii}^s & 0 & 0 \\ 0 & 0 & \mathbf{M}_{ii}^f & \mathbf{M}_{if} \\ 0 & 0 & \mathbf{M}_{fi} & \mathbf{M}_{ff} \end{bmatrix} \tag{5.16}$$

The damping matrix $\mathbf{C}$ and the stiffness matrix $\mathbf{K}$ are partitioned in the same way as $\mathbf{M}$ above, while $\tilde{\mathbf{M}}$ is of the form given in (5.9).

5.2 LOCALIZED NONLINEARITIES IN STRUCTURE-FOUNDATION INTERFACE

Natural base isolation is achieved if uplifting of the structure is allowed (Huckelbridge and Clough 1977). The generalization of such an isolation concept is the dynamic frictional contact which occurs at the structure-foundation interface, i.e. a structure can both uplift and slide. Here we first consider the case when only uplift of the structure is allowed.

The formulation for dynamic contact relies on the approach of Wilson (E. L. Wilson 1975), who introduced *relative dof* in order to avoid numerical sensitivity in contact problems. The relative dofs possibly should be defined in a rotated coordinate system (with axes normal and tangent to the contact boundary) so that the ill-conditioning of tangent operators does not occur. It should be emphasized that relative dofs fit naturally within the proposed split of the structure total motion field for an inelastic interface only (see (5.12)), since structure reference motion causes no contact stress (this follows from (5.11)).

The normal interface compliance is assumed in the form of a power law. This choice for the normal interface compliance plays a crucial role in modeling a more general case frictional contact phenomena (Ibrahimbegović and Wilson 1992). Normal interface compliance is obtained as the limit for the summation of elastic compliances of the individual asperities; different distributions of asperity heights result in different powers in the constitutive law for the normal interface. To account for energy dissipation that occurs in impact, a simple nonlinear viscoelastic model is suggested as a supplement to the elastic power law for the normal penetration interface. It can be considered as a generalization of the dissipative model of Hunt and Crossley (1975), which uses the restitution coefficient for different materials. The normal interface stress σ_n is given as:

$$\sigma_n = -C_n \langle u_n \rangle^{m_n} + B_n \langle u_n \rangle^{l_n} \dot{u}_n \tag{5.17}$$

where $\langle \cdot \rangle$ denotes a Macaulay bracket (i.e. $\langle u_n \rangle = u_n$ if $u_n > 0$ and $\langle u_n \rangle = 0$ if $u_n \leq 0$), and u_n is the normal interface deformation (approach). Normal interface deformation u_n is determined as the difference between the relative normal displacement Δ and the initial normal gap g (see Figure 5.3). In (5.17), C_n, B_n, m_n, l_n are the constitutive coefficients.

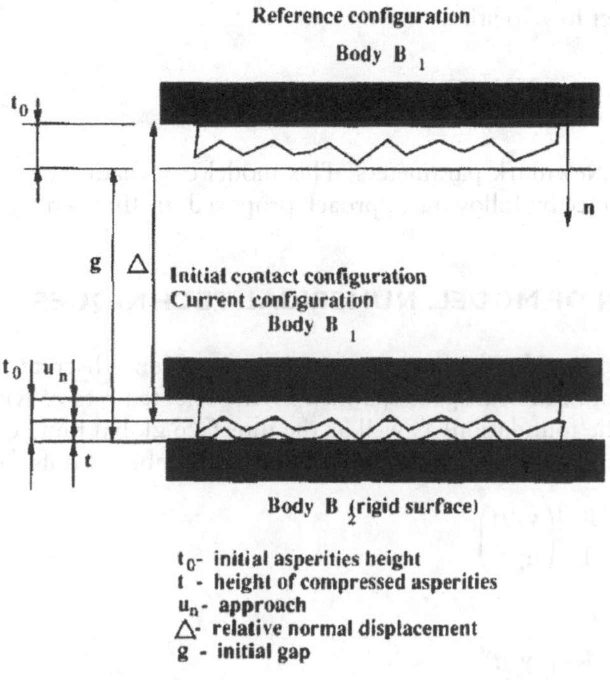

Figure 5.3 Initial gap, normal displacement and penetrating approach.

A family of isoparametric finite elements is developed to incorporate the model for dynamic contact presented herein. Contact elements which can be used for the modeling of contact boundary Γ_c in the form of a point, a line or a surface can be developed in a unified manner. The interpolation of contact boundary geometry **x** and displacement field **u** in each element is:

$$\mathbf{x} = \sum_{I=1}^{np} N_I \mathbf{x}_I ; \mathbf{u} = \sum_{I=1}^{np} N_I \mathbf{u}_I \tag{5.18}$$

where N_I is the standard isoparametric shape function, while $\mathbf{x}_I$ are nodal coordinates and $\mathbf{u}_I$ are relative nodal displacements.

If the unit's normal vector is denoted on the contact boundary Γ_c as $\mathbf{n} = \langle n_1, n_2, n_3 \rangle^T$ and defines vector **N**,

$$\mathbf{N} = \langle (n_1 N_1); (n_2 N_2); (n_3 N_3); \ldots; (n_1 N_{np}); (n_2 N_{np}); (n_3 N_{np}) \rangle^T \tag{5.19}$$

the contact element (e) contribution to the interface residual $\mathbf{r}_i(\mathbf{u}(t), \dot{\mathbf{u}}(t))$ in (5.15) is:

$$\mathbf{r}_i^{(e)}(\mathbf{u}(t), \dot{\mathbf{u}}(t)) = \int_{\Gamma_c^{(e)}} \mathbf{N} \sigma_n d\Gamma \tag{5.20}$$

Utilizing the consistent linearization procedure (Hughes and Pister 1978), for the contact interface residual (5.20), the tangent stiffness matrix can be written:

$$\mathbf{K}_{ii}^{(e)} = \int_{\Gamma_c^{(e)}} C_n m_n \langle u_n \rangle^{m_n - 1} d\Gamma + \int_{\Gamma_c^{(e)}} B_n{}^{l_n} \langle u_n \rangle^{l_n - 1} \dot{u}_n \sigma_n \mathbf{N}\mathbf{N}^T d\Gamma \tag{5.21}$$

and the tangent damping matrix resulting from the linearization of the normal interface residual with respect to velocities:

$$\mathbf{C}_{ii}^{(e)} = \frac{\gamma}{\beta \Delta t} \int_{\Gamma_c^{(e)}} B_n \langle u_n \rangle^{l_n} \mathbf{N}\mathbf{N}^T d\Gamma \tag{5.22}$$

where β and γ are Newmark parameters. This model can readily be extended to the frictional contact interface by following approach proposed in (Ibrahimbegovic, Wilson 1992).

5.3 REDUCTION OF MODEL: NUMERICAL TECHNIQUES

To enhance computational efficiency, the complete set of semi-discrete equations of motion (5.14) is projected onto the subspace spanned by the selected sets of Ritz vectors (for both the structure and the foundation) as well as the interface global finite element coordinates. Modal transformation are used in order to reduce the number of dofs for the structure:

$$\begin{pmatrix} \mathbf{u}_s(t) \\ \mathbf{u}_i^s(t) \end{pmatrix} = \begin{bmatrix} \mathbf{\Psi}_s & \mathbf{R}_s \\ 0 & \mathbf{I}_i \end{bmatrix} \begin{pmatrix} \mathbf{y}_s(t) \\ \mathbf{u}_i^s(t) \end{pmatrix} \tag{5.23}$$

and the foundation:

$$\begin{pmatrix} \mathbf{u}_f(t) \\ \mathbf{u}_i^f(t) \end{pmatrix} = \begin{bmatrix} \mathbf{\Psi}_f & \mathbf{R}_f \\ 0 & \mathbf{I}_i \end{bmatrix} \begin{pmatrix} \mathbf{y}_f(t) \\ \mathbf{u}_i^f(t) \end{pmatrix} \tag{5.24}$$

where:

$\boldsymbol{\psi}_s$ and $\boldsymbol{\psi}_f$ are the sets of Ritz vectors (subspace basis) for the structure and the foundation, respectively;

$\mathbf{R}_s$ and $\mathbf{R}_f$ are the corresponding pseudo-static transformations (following from (5.23) and (5.24));

$\mathbf{I}_i$ is the identity matrix of dimension equal to the number of interface dofs; and

$\mathbf{y}_s(t)$ and $\mathbf{y}_f(t)$ are sets of generalized Ritz coordinates for the structure and the foundation.

The sufficient number of Ritz vectors for representing the structure and the foundation can be determined by monitoring truncation criteria for the Ritz vector basis (Ibrahimbegović and Wilson 1990).

In the case of the paraxial boundary, the discretized damping matrix **C** has a *nonproportional form*. If lumping techniques (Chow 1985) for the evaluation of paraxial boundary damping matrices are used, the projection of the nonproportional damping matrix on the Ritz vector subspace can be easily obtained as a summation of rank-one matrices which affects only far-field boundary dofs. Material damping can also be dealt with only at the level of the subspace either by giving to it a Rayleigh damping form or directly specifying the damping ratios. An algorithm that accommodates nonproportional damping within the real Ritz vector subspace is selected, since it was demonstrated (Ibrahimbegović, Chen et al. 1990) to be more efficient than the adequate one that employs the complex Ritz vector subspace.

Introducing transformations (5.23) and (5.24) into the weak form of the equations of motion (5.14) and utilizing the additive split of the modal nonproportional damping matrix for the foundation substructure $\mathbf{C}_f = \mathrm{diag}(2\xi\omega_f) + \hat{\mathbf{C}}_f$, the new form of the equations of motion is obtained:

$$\mathbf{M}^*\ddot{\mathbf{u}}^*(t) + \mathbf{C}^*\dot{\mathbf{u}}^*(t) + \mathbf{K}^*\mathbf{u}^*(t) + \mathbf{r}(\mathbf{u}(t), \dot{\mathbf{u}}(t)) = \mathbf{T}^T\tilde{\mathbf{M}}\ddot{\mathbf{v}}_i^s(t) - \mathbf{C}_f\dot{\mathbf{y}}_f(t) \tag{5.25}$$

The form of the internal force vector $\mathbf{r}(\mathbf{u}(t), \dot{\mathbf{u}}(t))$ is preserved under modal transformation. The displacement vector in (5.25) has the form:

$$\mathbf{u}^*(t) = \begin{pmatrix} \mathbf{y}_s(t) \\ \mathbf{y}_f(t) \\ \mathbf{u}_i^s(t) \\ \mathbf{u}_i^f(t) \end{pmatrix} \tag{5.26}$$

which is obtained by the reduction of the form $\mathbf{u}^*(t) = \mathbf{T}\mathbf{u}(t)$. Now, the mass matrix of the complete system in an expanded form is given as:

$$\mathbf{M} = \begin{bmatrix} \mathbf{I}_s & 0 & \hat{\mathbf{M}}_{si} & 0 \\ 0 & \mathbf{I}_f & 0 & \hat{\mathbf{M}}_{fi} \\ \hat{\mathbf{M}}_{si}^T & 0 & \hat{\mathbf{M}}_{ii}^s & 0 \\ 0 & \hat{\mathbf{M}}_{fi}^T & 0 & \hat{\mathbf{M}}_{ii}^f \end{bmatrix} \tag{5.27}$$

where:

$$\hat{\mathbf{M}}_{si} = \boldsymbol{\Psi}_s^T\mathbf{M}_{ss}\mathbf{R}_s + \boldsymbol{\Psi}_s^T\mathbf{M}_{si}; \hat{\mathbf{M}}_{ii}^s = \mathbf{R}_s^T\mathbf{M}_{ss}\mathbf{R}_s + \mathbf{M}_{si}\mathbf{R}_s + \mathbf{R}_s^T\mathbf{M}_{is} + \mathbf{M}_{ii}^s \tag{5.28}$$

Similar partitioned forms can be written for the damping matrix $\mathbf{C}^*$ and stiffness matrix $\mathbf{K}^*$, except that the upper left blocks of those matrices have the diagonal form with elements $2\xi_i\omega_i$ and ω_i^2 instead of the identity matrices $\mathbf{I}_s$ and $\mathbf{I}_f$ (ξ_i are modal damping ratios and ω_i mi are approximate modal frequencies).

The semi-discrete equations of motion (5.25) can be further solved by utilizing step-by-step integration methods, i.e. fully discretizing them. For this purpose the implicit Newmark algorithm family' can be used. This requires the tangent operators (5.21) and (5.22) for all contact elements and the solution of the nonlinear set of equations at each time step. The task of solving a set of nonlinear equations is greatly simplified by first reducing the size of the linear part of the complete system to a small set of generalized Ritz coordinates, and second by reducing the linear part to its Schur complement (Duff, Erisman and Reid 1989), i.e. prior to the solution of the nonlinear equations set the static condensation is performed on the effective stiffness matrix $\hat{\mathbf{K}}$:

$$\begin{bmatrix} \hat{\mathbf{K}}_{11} & \hat{\mathbf{K}}_{12} \\ \hat{\mathbf{K}}_{21} & \hat{\mathbf{K}}_{22} \end{bmatrix} \rightarrow \left[\hat{\mathbf{K}}_{22} - \hat{\mathbf{K}}_{21}\hat{\mathbf{K}}_{11}{}^{-1}\hat{\mathbf{K}}_{12}\right] \tag{5.29}$$

where the static condensation is made trivial by the diagonal form of $\hat{\mathbf{K}}_{11}$ for all the matrices in (5.25), and consequently for the effective stiffness matrix. For the step-by-step algorithm with variable step size (a requirement of efficient solution for dynamic frictional contact), the linear part of the effective stiffness matrix has to be reformed and refactorized quite often. Computation of the Schur complement (5.29) of the linear part is much more efficient for the equations of the form (5.25) than for the complete set (5.14) where the linear part is represented in the finite element coordinates (Bathe and Gracewski 1981). Hence, the initial cost of formulating the Ritz vector subspace for both the structure and the foundation is likely to be compensated for, especially for loading of long duration (e.g. earthquake loading). In addition, the number of operations used to form the effective load vector is also significantly reduced, since $\dim(\mathbf{y}_s) \ll \dim(\mathbf{u}_s)$ and $\dim(\mathbf{y}_f) \ll \dim(\mathbf{u}_f)$.

5.4 CASE STUDIES OF STRUCTURE-FOUNDATION INTERACTION PROBLEMS

The analysis of the dam-foundation model presented in Figure 5.4 is performed when the uplifting of the dam occurs. The final selection for the foundation model is done after extensive parametric studies of radiation effects (Ibrahimbegović and Wilson 1989a). The model of the dam is very similar to the Pine Flat Dam analyzed by Chakrabarti and Chopra (1973) for hydrodynamic effects. Only the uplifting effect is considered; the tangential sliding is assumed to be restricted either by embedment or otherwise. Material properties for the dam and the foundations are given in the Table 5.1.

Material damping is given the form of Rayleigh damping with control frequencies equal to 13.27 and 21.39 rad/sec and damping ratio $\xi = 5\%$, while the nonproportional damping arises from paraxial boundaries used to represent the radiation condition. To model a maximum credible earthquake, the 1952 Taft earthquake (record S96E with peak ground acceleration 0.17949 at 3.74 sec) is arbitrarily scaled by 3. Since a nonlinear problem is elaborated, hydrostatic water pressure and dead load are considered simultaneously with the earthquake excitation. Hydrodynamic water pressure is completely disregarded. The initial conditions are represented by the deformed configuration under the influence of static loads.

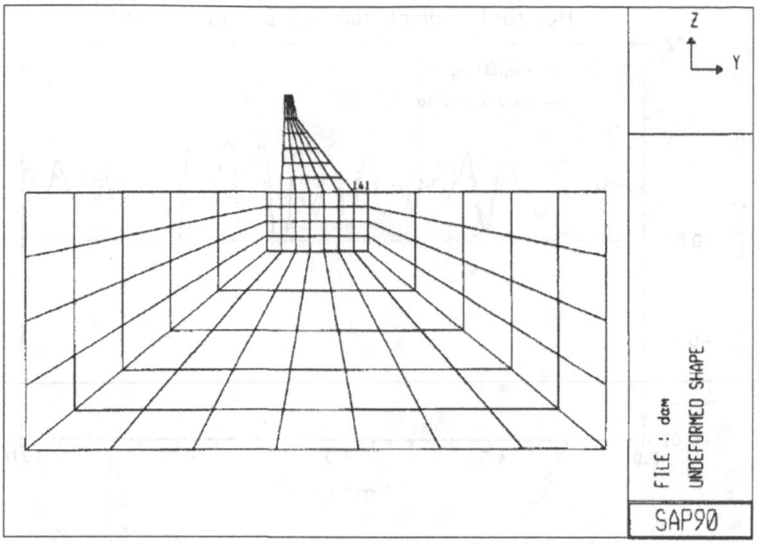

Figure 5.4 Finite element models of dam-foundation system.

Table 5.1 Material Properties for the Dam and the Foundations, and Contact Boundary

Dam and Foundations		*Contact Boundary*	
Young modulus	22150 MN/m^2	C_n	10^6 MN/m^2
Mass density	2500 kg/m^3	m_n	2
Poisson's ratio	0.25		

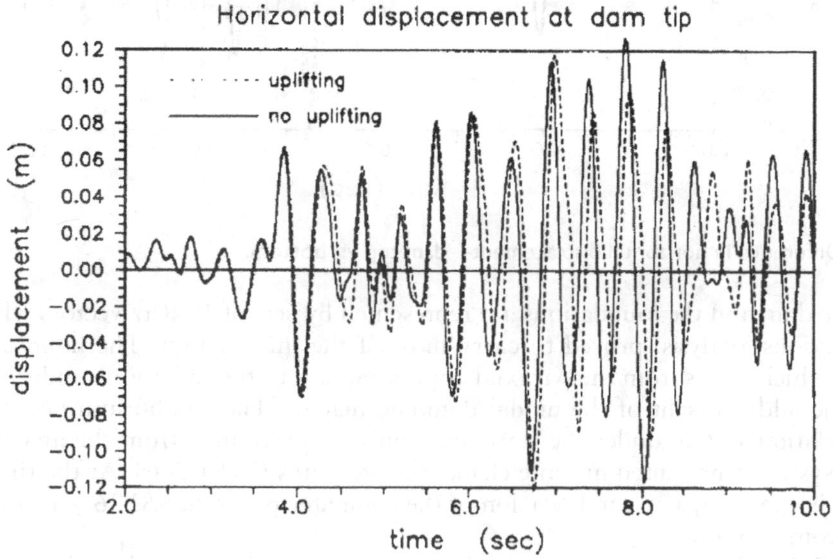

Figure 5.5 Horizontal displacement at dam tip.

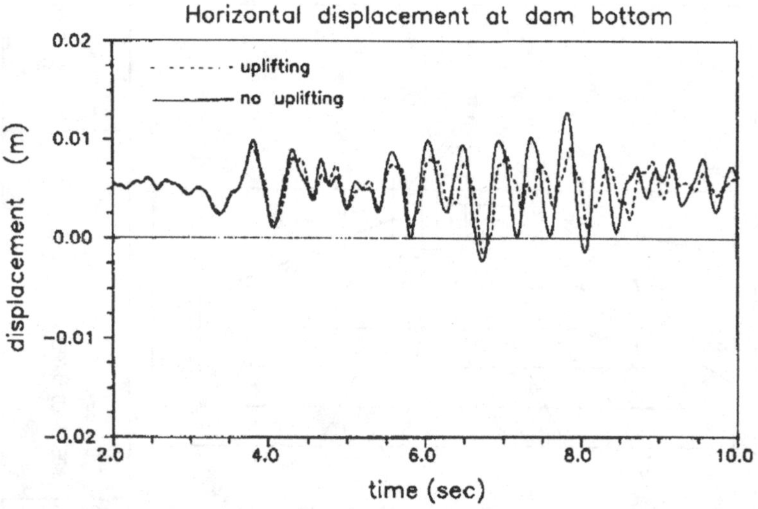

Figure 5.6 Horizontal displacement at dam bottom.

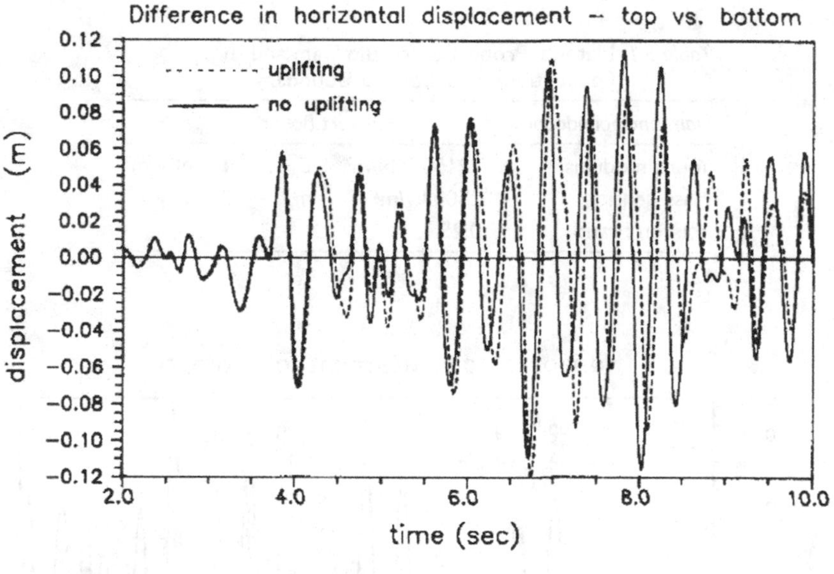

Figure 5.7 Difference in horizontal displacement—dam top vs. bottom.

Both the dam and the foundations are represented by sets of 10 Ritz vectors which have, by the previous analysis, proved to carry most of the information. The nonproportional damping, which arises from the paraxial approximation to the radiation condition, is handled by the additive split of the modal damping matrix. The equilibrium position of the dam-foundation system under the static load only is determined from the analysis of the complete system represented in finite element coordinates (I 154 dofs). At the time of conducting this example, a research version of the computer program SAP26 was used to perform the computations.

Figures 5.5 and 5.6 illustrates the displacements at the dam top and the bottom, respectively, for both cases where the uplifting of the dam is permitted and prevented. Over a

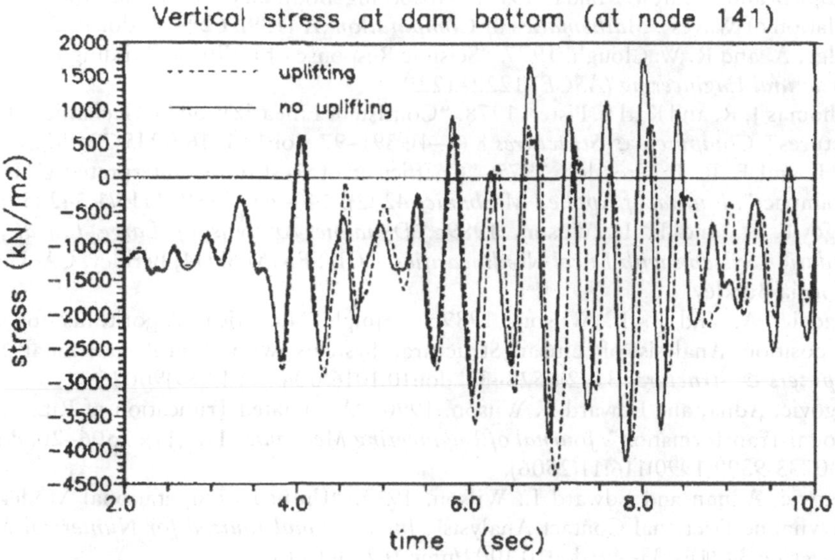

Figure 5.8 Vertical stress at dam bottom.

certain range of frequencies response amplification occurs due to the uplifting, but in the part of the strong shaking the beneficial isolation effects of the uplifting are evident. To assess the stress field in the dam, the difference of the horizontal displacement at the dam top vs. the bottom, as well as the vertical stress time-history at the dam bottom, are plotted in Figures 5.7 and 5.8, respectively. As it is assumed that the dam is made of low quality concrete, say C15/20 (which is consistent with the choice of Young's modulus value for the dam material), then the allowable compressive stress is 8 MPa while the allowable tensile stress is 0.8 MPa. Hence, if the uplifting of the dam is allowed for, no cracking of the dam body would occur. Figure 5.8 shows that if the uplifting is prevented this will not be the case.

BIBLIOGRAPHY

Bathe, Klaus-Jürgen and Sheryl Gracewski. 1981. "On Nonlinear Dynamic Analysis Using Substructuring and Mode Superposition". *Computers & Structures* 13 (5–6): 699–707. doi:10.1016/0045-7949(81)90032-8.

Chakrabarti, P. and Anil K. Chopra. 1973. "Earthquake Analysis of Gravity Dams Including Hydrodynamic Interaction". *Earthquake Engineering & Structural Dynamics* 2 (2): 143–60. doi:10.1002/eqe.4290020205.

Chow, Y. K. 1985. "Accuracy of Consistent and Lumped Viscous Dampers in Wave Propagation Problems". *International Journal for Numerical Methods in Engineering* 21 (4): 723–32. doi:10.1002/nme.1620210411.

Clough, Ray W. and Edward L. Wilson. 1979. "Dynamic Analysis of Large Structural Systems with Local Nonlinearities". *Computer Methods in Applied Mechanics and Engineering* (17–18): 107–29. doi:10.1016/0045-7825(79)90084-7.

Clough, Ray W. and Joseph Penzien. 1993. *Dynamics of Structures*. Edited by 2nd. New York: McGraw-Hill Education (ISE Editions).

Cohen, M. and P. C. Jennings. 1983. *Silent Boundary Methods for Transient Analysis*. Edited by T. Belytschko and T. J. R. Hughes. New York: Elsevier Science Publications.

Duff, I. S., A. M. Erisman and J. K. Reid. 1989. *Direct Methods for Sparse Matrices*. Oxford: Oxford University Press.

Engquist, Bjorn and Andrew Majda. 1977. "Absorbing Boundary Conditions for the Numerical Simulation of Waves". *Mathematics of Computation* 31 (139): 629–51. doi:10.2307/2005997.

Huckelbridge, A. and R. W. Clough. 1977. "Seismic Response of Uplifting Building Frame". *Journal of Structural Engineering (ASCE)* 1222–1229.

Hughes, Thomas J. R. and Karl S. Pister. 1978. "Consistent Linearization in Mechanics of Solids and Structures". *Computers & Structures* 8 (3–4): 391–97. doi:10.1016/0045-7949(78)90183-9.

Hunt, K. H. and F. R. E. Crossley. 1975. "Coefficient of Restitution Interpreted as Damping in Vibroimpact". *Journal of Applied Mechanics* 42 (2): 440–45. doi:10.1115/1.3423596.

Ibrahimbegović, A. and E. L. Wilson. 1989a. *Dynamic Analysis of Large Linear Structure-Foundation Systems with Local Nonlinearities*. UCB/SEMM 89/14, Berkley, CA: University of California Berkley.

Ibrahimbegović, A. and E. L. Wilson. 1989b. "Simple Numerical Algorithms for the Mode Superposition Analysis of Linear Structural Systems with Non-Proportional Damping". *Computers & Structures* 33 (2): 523–31. doi:10.1016/0045-7949(89)90026-6.

Ibrahimbegović, Adna, and Edward L. Wilson. 1990. "Automated Truncation of Ritz Vector Basis in Modal Transformation". *Journal of Engineering Mechanics* 116 (11): 2506–20. doi:10.1061/(asce)0733-9399(1990)116:11(2506).

Ibrahimbegović, Adnan and Edward L. Wilson. 1992. "Unified Computational Model for Static and Dynamic Frictional Contact Analysis". *International Journal for Numerical Methods in Engineering* 34 (1): 233–47. doi:10.1002/nme.1620340115.

Ibrahimbegović, Adnan, Harc C. Chen, Edward L. Wilson and Robert L. Taylor. 1990. "Ritz Method for Dynamic Analysis of Large Discrete Linear Systems with Non-Proportional Damping". *Earthquake Engineering & Structural Dynamics* 19 (6): 877–89. doi:10.1002/eqe.4290190608.

Kausel, Eduardo. 1988. "Local Transmitting Boundaries". *Journal of Engineering Mechanics* 114 (6): 1011–27. doi:10.1061/(asce)0733-9399(1988)114:6(1011).

Lin, Y. K. 1976. *Probabilistic Theory of Structural Dynamics*. Huntington, NY: R. E. Krieger Pub. Co.

Roesset, J. M. and J. S. Kim. 1987. "Specification of Control Point for Embedded Foundations". *5th Canadian Conference on Earthquake Engineering*. 63–86.

Wilson, E. L. 1975. "Finite Elements for Foundations, Joints and Fluids". *Conf. on Numerical Methods Soil Roch Mechanics*, September 15–19. Karlsruhe, Germany.

Wilson, Edward L. and Eduardo P. Bayo. 1986. "Use of Special Ritz Vectors in Dynamic Substructure Analysis". *Journal of Structural Engineering* 112 (8): 1944–54. doi:10.1061/(asce)0733-9445(1986)112:8(1944).

Wolf, John P. 1986. "A Comparison of Time-Domain Transmitting Boundaries". *Earthquake Engineering & Structural Dynamics* 14 (4): 655–73. doi:10.1002/eqe.4290140412.

Chapter 6

The dynamics of extreme impact loads in an airplane crash

Present day challenges in fighting man-made and natural hazards have brought about new issues in dealing with extreme transient conditions for engineering structures. As already indicated, the main advantage of the multi-scale approach is that it has by far the greatest capabilities for providing a basis for constructing a sufficiently predictive model. Namely, this kind of approach provides refined prediction capabilities of the mechanics model with respect to the smeared models of plasticity or damage (Lemaitre and Chaboche 1988; Lubliner 1990), as well as more detailed information regarding inelastic mechanisms rather than just the total inelastic dissipation. It was only at the beginning of the 21st century that models for predicting detailed crack-spacing and opening (fracture process zones developed in more detail) were developed. Application to metals was performed by Ibrahimbegović and Brancherie (2003) and to concrete by Brancherie et al. (2005) and Ibrahimbegović (2006) or to reinforced concrete by Dominguez et al. (2005).These models are used as the basis of models for massive structures exposed to dynamic extreme airplane actions, as a man-made hazard for which most engineering structures have not been designed. From the standpoint of nonlinear analysis, the main difficulty in such a problem pertains to the significant high-frequency content typical of impact phenomena, as well as the likely presence of extensive structural damage. In this respect, first of all, it is necessary to develop models that will be able to predict inelastic behavior and damage for both the airplane and the massive structure. Second, it is interesting to examine different impact scenarios in trying to quantify any potential reserve that might exist in the given design of engineering structures for taking on a higher level of risk. The complementary goal, which also of great importance, pertains to providing the best-reduced basis for carrying out the parametric studies that are needed for a sound design procedure.

The idea for such analysis was born after the September 11, 2001, attacks on the World Trade Center (Figure 6.1).

In order to solve such a complex problem, it is necessary to incorporate the significant advances in nonlinear analysis within the design procedure of a complex engineering structure under an equally complex dynamic loading. One possible example of this would be the impact by an airplane on a massive structure, such as a nuclear power plant (Figure 6.2).

It is important to note that traditional design procedures are not applicable to these types of complex structures, and even less so to a complex, nonproportional loading program such as the one produced by the frictional contact of the airplane with the structure. At the moment, some empirical expressions are used, which are based on rather simplified interpretations of experimental results of projectile tests perforating different plate-like structures, and this is very difficult to apply on complex structures.

This all shows that there was and still is the need for a novel approach to the nonlinear analysis of these kinds of problems.

One of the elements that has a major influence on structure damage, as well as the structure–projectile interaction, is the true nature of the impact. In this respect, it is necessary to

Figure 6.1 World Trade Center attack, 9/11/2001 (www.telegraph.co.uk).

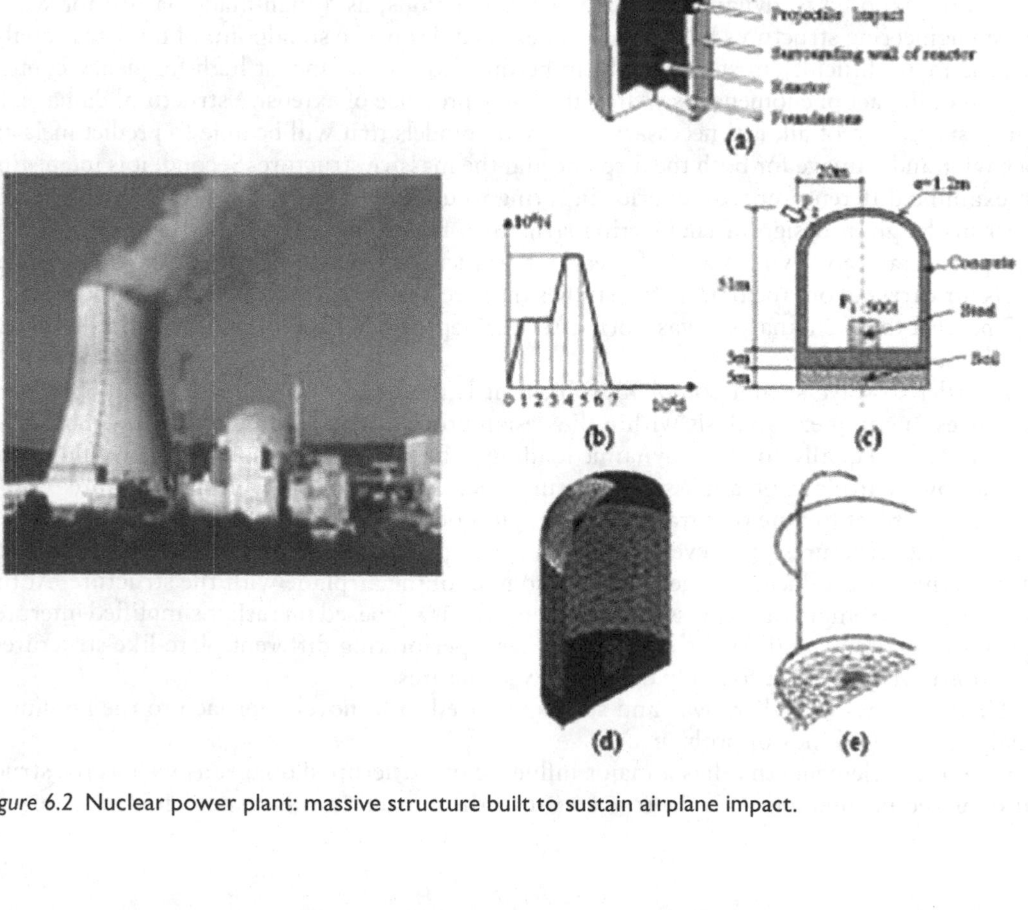

Figure 6.2 Nuclear power plant: massive structure built to sustain airplane impact.

define the type of impact action, either hard or soft impact. In the case of hard impact, the projectile is much stiffer than the target; whereas in the case of the soft impact, the target is at least as stiff as the projectile (Figure 6.3). Additionally, a great diversity in damage modes creates another problem that has to be taken into account and elaborated in detail, including perforation, cauterization or spalling (Figure 6.3).

In order to solve this kind of complex problem, it is necessary to split the analysis into two phases, local and global. The local phase deals with the analysis of a single structural component (e.g. an impacted area of a nuclear power plant or a concrete slab in a facility, see Figure 6.4); whereas the global analysis deals with the complete (complex) structure. The result of the local analysis will be incorporated into the final stage of the global analysis in the sufficiently representative manner. This kind of formulation will provide a much higher efficiency of the computations in the global phase. In this way, all the parametric studies, needed in the design phase, can be performed in a very efficient manner, which is the main advantage of this kind of approach (Figure 6.4).

6.1 CENTRAL DIFFERENCE SCHEME COMPUTATION OF IMPACT PROBLEMS

The local phase of the analysis is carried out first using a detailed local model of a single component corresponding to the impacted zone, which is exposed to an impact action of short duration. Analysis of this kind considers the finite element model of the projectile (for example an airplane), which can account for the large plastic deformations developing at the impact and the frictional contact with potentially large frictional sliding. The explicit, central difference scheme (Bathe and Wilson 1976; Owen and Hinton 1980) is most appropriate for such an analysis. The state variable values over a typical increment can be computed according to:

$$\begin{aligned}
&\mathbf{d}_{n+1} = \mathbf{d}_n + \Delta t \mathbf{v}_n + \Delta t^2 \mathbf{a}_n, \\
&\mathbf{M}\mathbf{a}_{n+1} = \mathbf{r}_{\text{contact},n+1} - \hat{\mathbf{f}}^{\text{int}}\left(\mathbf{d}_{n+1}\left[\varepsilon^{vp}, \mathcal{D}, \ldots\right]_{\text{GNP}}\right), \qquad (6.1) \\
&\mathbf{v}_{n+1} = \mathbf{v}_n \frac{\Delta t}{2}\left(\mathbf{a}_n + \mathbf{a}_{n+1}\right)
\end{aligned}$$

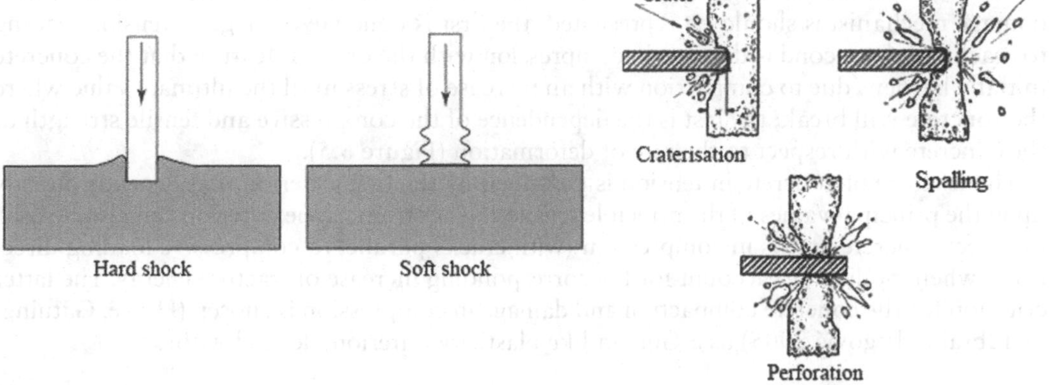

Figure 6.3 Hard vs. soft impact and damage modes.

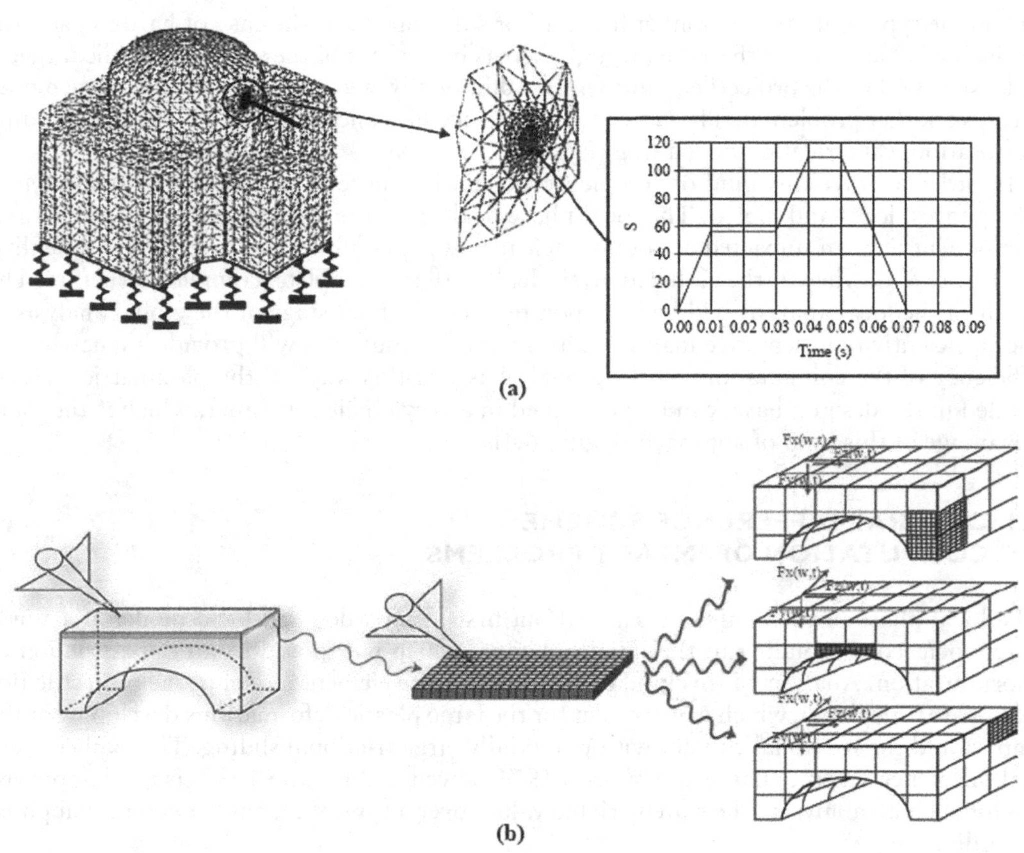

Figure 6.4 An analysis-design procedure based on the split between the local phase and the global phase.

where **d**, **v** and **a** are, respectively, displacements, velocities and accelerations, and **M** is the mass matrix in a diagonal form for computational efficiency. The main sources of nonlinearity belong to the impact and frictional contact of the airplane with contact forces denoted as $\mathbf{r}_{contact}$, as well as to the plastic and damage deformations of the impacted component of the structure, which are all stored in internal force term $\hat{\mathbf{f}}^{int}$.

A reliable representation of damage in the impacted structural component made of reinforced concrete has to be given, which can then be used for the structural design. Several damage mechanisms should be represented: the first is concrete damage in tension leading to cracking; the second is damage in compression with the original feature that the concrete initially hardens due to compaction with an increase of stress until the ultimate value where the concrete will break; the last is the dependence of the compressive and tensile strength of the concrete with respect to the rate of deformation (Figure 6.5).

The fracture of concrete in tension is described by the first criterion that depends directly upon the principal values of the principle tensile elastic strain. Same criterion can also be used to detect concrete cracks in compression (with cracks parallel to compressive loading direction), when modified to account for the corresponding increase of fracture energy. The latter criterion for the concrete compaction and damage in compression is chosen (Hervé, Gatuingt and Ibrahimbegović 2005) as a Gurson-like plasticity criterion, defined with:

$$\phi\left(\sigma_{ij},\sigma_m,f^*\right)=\frac{3J_2}{\sigma_m^2}+2q_1f^*\cosh\left(q_2\frac{I_1}{2\sigma_m}\right)-\left(1+\left(q_3f^*\right)^2\right)\leq 0 \tag{6.2}$$

where:

σ_{ij}	are the components of the nominal stress tensor
σ_m	is the reference value of the stress in the matrix
f^*	is the concrete porosity and
q_1, q_2, q_3	are the chosen coefficients.

The graphic illustration of this criterion for both compressive and tensile stress is shown in Figure 6.5.

The constitutive model also features porosity-dependent hardening that evolves as a function of the equivalent compressive plastic deformation. The same parameter defines the threshold for element erosion, defining the stage where the element is damaged to such an extent that it ought to be completely removed from the mesh.

The evolution equations for plastic deformation in compression and damage deformation in tension are chosen as rate-dependent in order to account for the effects of rate of deformation, which are quite noticeable for this class of problem (Figure 6.5). The numerical implementation of the proposed model is incorporated within the proposed computational framework in (6.1). However, contrary to the explicit scheme computations for the global momentum balance equations, the integration of the evolution equations for internal variables, such as plastic strains, damage compliance and hardening/softening variables, is carried out by an implicit scheme. The latter provides at each step the admissible values of stress with respect to the chosen criteria and leads to a robust numerical implementation.

The tests for simulating impact were done in Sandia Laboratory, and the reliability of the proposed constitutive model was checked. This was done using different sized missiles, which simulated aircraft engines, and their impact on the reinforced concrete slabs. In the

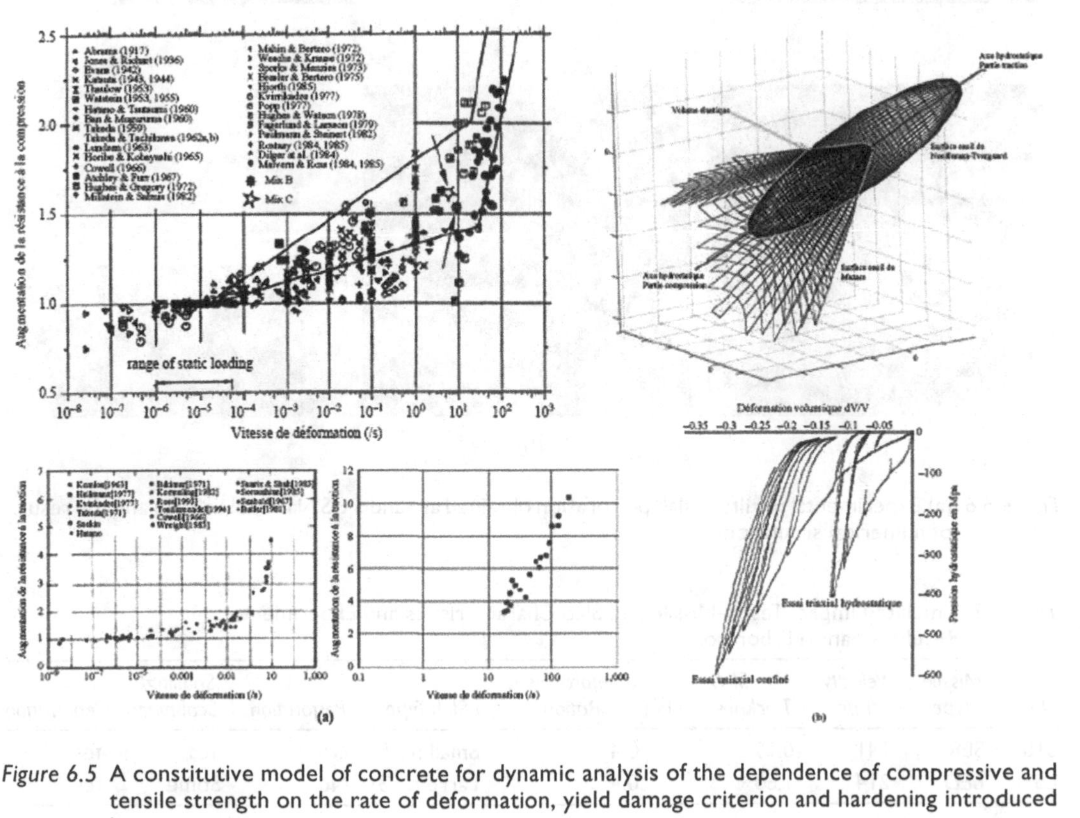

Figure 6.5 A constitutive model of concrete for dynamic analysis of the dependence of compressive and tensile strength on the rate of deformation, yield damage criterion and hardening introduced by compaction.

simulation process, it was seen as best to simulate both rigid and soft missiles. In this respect, simulations for large size equivalent and deformable missiles, medium size equivalent and deformable missiles and small size equivalent rigid missiles (SER) were done (see Figure 6.6 for details). The results of two of the conducted experiments are given in Table 6.1.

The missiles where modeled using the shell finite element models, except for the SER missiles where 3D solid elements were used. The concrete slabs were modeled using under-integrated 3D solid elements with the proposed constitutive model, and the reinforcement was modeled with truss-bar elastoplastic elements. In order to define the different types of damage, it was necessary to distinguish between penetration, perforation and scabbing. Penetration denotes that the missile penetrated the slab without having gone through it;

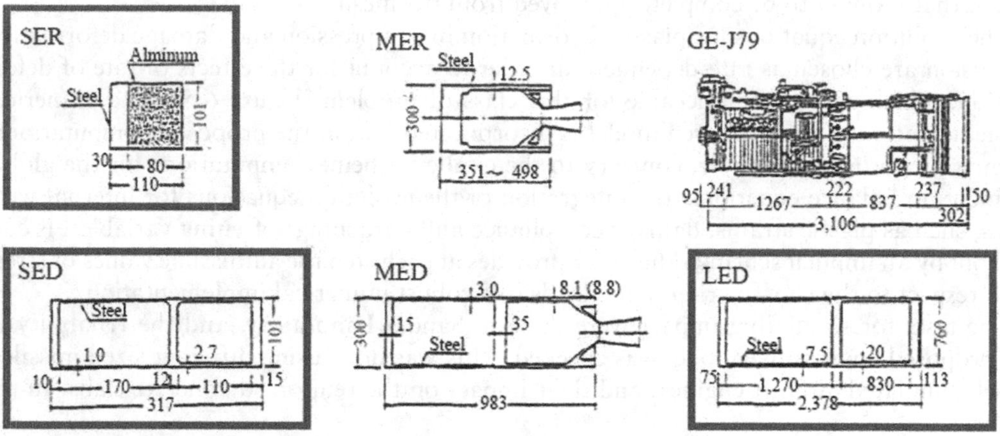

(a)

(b)

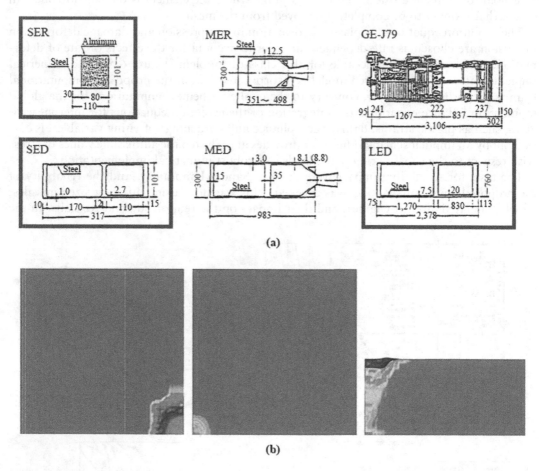

Figure 6.6 (a) Experimental results of slab perforation obtained at Sandia US National Lab, (b) and the results of numerical simulations.

Table 6.1 Simulated Impact Test—Missile and Slab Characteristics and Experimental Results—Sandia Laboratory

No	*Missile Type*	*Velocity (m/s)*	*Slab Thickness (M)*	*Reinforcement Ration*	*Slab Type*	*Perforation*	*Spalling/ Scabbing*	*Penetration*
S10	SER	141	0.15	0.4	Small no.1	No	Yes	Yes
L5	LED	214	1.60	0.4	Large no.3	No	Some	Yes

perforation denotes that the missile went through; and scabbing denotes that the impact generated a scab on the rear face of the slab.

Figure 6.6b shows the numerical simulations for the selected examples. It is clear that quite similar results were obtained. Once the velocity of the missile stopped decreasing, which implied the missile was finally stopped by the concrete, the computations where stopped. Numerical simulation of example S10 (thin slab) showed damage of the slab, spalling occurred on the rear face, but there was no penetration of the missile. The second example was a thick slab (L5), where the damage remained localized in the impact area with a large undamaged volume separating damage on the front face and damage on the rear face. The latter implies the presence of only some small cracks. A very good correlation between the model and the tests were obtained.

6.2 LOCAL MODEL EXTENSIONS

One phenomena that needs to be elaborated further is spalling, where a damaged piece from the main structure can detach and fly away. This represents an additional issue that has to be modeled. Damage of the interior equipment or storage material depends on the size of the detached piece, more precisely its mass, and its velocity. If the velocity is too high and/or if the piece of detached material is too big, it can cause significant damage to the equipment or storage material without causing larger damage to the structure. However, this can cause significant damage, explosion or fire, and lead to structural damage and even collapse.

The best way to represent the spalling phenomena is through a three-point bending test in dynamics. This test is performed in the manner that a weight is dropped from a certain height. In this way, an impact is caused on the fallen structure producing a compressive wave that travels from the upper surface in a downward direction toward the bottom surface of the specimen. When this wave reaches the lower surface of the specimen, it will reflect, double in size and turn into a tensile wave. This is the moment when the damage can be introduced in tension sensitive material, with pieces that can detach and freely move away from the main structure.

The results of the performed numerical simulation are illustrated in Figure 6.7.

This damage-plasticity continuum model is capable of predicting the extent of the damage zone associated with a large value of the damage variables or even eliminate all the elements which are extensively damaged by activating the erosion criterion. The continuum model enables one to eliminate the pieces of the damaged structure that have detached during the impact; however, it is not possible to follow the motion of these detached pieces. In order to grasp these phenomena thoroughly, it is necessary to use a discrete model. In this case, a model based on the Vornoi cell representation of the specimen was developed by Ibrahimbegović and Delaplace (2003), where the cohesive forces between the adjacent cells are represented by a geometrically exact Reissner beam model (Ibrahimbegović and Taylor 2002). This particular feature of the beam model with its capability for representing the overall large motion (accompanied by small strains) is crucial for the present application, where the detached piece is represented by several Vornoi cells with preserved cohesive forces (Figure 6.7). There is a very delicate question here concerning the implementation of appropriate time-integration schemes (Delaplace and Ibrahimbegović 2006) capable of controlling the high-frequency content of motion (Ibrahimbegović and Mamouri 2002) and thus minimizing the risk of the spurious stress oscillation introduced by brittle fracture.

An example of an anisotropic damage model is given; which can be generalized to even a complex model. Namely, a coupled damage-plasticity model can be obtained, which is capable of representing the inelastic behavior of porous metals (Ibrahimbegović, Markovic

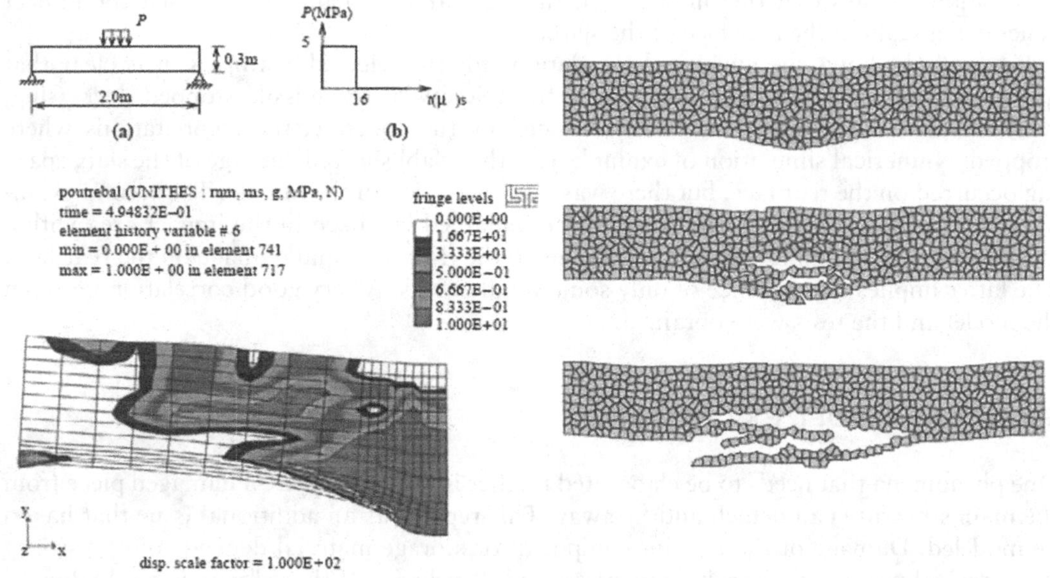

Figure 6.7 The numerical simulation of three-point bending testing: A continuum model with element erosion can completely remove damage elements, and the discrete model can also represent spalling as well as the motion of any detached piece of the specimen.

and Gatuingt 2003) or concrete under compaction (Hervé, Gatuingt and Ibrahimbegović 2005). The latter is used in the application of a commercial airplane impact problem on a massive structure. Finally, the same kind of development can be carried out for an even more general framework of the reinforced concrete model, which is assembled from the anisotropic damage model for concrete, the elastoplastic model for steel and a coupled damage-plasticity model for bond-slip (Dominguez et al. 2005). The latter is implemented within the zero-thickness bond element with normal and tangential degrees of freedom, which is constructed in the same manner as the contact element described by Ibrahimbegović and Wilson (1992). An increase in slip resistance due to confinement pressure can be easily represented. Figure 6.8 illustrates the predictive capabilities with respect to crack-spacing and opening for this kind of model, where the important role played by a bond-slip element in redistributing the stress transfer mechanism along the steel bar more evenly was demonstrated, as well as an excellent correlation between the experimental and numerical results, even in the case of the dispersion of slip resistance (Dominguez et al. 2005).

6.3 FIELD TRANSFER STRATEGY FOR COMPLEX STRUCTURE IMPACT LOADING

Once the local analysis phase of the airplane impact on a single structural component (or impacted zone) has been completed, the global phase can start. The main goal of this phase is to check the integrity of the whole structural assembly and perform any eventual parametric studies where different design possibilities are proposed. Efficiency is ensured by the chosen coarse mesh analysis at the structural scale, which incorporates the results from the structural component which were considered in the local phase. Global analysis can directly integrate the results obtained in the local analysis phase, providing a couple of possible simplifications. First of all, the fine mesh used in the local analysis phase is replaced by a coarse

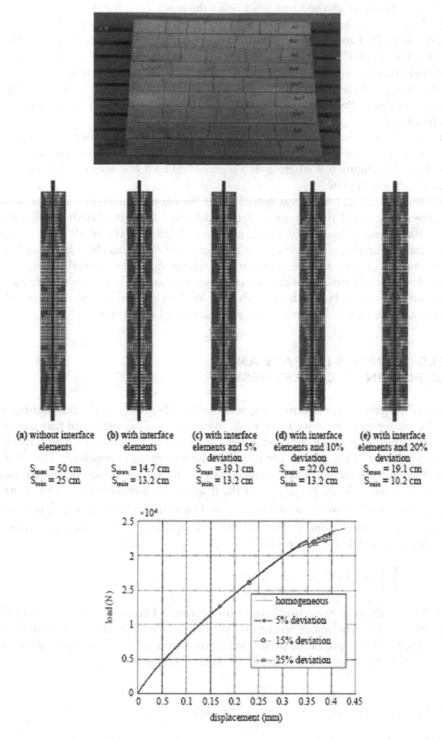

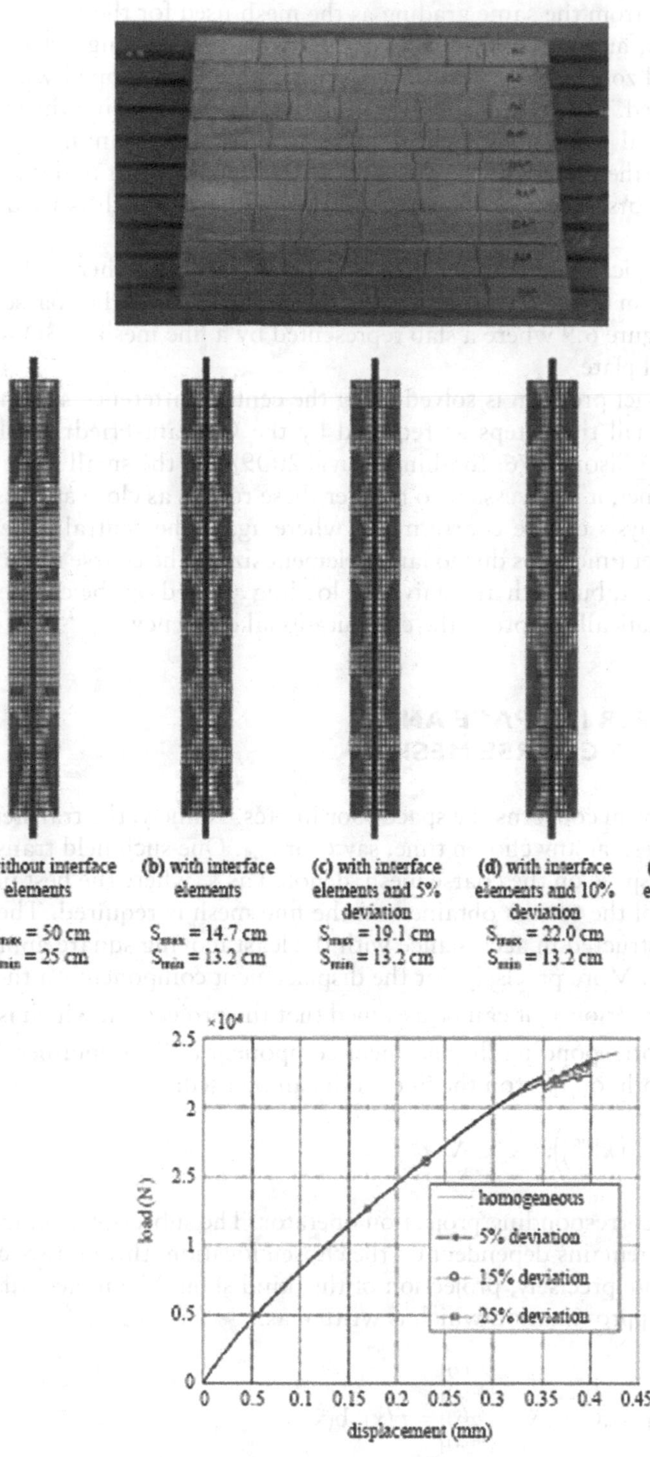

Figure 6.8 Pull-out test for a reinforced concrete beam: Experimental results, numerical results (with no bond-slip, with constant bond-slip and with variable bond-slip resistance using the standard variation of 0.05, 0.10 and 0.20 and the resulting force–displacement diagram).

mesh which is chosen from the same grading as the mesh used for the rest of the structural assembly. In this way, any potential risk that the waves propagating across the interface between the impacted zone and the rest of the structure remain trapped within the refined mesh zone is eliminated. The second simplification consists of replacing the true loading on the impacted structural component, which stems from time-consuming computations of frictional contact and the impact of the airplane, with the equivalent nodal loading applied at the nodes of the coarse mesh of the structural component, which is used in the global analysis phase.

This represents a typical field transfer problem. The biggest issue here is how, in the best possible way, to perform the field transfer task between the fine and the coarse mesh. This is easily explained in Figure 6.9 where a slab represented by a fine mesh of 3D solid elements, is impacted by a small plate.

This nonlinear impact problem is solved using the central difference scheme which typically employs very small time steps as required by the Courant-Friedrichs-Levy stability condition (Bathe and Wilson 1976; Ibrahimbegović 2009) and the small finite element sizes in the fine mesh. Further, it is necessary to recover these results as close as possible from the nonlinear impact analysis on the coarse mesh, where again the central difference scheme will be used with larger time steps due to larger element sizes. The coarse mesh analysis is no longer driven by contact, but with an equivalent loading applied on the coarse mesh, which can further and dramatically improve the computational efficiency.

6.4 FIELD TRANSFER IN SPACE AND PROJECTION TO COARSE MESH

The first transfer problem concerns the space coordinates. Namely, the transfer between the fine and the coarse mesh at any chosen time, say t_n or t_{n+1}. One such field transfer should be done for any point in space on the coarse mesh, denoted as $\mathbf{x}$, where the best possible coarse mesh representation of the results obtained on the fine mesh is required. The coarse mesh representation is constructed in accordance with the least moving square approximation of the fine mesh transfer. More precisely, for the displacement component on the coarse mesh $d_i^{\text{coarse}}(\mathbf{x})$ at a chosen location $\mathbf{x}$, it can be assumed that the projection, which is based on the nodal values of the corresponding displacement component $d_i^{\text{fine}}(\mathbf{x})$ obtained from solution to (6.1) in the neighborhood of $\mathbf{x}$ on the fine mesh can be used:

$$d_i^{\text{coarse}}(\mathbf{x}) = \prod_x \left(d_i^{\text{fine}}\left(\mathbf{x}_j^{\text{fine}}\right)\right); \forall \mathbf{x}_j \in \mathcal{N}(\mathbf{x}) \tag{6.3}$$

where Π denotes the corresponding projection operator. The subscript x indicates that such a projection operator remains dependent on the chosen location, through space dependence of its coefficients. More precisely, projection of this kind should be in accordance with the moving least square approximation which is written as:

$$d_i^{\text{coarse}}(\overline{\mathbf{x}}) = \begin{bmatrix} 1 & x_1 & x_2 & x_3 \end{bmatrix}^T \begin{bmatrix} b_0(\bar{x}) \\ b_1(\bar{x}) \\ b_2(\bar{x}) \\ b_3(\bar{x}) \end{bmatrix} = \mathbf{p}(\mathbf{x})^T \mathbf{b}(\overline{\mathbf{x}}) \tag{6.4}$$

where $\mathbf{b}(\overline{\mathbf{x}})$ are the approximation parameters to be determined from the least square fit at each desired location $\overline{\mathbf{x}}$. This location is typically selected in accordance with the coarse mesh, either as the nodal point of the numerical integration point, with a known coordinates values $\overline{\mathbf{x}}$. Projection to the coarse mesh is constructed using the complete linear polynomials $\mathbf{p}(\mathbf{x})$.

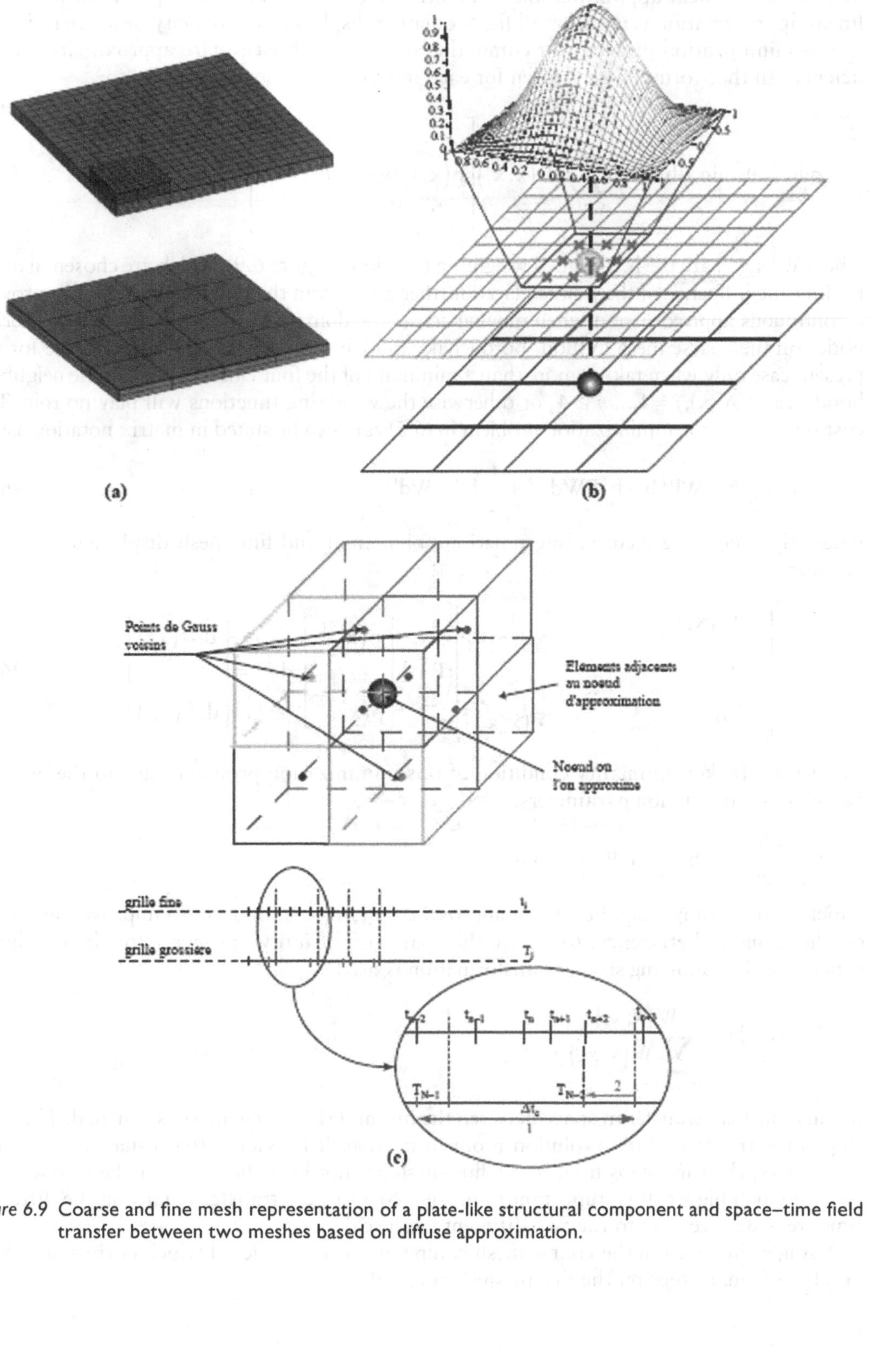

Figure 6.9 Coarse and fine mesh representation of a plate-like structural component and space–time field transfer between two meshes based on diffuse approximation.

The choice of linear approximation is conditioned by coarse mesh interpolation, where the linear approximation is used for all fields of either displacement, velocity or acceleration.

The minimization problem for computing the moving least square approximation coefficients can thus formally be written for each motion component according to:

$$\min_b J(\mathbf{b}); \min_b J(\mathbf{b}) := \frac{1}{2} \sum_{j \in \mathcal{N}(x)} W(\bar{\mathbf{x}}, \mathbf{x}_j) \left[\underbrace{\mathbf{p}(\mathbf{x}_j)^T \mathbf{b}(\bar{\mathbf{x}})}_{d_i^{\text{coarse}}(x_j)} - d_i^{\text{fine}}(\mathbf{x}_j) \right]^2 \tag{6.5}$$

where $W(\bar{\mathbf{x}}, \mathbf{x}_j)$ are the bell-shaped weighting functions (Figure 6.9), which are chosen in order to limit the influence of the points placed farther away from the point $\mathbf{x}$, as well as to provide a continuous approximation when moving across the domain of influence of the neighboring nodes on the coarse mesh (Villon, Borouchaki and Khemais 2002). The latter is true for the present case only when taking more than a minimum of the four closest points in the neighborhood $\bar{\mathbf{x}}, \mathbf{x}_j \in \mathcal{N}(\mathbf{x}), j = 1, \ldots m \geq 4$; or otherwise the weighting functions will play no role. The cost function in the minimization problem in (6.5) can also be stated in matrix notation as:

$$J(\mathbf{b}) = \frac{1}{2} \mathbf{b}^T \mathbf{P} \mathbf{W} \mathbf{P}^T \mathbf{b} - \mathbf{b}^T \mathbf{P} \mathbf{W} \tilde{\mathbf{d}}^{\text{fine}} + \frac{1}{2} \tilde{\mathbf{d}}^{\text{fine},T} \mathbf{W} \tilde{\mathbf{d}}^{\text{fine}} \tag{6.6}$$

where the weighting factors, interpolation polynomial and fine mesh displacement values are stored as:

$$\mathbf{W} = \begin{bmatrix} W(\bar{\mathbf{x}}, \mathbf{x}_1) & .. & 0 \\ & .. & \\ & .. & \\ 0 & \ldots & W(\bar{\mathbf{x}}, \mathbf{x}_m) \end{bmatrix}; \mathbf{P}^T = \begin{bmatrix} \mathbf{p}(\mathbf{x}_1)^T \\ \\ \mathbf{p}(\mathbf{x}_m)^T \end{bmatrix}; \tilde{\mathbf{d}}^{\text{fine}} = \begin{bmatrix} \mathbf{d}_i^{\text{fine}}(\mathbf{x}_1) \\ \\ \mathbf{d}_i^{\text{fine}}(\mathbf{x}_m) \end{bmatrix} \tag{6.7}$$

The Kuhn–Tucker optimality condition of this minimization problem leads to the optimal value of approximation parameters:

$$0 = \frac{\partial J(\mathbf{b})}{\partial \mathbf{b}} \Rightarrow \mathbf{b} = \left[\mathbf{P} \mathbf{W} \mathbf{P}^T \right]^{-1} \mathbf{P} \mathbf{W} \tilde{\mathbf{d}}^{\text{fine}} \tag{6.8}$$

which defines completely the chosen approximation in (6.3). In order to improve the system conditioning and efficiency, to ensure the "partition of unity" property for the weighting functions, the following simple transformation is used:

$$W(\bar{\mathbf{x}}, \mathbf{x}_i) \leftarrow \frac{W(\bar{\mathbf{x}}, \mathbf{x}_i)}{\sum_{j \in \mathcal{N}(\bar{\mathbf{x}})} W(\bar{\mathbf{x}}, \mathbf{x}_j)} \tag{6.9}$$

By this, the field transfer in space between the fine and the coarse mesh is clarified. The next step is the transfer of the evolution problem in time. It is evident that usually, and in the most cases, the time steps used in the fine mesh are not kept the same for the coarse mesh but are much bigger. It is important to make a correct field transfer regarding the different time steps with respect to the two different mesh scales.

A single time step in the coarse mesh computation is considered which corresponds to a number of small steps on the fine mesh (Figure 6.9).

There are two possible ways for this field transfer. The first possible way is by first carrying out the computations on the fine mesh using the central difference scheme in (6.1) and then use the projection in (6.3) of computed result on the coarse mesh. This kind of the field transfer is carried out for each component of the displacement, velocity and acceleration vectors, which can formally be written as:

$$\begin{aligned}
&\mathbf{d}_{n+1}^{\text{coarse}}(\mathbf{x}) := \prod_x \left(\mathbf{d}_{n+1}^{\text{fine}}\left(\mathbf{x}_j^{\text{fine}}\right)\right); \forall \mathbf{x}_j \in \mathcal{N}(\mathbf{x})\\
&\mathbf{v}_{n+1}^{\text{coarse}}(\mathbf{x}) := \prod_x \left(\mathbf{v}_{n+1}^{\text{fine}}\left(\mathbf{x}_j^{\text{fine}}\right)\right); \forall x_j \in \mathcal{N}(\mathbf{x})\\
&\mathbf{a}_{n+1}^{\text{coarse}}(\mathbf{x}) := \prod_x \left(\mathbf{a}_{n+1}^{\text{fine}}\left(\mathbf{x}_j^{\text{fine}}\right)\right); \forall x_j \in \mathcal{N}(\mathbf{x})\\
&\mathbf{f}_{n+1}^{\text{coarse}}(\mathbf{x}) := \prod_x \left(\mathbf{r}_{\text{contact},\,n+1}^{\text{fine}}\left(\mathbf{x}_j^{\text{fine}}\right)\right); \forall \mathbf{x}_j \in \mathcal{N}(\mathbf{x})
\end{aligned} \tag{6.10}$$

where $\mathbf{d}_{n+1}^{\text{coarse}}$, $\mathbf{v}_{n+1}^{\text{coarse}}$,$\mathbf{a}_{n+1}^{\text{coarse}}$ and $\mathbf{f}_{n+1}^{\text{coarse}}$are, respectively, the computed vectors of displacements, velocities and acceleration at time t_{n+1}, while $\mathbf{r}_{\text{contact},\,n+1}^{\text{fine}}$ are the nodal forces computed from the contact problem on the fine mesh. The last expression states that any time we need the corresponding values on the coarse mesh (typically, either at nodes or at numerical integration points), they can be obtained by projection, which completely eliminates the need for central difference scheme computations on the coarse mesh.

6.5 FIELD TRANSFER IN TIME AND MINIMIZATION PROBLEM FOR OPTIMAL TRANSFER PROCEDURE

The second way that transfer is possible is where the computations are kept on the coarse mesh. In this procedure, the results obtained on the fine mesh are transferred, and then computations are done on the coarse mesh. The latter can formally be written according to:

$$\begin{aligned}
&\mathbf{d}_{n+1}^{\text{coarse}}(\mathbf{x}) = \prod_x \left(\mathbf{d}_n^{\text{fine}}\left(\mathbf{x}_j^{\text{fine}}\right)\right) + \Delta t \prod_x \left(\mathbf{v}_n^{\text{fine}}\left(\mathbf{x}_j^{\text{fine}}\right)\right) + \Delta t^2 \prod_x \left(\mathbf{a}_n^{\text{fine}}\left(\mathbf{x}_j^{\text{fine}}\right)\right)\\
&\mathbf{M}\mathbf{a}_{n+1}^{\text{coarse}}(\mathbf{x}) = \mathbf{g}_{n+1}(\mathbf{x}) - \hat{\mathbf{f}}^{\text{int}}\left(\mathbf{d}_{n+1}^{\text{coarse}}(\mathbf{x}),\ldots\right)\mathbf{v}_{n+1}^{\text{coarse}}(\mathbf{x}) =\\
&= \mathbf{v}_n^{\text{coarse}}(\mathbf{x}) + \frac{\Delta t}{2}\left(\prod_x \left(\mathbf{a}_n^{\text{fine}}\left(\mathbf{x}_j^{\text{fine}}\right)\right) + \mathbf{a}_{n+1}^{\text{coarse}}(\mathbf{x})\right)
\end{aligned} \tag{6.11}$$

When referring to computational efficiency, there is one crucial difference between the computational procedure on fine and on coarse mesh. The computation on the fine mesh is motivated by impact and frictional contact while the latter is driven by equivalent nodal loading g_{n+1}. This equivalent nodal loading is obtained as the solution of the minimization problem seeking to render the results of two field transfer procedures as close as possible to each other. Indicating that:

$$\min_{g_{n+1}} F(\mathbf{g}_{n+1}) \tag{6.12}$$

having the explicit form of the cost function defined by:

$$F(\mathbf{g}_{n+1}) = \frac{1}{2} \int_{\Omega\text{coarse}} \left\{ \left(\mathbf{d}_{n+1}^{\text{coarse}}(\mathbf{x}) - \prod_x \left(\mathbf{d}_{n+1}^{\text{fine}}(\mathbf{x}) \right) \right)^T \times \right.$$

$$\times \left(\mathbf{d}_{n+1}^{\text{coarse}}(\mathbf{x}) - \prod_x \left(\mathbf{d}_{n+1}^{\text{fine}}(\mathbf{x}) \right) \right)$$

$$+ \left(\mathbf{v}_{n+1}^{\text{coarse}}(\mathbf{x}) - \prod_x \left(\mathbf{v}_{n+1}^{\text{fine}}(\mathbf{x}) \right) \right)^T \times \left(\mathbf{v}_{n+1}^{\text{coarse}}(\mathbf{x}) - \prod_x \left(\mathbf{v}_{n+1}^{\text{fine}}(\mathbf{x}) \right) \right) + \tag{6.13}$$

$$\left. + \left(\mathbf{a}_{n+1}^{\text{coarse}}(\mathbf{x}) - \prod_x \left(\mathbf{a}_{n+1}^{\text{fine}}(\mathbf{x}) \right) \right)^T \times \left(\mathbf{a}_{n+1}^{\text{coarse}}(\mathbf{x}) - \prod_x \left(\mathbf{a}_{n+1}^{\text{fine}}(\mathbf{x}) \right) \right) \right\} dV$$

The dependence of such a cost function on the equivalent load vector acting in the coarse mesh $\mathbf{g}_{n+1}$ is defined through the corresponding central difference equation (6.11). A different weighting factor for enforcing the equivalence of the displacement, velocity or acceleration fields between the coarse and the fine meshes can be selected, and in this way the particular results could be enhanced.

Further improvement is possible if the work of the contact forces on the fine mesh are matched as close as possible over any time step of the coarse mesh computations through the work of equivalent loads on the coarse mesh. The latter can be obtained using the trapezoidal rule for computing the work of equivalent loads on the coarse mesh, which is in accordance with the accuracy of the central difference scheme. It can be stated that:

$$C(\mathbf{g}_{n+1}) = \int_{\Omega\text{coarse}} \left\{ \frac{t_{n+1} - t_n}{2} \left(\mathbf{g}_{n+1}^T \mathbf{v}_{n+1}^{\text{coarse}} + \mathbf{g}_n^T \mathbf{v}_n^T \right) - W_{\text{contact}}^{\text{fine}} \right\} dV \tag{6.14}$$

where $W_{\text{contact}}^{\text{fine}}$ is the work of the contact forces which is computed on the fine mesh during the same time step. The optimization problem in (6.12) can be customized by adding the last condition as the constraint $C(\mathbf{g}_{n+1}) = 0$. This kind of problem can be solved using the well-known method of Lagrange multipliers (Strang 1986; Luenberger 1984) to define the corresponding Lagrangian:

$$\min_{\mathbf{g}_{n+1}} \max_{\lambda_{n+1}} L(\mathbf{g}_{n+1}, \lambda_{n+1}); L(\mathbf{g}_{n+1}, \lambda_{n+1}) = F(\mathbf{g}_{n+1}) + \lambda_{n+1} C(\mathbf{g}_{n+1}) \tag{6.15}$$

The Kuhn–Tucker equations for this most general case of constrained minimization is written in the form:

$$0 = \frac{\partial L}{\partial \mathbf{g}_{n+1}} := \frac{\partial F}{\partial \mathbf{g}_{n+1}} + \frac{\partial C}{\partial \mathbf{g}_{n+1}} \tag{6.16}$$

$$0 = \frac{\partial L}{\partial \lambda_{n+1}} := C(g_{n+1}) \tag{6.17}$$

Four different implementations of the proposed field transfer method were developed and tested, considering different bases of the minimization problems presented herein.

The first considered the direct transfer of the nodal values computed on the fine mesh. The second took the same cost function and added the work conservation constraint on top. The third method seeks to improve upon the computation of the cost function by using patch-like computations. Finally, the fourth method was using the direct transfer of the values at the Gauss numerical points of the coarse mesh, leading in general to the highest precision of results.

6.6 NUMERICAL EXAMPLES

6.6.1 Example 1 — Element Level

Computations with these four methods were performed for the simply supported thick plate component impacted with a square projectile (see Figure 6.9). The results representing

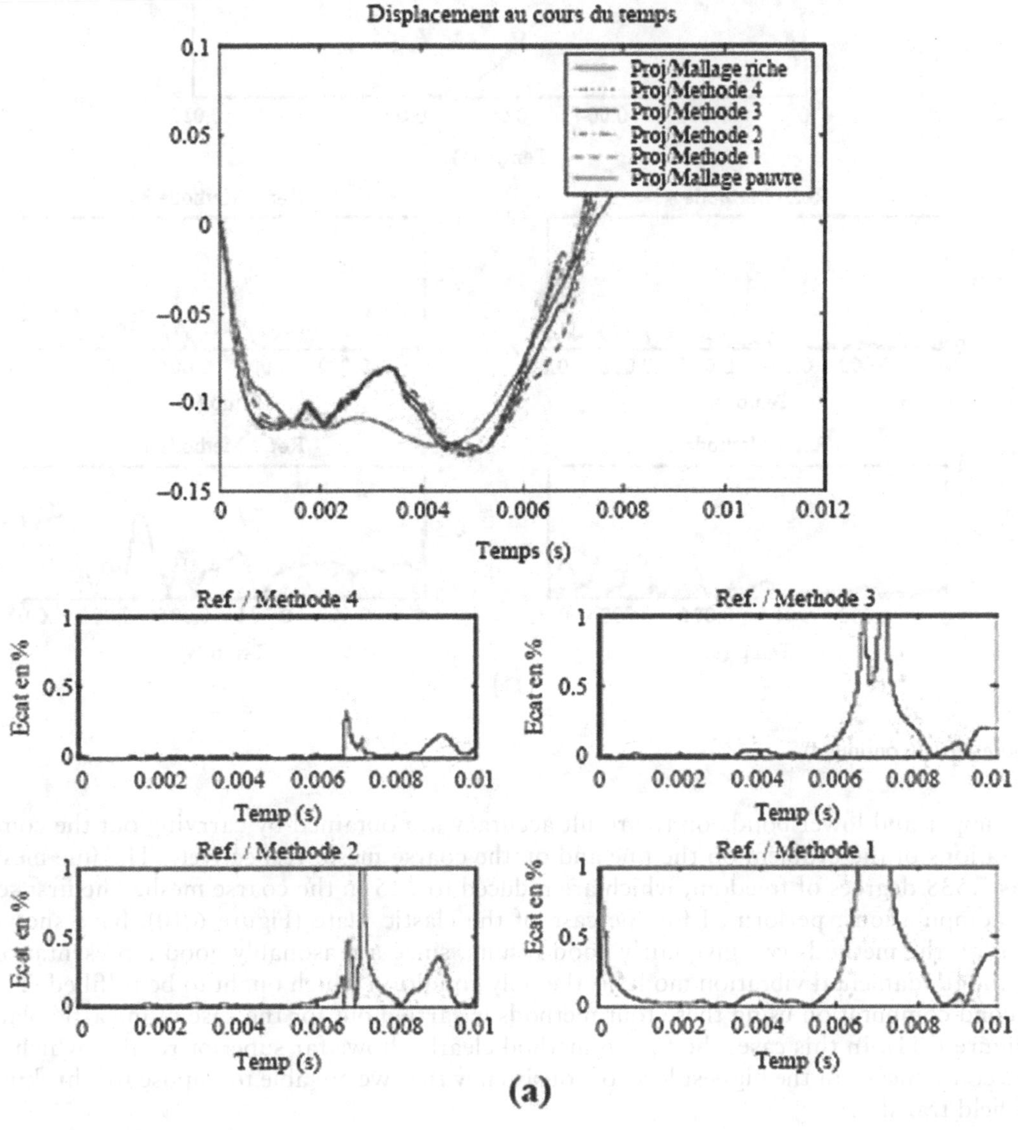

Figure 6.10 (a) Computed displacement and energy with fine and coarse mesh and (b) four different project methods—elastic case.

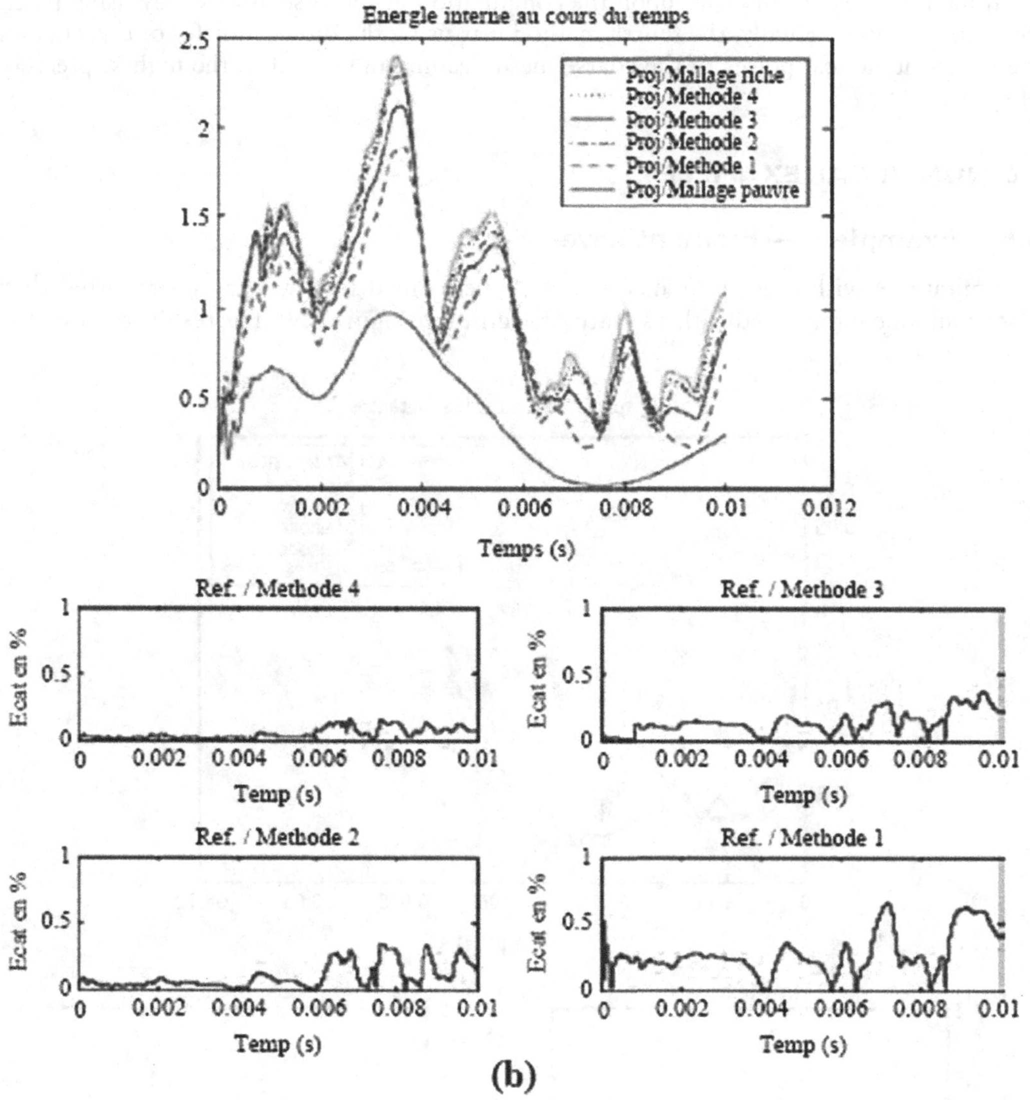

Figure 6.10 (Continued)

the upper and lower bonds on the result accuracy are obtained by carrying out the computations of the problem on the fine and on the coarse mesh, respectively. The fine mesh has 7,538 degrees of freedom, which are reduced to 225 in the coarse mesh. The first set of computations, performed for the case of the elastic plate (Figure 6.10), have shown that all the methods can give fairly good results, since a reasonably good representation of the fundamental vibration modes is the only condition which ought to be fulfilled. The second computation using these four methods is carried out for the case of inelastic plate (Figure 6.11). In this case, the fourth method clearly shows far superior results, which is the consequence of the highest level of consistency that we are able to impose for this kind of field transfer.

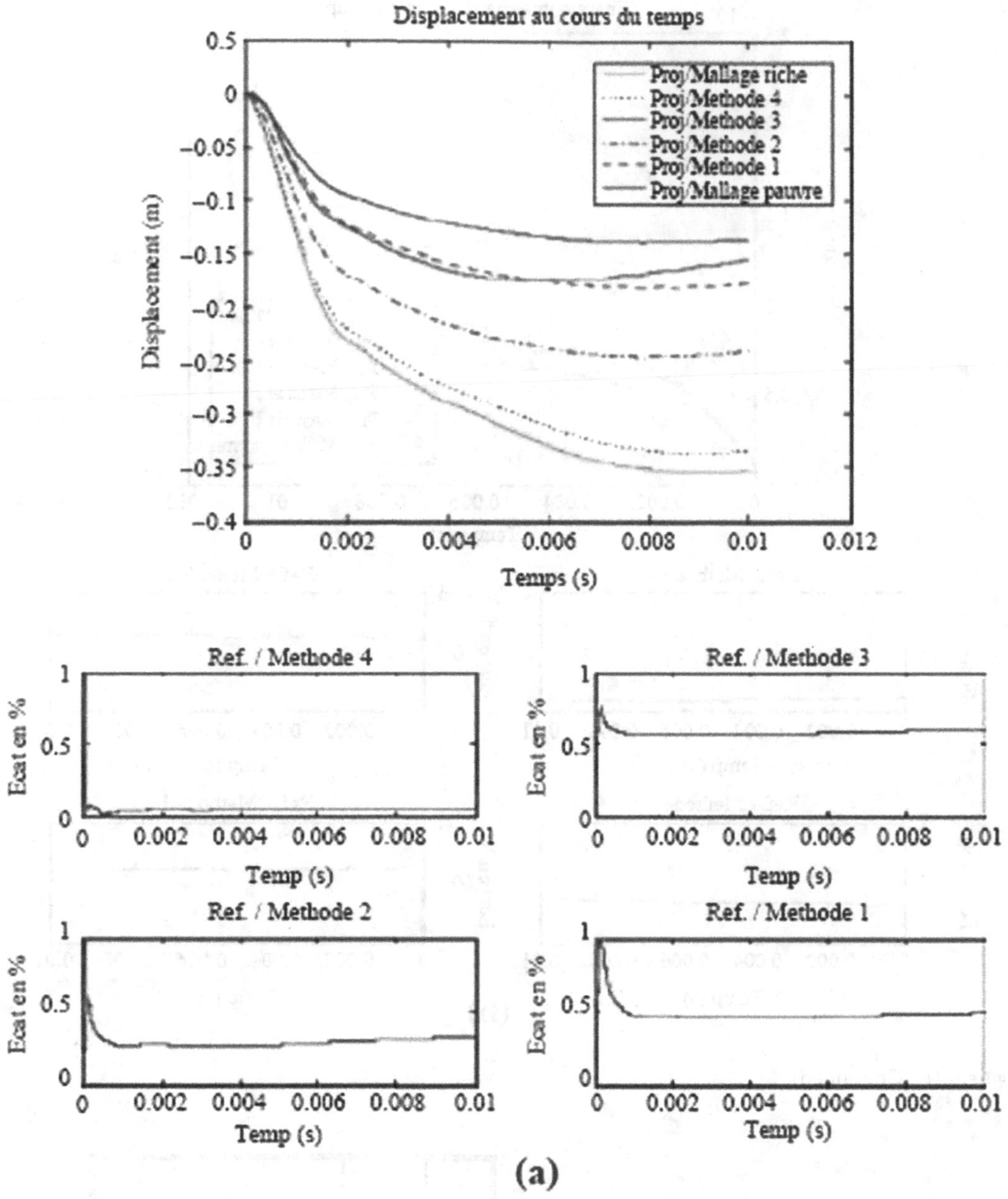

Figure 6.11 (a) Computed displacement and energy with fine and coarse mesh and (b) four different project methods—inelastic case.

6.6.2 Example 2 — Structure Level

The second example concerns the application of this strategy to a more realistic case of application. Due to the confidential nature of the study, detailed data are not provided (Figure 6.12), but the order of magnitude of the obtained results correspond to a quite realistic case. The roof slab was placed on the walls with simple supports, in a manner which

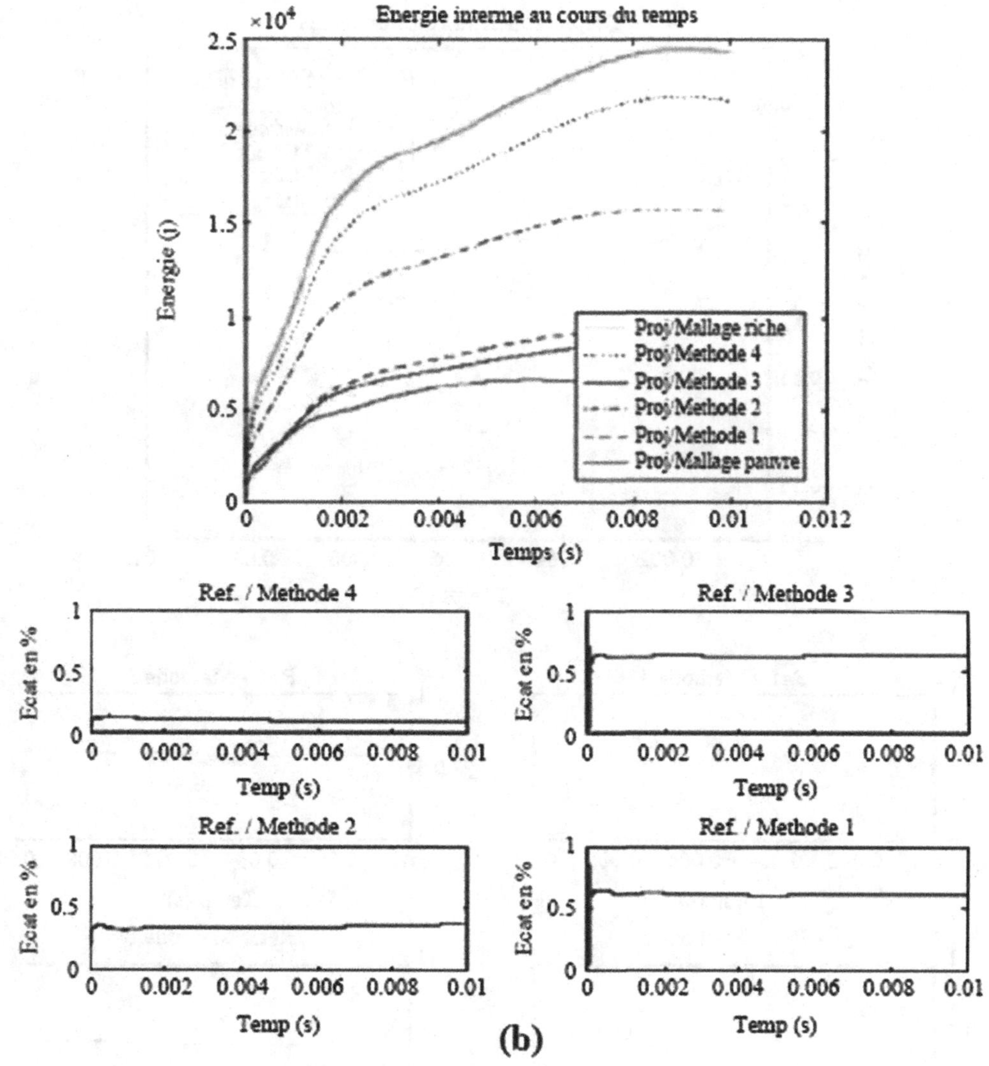

Figure 6.11 (Continued)

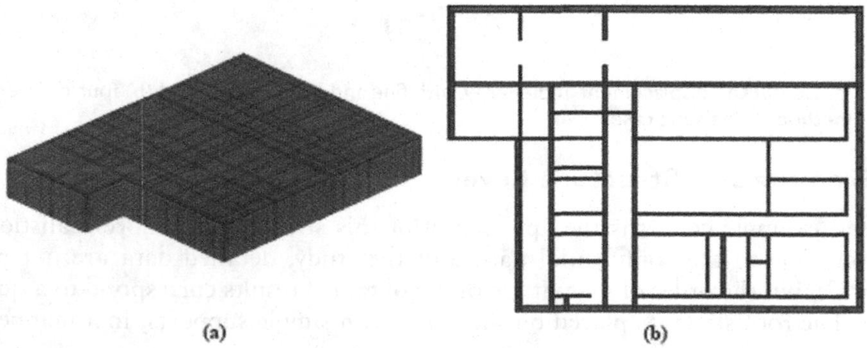

Figure 6.12 A structure under the impact of a large airplane (a) 3D model; (b) plane view.

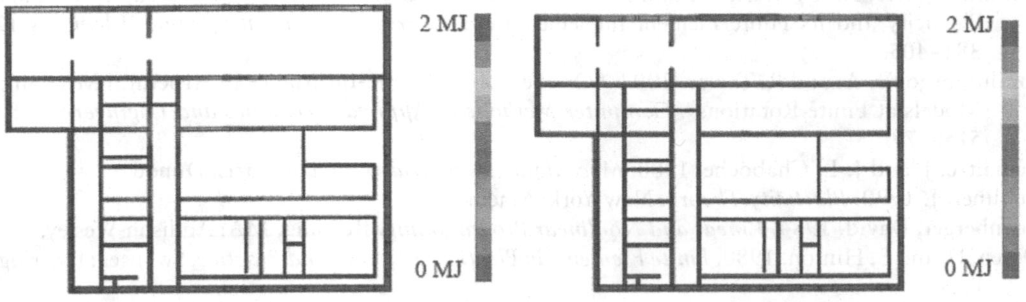

Figure 6.13 Computed energy under the impact of a large airplane (a) after 350 ms; (b) after 1,000 ms.

allowed the slab to uplift and reduce negative effects in the case of an internal explosion. The local problem of the impact on the roof slab was solved with a fine mesh including roughly 75,000 degrees of freedom. In the global computational phase, the roof slab finite element representation was reduced in order of magnitude to close to 2,500 degrees of freedom, with as many degrees of freedom used for the remaining part of the structure. The corresponding strain energy plots computed at 350 ms and 1,000 ms of motion are shown in Figure 6.13.

REFERENCES

Bathe, K.-J. and E. Wilson. 1976. *Numerical Methods in Finite Element Analysis*. Englewood Cliffs, NJ: Prentice-Hall.

Brancherie, D., P. Villon, A. Ibrahimbegović, A. Rassineux and P. Breitkopf. 2005. "Transfer Operator Based on Diffuse Interpolation and Energy Conservation for Damage Materials". Edited by Bathe K.-J. *International Journal of Heat and Mass Transfer*. Amsterdam: Elsevier. doi:10.1016/j.ijheatmasstransfer.2004.11.006.

Delaplace, Arnaud and Adnan Ibrahimbegović. 2006. "Performance of Time-Stepping Schemes for Discrete Models in Fracture Dynamic Analysis". *International Journal for Numerical Methods in Engineering* 65 (9): 1527–44. doi:10.1002/nme.1509.

Dominguez, Norberto, Delphine Brancherie, Luc Davenne and Adnan Ibrahimbegović. 2005. "Prediction of Crack Pattern Distribution in Reinforced Concrete by Coupling a -Strong Discontinuity Model of Concrete Cracking and a Bond-Slip of Reinforcement Model". *Engineering Computations* 22 (5/6): 558–82. doi:10.1108/02644400510603014.

Hervé, Guillaume, F. Gatuingt and Adnan Ibrahimbegović. 2005. "On Numerical Implementation of a Coupled Rate Dependent Damage-Plasticity Constitutive Model for Concrete in Application to High-rate Dynamics". *Engineering Computations* 22 (5/6).

Ibrahimbegović, Adnan. 2009. *Nonlinear Mechanics of Deformable Solids: Theoretical Formulation and Finite Element Implementation (in French)*. Paris: Hermes Science – Lavoisier.

Ibrahimbegović, A. and D. Brancherie. 2003. "Combined Hardening and Softening Constitutive Model of Plasticity: Precursor to Shear Slip Line Failure". *Computational Mechanics* 31 (1–2): 88–100. doi:10.1007/s00466-002-0396-x.

Ibrahimbegović, A. and A. Delaplace. 2003. "Microscale and Mesoscale Discrete Models for Dynamics Fracture of Structures Built of Brittle Material". *Computers & Structures* 81: 1255–65.

Ibrahimbegović, Adnan and Edward L. Wilson. 1992. "Unified Computational Model for Static and Dynamic Frictional Contact Analysis". *International Journal for Numerical Methods in Engineering* 34 (1): 233–47. doi:10.1002/nme.1620340115.

Ibrahimbegović, A. and S. Mamouri. 2002. "Energy Conserving/Decaying Implicit Time-Stepping Scheme for Nonlinear Dynamics of Three-Dimensional Beams Undergoing Finite Rotation". *Computer Methods in Applied Mechanics and Engineering* 191: 4241–5.

Ibrahimbegović, A., D. Markovic and F. Gatuingt. 2003. "Constitutive Model of Coupled Damage-Plasticity and Its Finite Element Implementation". *European Journal of Finite Elements* 12: 381–405.

Ibrahimbegović, A. and R. Taylor. 2002. "On the Role of Frame-Invariance in Structural Mechanics Models at Finite Rotations". *Computer Methods in Applied Mechanics and Engineering* 191: 5159–79.

Lemaitre, J. and J.-L. Chaboche. 1988. *Mecanique des matriaux solides*. Paris: Dunod.

Lubliner, J. 1990. *Plasticity Theory*. New York: Macmillan.

Luenberger, David. 1984. *Linear and Nonlinear Programming*. Reading, MA: Addison-Wesley.

Owen, D. and E. Hinton. 1980. *Finite Elements in Plasticity: Theory and Practice*. Swansea: Pineridge Press.

Strang, G. 1986. *Introduction to Applied Mathematics*. Cambridge: Wellesley-Cambridge Press.

Villon, Pierre, Houman Borouchaki and Saanouni Khemais. 2002. "Transfert de Champs Plastiquement Admissibles". *Comptes Rendus Mécanique* 330 (5): 313–18. doi:10.1016/s1631-0721(02)01457-2.

Chapter 7

Fire-induced extreme loads

How to determine the inelastic behavior of a structure subjected to mechanical and thermal loads jointly applied is an important task in civil and nuclear engineering, especially in the case of accidental loading scenarios and/or fire resistance. A vast number of cellular structures are jointly subjected to mechanical and thermal loads, especially in the case of accidental loading scenarios and/or fire resistance, and in this respect the development of predictive models capable of describing inelastic behavior is needed. Most of them are built either of folded plates and/or non-smooth shells. Heat transfer problems are dealt with and solved in a satisfactory way for solid bodies (e.g. Armero and Simo 1992; Coleman and Gurtin 1967; Ibrahimbegović, Lotfi Chorf and Gharzeddine 2001; Lewis et al. 1996; Simo and Miehe 1992; Simmonds 2001; Stabler and Baker 2000), including the pertinent aspects of thermomechanical coupling; however, for cellular structures built of clay or concrete hollow blocks, several novel issues arise (e.g. Bischoff and Armero 2001; Simo and Kennedy 1992; Simmonds 2001) both in terms of describing pertinent heat transfer phenomena and of accounting for thermomechanical coupling. It is envisaged that this model should initially supplement and then eventually replace the standard testing procedure for evaluating the fire resistance of the clay or concrete hollow blocks cellular structures. The model should be able to account for a number of complex phenomena of heat conduction and radiation, as well as the inelastic behavior of material with thermomechanical coupling.

Due to the complexity of the problem here only cellular structures will be treated. The classical shell model (e.g. Naghdi 1972; Libai and Simmonds 2005; Simo, Fox and Rifai 1989) is not applicable for the modeling of folded plates or non-smooth shells. The main difficulty in this sense is the lack of compatibility between the displacement degrees-of-freedom-only, which one uses to describe the membrane deformation field, and the displacement combined with rotations, which are necessary to describe the bending deformations. This is solved using the application of shell model with so-called drilling rotations (e.g. Hughes and Brezzi 1989; Ibrahimbegović and Frey 1994; Gruttmann, Wagner and Wriggers 1992). From the standpoint of the finite element implementation, the most convenient format of the shell theory with drilling rotations is for shallow shells (e.g. Ibrahimbegović and Frey 1994). The latter combines the kinematics hypothesis appropriate for a particular shell model, such as the Kirchhoff hypothesis where the shear deformation is neglected or the Reissner–Mindlin hypothesis including constant shear deformation, with the hypothesis of Marguerre on shallow shell geometry, which allows one to work with the projected form of the shallow shell.

7.1 TRANSIENT HEAT TRANSFER COMPUTATIONS

The 3D classical form of the energy balance equation when the mechanical part is ignored is described as:

$$c\frac{\partial\theta}{\partial t}+\frac{\partial q}{\partial\alpha}=r;q=-k\frac{\partial\theta}{\partial\alpha}\Rightarrow c\frac{\partial\theta}{\partial t}=-q_{\alpha,\alpha}+r \tag{7.1}$$

where c is the heat capacity coefficient, θ is temperature, r is heart source, $q_{\alpha,\alpha}$ stands for the heat flux and k is the thermal conductivity coefficient. The non-stationary heat transfer is described by a partial differential equation featuring the temperature field θ which is a function of space and time, as well as its partial derivatives with respect to space and with time. Besides the boundary conditions, in order to acquire unique solutions, it is necessary to obtain the initial conditions as well. In this respect, the initial conditions, specifying the temperature field at the time where the heat transfer starts:

$$\theta(x,0)=\theta_0(x) \text{ in } \Omega \tag{7.2}$$

and the boundary conditions, specifying the value of temperature or its space derivative on the domain boundary as:

$$\theta|_{\partial\Omega_\theta}=\bar{\theta} \text{ or } q_\alpha n_\alpha|_{\partial\Omega_q}=\bar{q}_n \tag{7.3}$$

In order to develop the shell-like formulation, representation of the shell 3D domain has to be split into mid-surface $\bar{A}$ and thickness direction, and further assume (in analogy with the mechanics part) a linear variation of the weighting temperature field:

$$\theta^*(x_\alpha,\zeta)=\vartheta^*(x_\alpha)+\zeta\varphi^*(x_\alpha) \tag{7.4}$$

where ϑ^* is the mid-surface temperature and φ^* is the through-the-thickness gradient. The weak form (the advantage is the reduction of the solutions regularity) of the energy balance equation (7.1) in a shell-like domain can be written as:

$$\begin{aligned}0=G_\theta(\vartheta,\varphi):=&\int_{\bar{A}}\left(ct\vartheta^*\dot{\vartheta}+c\frac{t^3}{12}\varphi^*\dot{\varphi}\right)dA-\int_{\bar{A}}\left(\vartheta^*_{,\alpha}p_\alpha+\varphi^*_{,\alpha}r_\alpha+\varphi^*p_3\right)dA\\&+\int_{\bar{A}}\left(\left(\vartheta^*+\frac{t}{2}\varphi^*\right)q_n^++\left(\vartheta^*-\frac{t}{2}\varphi^*\right)q_n^-\right)dA\end{aligned} \tag{7.5}$$

and by q_n^+ and q_n^- the heat fluxes on the upper and lower surface, respectively.

The resultant heat fluxes can be related to temperature and temperature gradient through a set of constitutive relations. For example, the choice made to assure compatibility with the resultant stress variations leads to the following form:

$$p_\alpha(x_\alpha)=-kt\vartheta_{,\alpha};p_3=\varphi e_3, \tag{7.6}$$

$$r_\alpha(x_\alpha)=-\frac{kt^3}{12}\varphi_{,\alpha} \tag{7.7}$$

The discrete finite element approximation is based upon the weak form. Implementation of the thermal part of the problem is carried out by the application of the same isoparametric interpolations. First of all, it is necessary to choose the finite element interpolation for the weighting functions ϑ and φ. For the chosen four node shell element the finite element interpolation is:

$$\vartheta \rightarrow \vartheta^h\big|_{\bar{A}^e} = \sum_{a=1}^{4} N_a(\xi,\eta)\,\vartheta_a, \underset{\sim}{\varphi} \rightarrow \underset{\sim}{\varphi}^h\big|_{\bar{A}^e} = \sum_{a=1}^{4} N_a(\xi,\eta)\,\underset{\sim}{\varphi}_a \tag{7.8}$$

ϑ_a and φ_a are the nodal values of temperature and temperature gradients, and N_a are the chosen shape functions for this type of element. By choosing the same kind of interpolations for weighting temperature, the weak form for the thermal part in (7.5) can be reduced to a set of algebraic equations (7.9). In order to integrate the temperature evolution over a given time step $\Delta t = t_{n+1} - t_n$, a backward Euler method was selected.

$$\underset{\sim}{f}^\theta = \underset{\sim}{0}, \underset{\sim}{f}^\theta = \left[f_a^\theta\right],$$
$$f_a^\theta = \int_A \left(ctN_a \frac{\vartheta_{a,n+1} - \vartheta_{a,n}}{\Delta t} + c\frac{t^3}{12} N_a \frac{\varphi_{a,n+1} - \varphi_{a,n}}{\Delta t} - N_{a,\alpha}\left(p_\alpha + r_\alpha + p_3\right)\right) dA \tag{7.9}$$

7.2 DAMAGE MECHANISMS REPRESENTATION UNDER INCREASED TEMPERATURE

The choice of a material model is of crucial importance for determining the most appropriate constitutive equations for the mechanics part of the shell problem. A constitutive model able to describe the behavior of brittle materials in terms of a generalized plasticity model based on a Saint-Venant or Rankine-like yield criterion has been chosen. It is well known that the von Mises plasticity criterion is used for metals, but this will not be further elaborated, as many authors have been and are working on this matter. This model is based on limiting the elastic domain using maximum mechanical strain values, which is in accordance with the experimentally observed behavior of brittle materials. Failure or brittle materials are primarily driven by extensions (positive strains) leading to cracking in the direction perpendicular to the principal tension stress or parallel to the principal compressive stress (e.g. Colliat, Ibrahimbegović and Davenne 2005b). Accordingly, the elastic domain is defined in terms of principal values by using a multisurface plasticity criterion of mechanical strain (e.g. Colliat, Ibrahimbegović and Davenne 2005 for details). This criterion can be recast in a stress-resultant space, leading to a multisurface yield criterion which consists of four surfaces intersecting in a non-smooth fashion of the form (Hughes and Brezzi 1989; Ibrahimbegović, Taylor and Wilson 1990; Ibrahimbegović and Wilson 1991; Ibrahimbegović and Frey 1993):

$$\Phi_1 = \frac{K + \frac{4\mu}{3}}{2\mu}\left|\hat{n}_{\alpha\beta} + \hat{m}_{\alpha\beta}\right|_I - \frac{K - \frac{2\mu}{3}}{2\mu}\left|\hat{n}_{\alpha\beta} + \hat{m}_{\alpha\beta}\right|_{II} - \left[\sigma_y(\theta) - q(\theta)\right] \tag{7.10}$$

$$\Phi_2 = \frac{K + \frac{4\mu}{3}}{2\mu}\left|\hat{n}_{\alpha\beta} + \hat{m}_{\alpha\beta}\right|_{II} - \frac{K - \frac{2\mu}{3}}{2\mu}\left|\hat{n}_{\alpha\beta} + \hat{m}_{\alpha\beta}\right|_I - \left[\sigma_y(\theta) - q(\theta)\right] \tag{7.11}$$

$$\Phi_3 = \frac{K + \frac{4\mu}{3}}{2\mu}\left|\hat{n}_{\alpha\beta} - \hat{m}_{\alpha\beta}\right|_I - \frac{K - \frac{2\mu}{3}}{2\mu}\left|\hat{n}_{\alpha\beta} + \hat{m}_{\alpha\beta}\right|_{II} - \left[\sigma_y(\theta) - q(\theta)\right] \tag{7.12}$$

$$\Phi_4 = \frac{K+\frac{4\mu}{3}}{2\mu}\left|\hat{n}_{\alpha\beta} - \hat{m}_{\alpha\beta}\right|_{II} - \frac{K-\frac{2\mu}{3}}{2\mu}\left|\hat{n}_{\alpha\beta} - \hat{m}_{\alpha\beta}\right|_{I} - \left[\sigma_y(\theta) - q(\theta)\right] \tag{7.13}$$

with the stress-resultant and couple-normalized values defined with:

$$\hat{n}_{\alpha\beta} = \frac{n_{\alpha\beta}}{t}, \hat{m}_{\alpha\beta} = \frac{m_{\alpha\beta}}{t^2} \tag{7.14}$$

and $|\cdot|_{I/II}$ denoting the principal values of the symmetric tensor, and K and μ denoting, respectively, the bulk and the shear moduli. It is for both the elastic limit $\sigma_y(\theta)$ and the variable which controls the evolution of the elastic domain $q(\theta)$ that temperature dependence is assumed. The latter is typically related to a stress softening branch which eventually drives the stress to zero (e.g. Ibrahimbegović and Brancherie 2003). A special provision is taken (e.g. Colliat, Ibrahimbegović and Davenne 2005) to incorporate different behaviors and corresponding fracture energies in compression and in tension. A plasticity model of this kind also features the standard additive split of generalized strain measures for both membrane and bending components with:

$$\varepsilon_{\alpha\beta} = \varepsilon^e_{\alpha\beta} + \varepsilon^p_{\alpha\beta}, \chi_{\alpha\beta} = \chi^e_{\alpha\beta} + \chi^p_{\alpha\beta} \tag{7.15}$$

The free energy can then be written in terms of the elastic strain components according to:

$$\Psi\left(\varepsilon^e_{\alpha\beta}, \chi^e_{\alpha\beta}, \xi\right) = \frac{1}{2}\left[\varepsilon^e_{\alpha\beta} t \tilde{C}_{\alpha\beta\gamma\delta} \varepsilon^e_{\gamma\delta} + \chi^e_{\alpha\beta} \frac{t^3}{12} \tilde{C}_{\alpha\beta\gamma\delta} \chi^e_{\gamma\delta}\right] + \mathcal{H}\left(\xi, \varepsilon^e_{\alpha\beta}\right) \tag{7.16}$$

where:

- $\tilde{C}_{\alpha\beta\gamma\delta}$ is the fourth-order elasticity tensor modified for plane stress case and
- ξ is the internal variable which controls hardening/softening response.

Further developments of this plasticity model's ingredients follow in the footsteps of the work of Simo, Kennedy and Govindjee (1998). In particular, the evolution equations for plastic components of the generalized strain measures are obtained by the Koiter rule; the latter can be written as:

$$\dot{\varepsilon}^p_{\alpha\beta} = \sum_{j\in J_{\text{act}}} \dot{\gamma}^j \frac{\partial \Phi_j}{\partial n_{\alpha\beta}}, \dot{\chi}^p_{\alpha\beta} = \sum_{j\in J_{\text{act}}} \dot{\gamma}^j \frac{\partial \Phi_j}{\partial m_{\alpha\beta}} \tag{7.17}$$

with $j \in J_{act}$ denoting the active yield surfaces and $\dot{\gamma}^j$ denoting the corresponding plastic multipliers. In the determination of J_{act}, special care has to be provided to the case in which:

$$\left(\frac{\partial \Phi_2}{\partial \hat{n}_{\alpha\beta}} \frac{\partial \Phi_2}{\partial \hat{m}_{\alpha\beta}}\right) \cdot \begin{bmatrix} tC_{\alpha\beta\gamma\delta} & 0 \\ 0 & \frac{t^3}{12} C_{\alpha\beta\gamma\delta} \end{bmatrix} \begin{pmatrix} \dot{\varepsilon}_{\gamma\delta} \\ \dot{\chi}_{\gamma\delta} \end{pmatrix} \leq 0 \tag{7.18}$$

and $\dot{\gamma} > 0$ at the same time (the same problem exists for $i=4$) (e.g. Colliat, Ibrahimbegović and Davenne 2005). The plastic multipliers are computed from the consistency condition of the given plastic state and consistent linearization provides the elastoplastic tangent moduli.

The time-integration of the constitutive response is carried out using the backward Euler scheme and the return mapping algorithm. The internal variables marked as $e^p = \left(\varepsilon^p_{\alpha\beta,n}, \chi^p_{\alpha\beta,n}\right)$ and ξ_n which are known at time step t_n need to be determined at the next time step t_{n+1}.

Before moving to the plastic step, the elastic trial step has to be assumed. At this moment all the values of internal variables at time t_n remain frozen, trial values for stress resultants and couples as well as the corresponding yield criterion values are being calculated employing the following:

$$n_{\alpha\beta,n+1}^{\text{trial}} = \frac{\partial \Psi}{\partial \varepsilon_{\alpha\beta,n+1}^{e,\text{trial}}}, m_{\alpha\beta,n+1}^{\text{trial}} = \frac{\partial \Psi}{\partial \chi_{\alpha\beta,n+1}^{e,\text{trial}}}, q_{n+1}^{\text{trial}} = \frac{\partial \Psi}{\partial \xi_{\alpha\beta,n+1}^{e,\text{trial}}} \tag{7.19}$$

giving:

$$\Phi_{i,n+1}^{\text{trial}} = \Phi_i\left(n_{\alpha\beta,n+1}^{\text{trial}}, m_{\alpha\beta,n+1}^{\text{trial}}, q_{n+1}^{\text{trial}}\right), i \in [1...4] \tag{7.20}$$

The elastic trial step is admissible only if:

$$\Phi_{i,n+1}^{\text{trial}} \leq 0, \forall i \in [1...4] \tag{7.21}$$

leading to:

$$\varepsilon_{\alpha\beta,n+1}^{p} = \varepsilon_{\alpha\beta,n}^{p}, \chi_{\alpha\beta,n+1}^{p} = \chi_{\alpha\beta,n}^{p}, \xi_{n+1} = \xi_n \tag{7.22}$$

If any of the yield surfaces is active, as indicated by a positive trial value, the elastic trial step values have to be corrected:

$$\exists i, \Phi_{i,n+1}^{\text{trial}} > 0 \Rightarrow \varepsilon_{\alpha\beta,n+1}^{p} \neq \varepsilon_{\alpha\beta,n}^{p}, \chi_{\alpha\beta,n+1}^{p} \neq \chi_{\alpha\beta,n}^{p}, \xi_{n+1} \neq \xi_n \tag{7.23}$$

It is extremely important to emphasize that if only a single yield surface is active, this will not necessarily imply that only a single Lagrange multiplier is non-zero, i.e.

$$\Phi_{1,n+1}^{\text{trial}} > 0 \text{ and } \Phi_{2,n+1}^{\text{trial}} \leq \not\Leftrightarrow \gamma_{1,n+1} > 0 \text{ and } \gamma_{2,n+1} = 0 \tag{7.24}$$

or:

$$\Phi_{3,n+1}^{\text{trial}} > 0 \text{ and } \Phi_{4,n+1}^{\text{trial}} \leq \not\Leftrightarrow \gamma_{3,n+1} > 0 \text{ and } \gamma_{4,n+1} = 0 \tag{7.25}$$

By taking into account these and other possibilities to obtain the number of unknown Lagrange multipliers, a complete set of nonlinear equations is formed with $\varepsilon_{\alpha\beta,n+1}^{p}$, $\chi_{\alpha\beta,n+1}^{p}$, ξ_{n+1} and $\lambda_{i,n+1} = \Delta t \gamma_{i,n+1}$ as the unknowns.

$$R_{n+1}(\cdot) = 0 \tag{7.26}$$

The elastoplastic tangent modulus needed for the incremental, iterative solution procedure via Newton's method can be acquired by application of the consistent linearization of the last expression and the systematic application of the static condensation (e.g. Ibrahimbegović, Gharzeddine and Chorfi 1998).

$$\text{Lin}\left[R_{n+1}^{i+1}\right] = R_{n+1}^{i} + DR_{n+1}^{i} \Rightarrow \underset{\sim}{\tilde{C}}_{n+1}^{\text{ep}} \tag{7.27}$$

Finally, the consistent tangent for the mechanics part of the problem can be written as:

$$\underset{\sim}{K}^{uu} = \int_{\bar{A}} \underset{\sim}{\bar{B}}^{T} \underset{\sim}{\tilde{C}}_{n+1}^{\text{ep}} \underset{\sim}{\bar{B}} dA \tag{7.28}$$

where $\underset{\sim}{B}$ is the strain–displacement matrix which contains all the corresponding sub-matrices for membrane and bending strain fields.

7.3 COUPLED THERMOMECHANICAL PROBLEM COMPUTATIONS

In this case, the thermomechanical coupling is taken into account for the shell model and the constitutive model should change. The starting point in defining the constitutive response can be selected in terms of the Helmholtz free energy which is now defined as:

$$\Psi\left(\varepsilon^e_{\alpha\beta},\chi^e_{\alpha\beta},\underset{\sim}{\xi},\vartheta,\varphi_\alpha\right)=\frac{1}{2}\left[\varepsilon^e_{\alpha\beta}t\tilde{C}_{\alpha\beta\gamma\delta}\varepsilon^e_{\gamma\delta}+\chi^e_{\alpha\beta}\frac{t^3}{12}\tilde{C}_{\alpha\beta\gamma\delta}\chi^e_{\gamma\delta}\right]+\mathcal{H}\left(\underset{\sim}{\xi},\varepsilon^e_{\alpha\beta},\vartheta,\varphi_\alpha\right)$$
$$-\frac{\rho c}{2\theta_{ref}}\left[t\vartheta^2+\frac{t^3}{12}\varphi^2\right]-\left[\varepsilon^e_{\alpha\beta}t\alpha\tilde{C}_{\alpha\beta\gamma\delta}\delta_{\gamma\delta}\vartheta+\chi^e_{\alpha\beta}\frac{t^3}{12}\alpha\tilde{C}_{\alpha\beta\gamma\delta}\delta_{\gamma\delta}\varphi\right] \tag{7.29}$$

where α a is the thermal expansion coefficient. In the last expression, the free energy is written by adding upon the mechanics part in the first term a more general temperature dependent form of the hardening potential in the second term; followed by the thermal potential in the third term and the thermomechanical coupling effect in the fourth term. With this choice of free energy, a modified form of the stress-resultant constitutive equation is acquired which contains the temperature dependent term according to:

$$n_{\alpha\beta}=\frac{\partial\Psi}{\partial\varepsilon^e_{\alpha\beta}}=n^{\text{meca}}_{\alpha\beta}+t\alpha\tilde{C}_{\alpha\beta\gamma\delta}\delta_{\gamma\delta}\vartheta \tag{7.30}$$

$$m_{\alpha\beta}=\frac{\partial\Psi}{\partial\chi^e_{\alpha\beta}}=m^{\text{meca}}_{\alpha\beta}+\frac{t^3}{12}\alpha\tilde{C}_{\alpha\beta\gamma\delta}\delta_{\gamma\delta}\varphi \tag{7.31}$$

For the case of thermomechanical coupling, the stress-resultant in (7.30) is replaced by the weak form of the equilibrium equations in the mechanical part.

On the other hand, the weak form of the thermal balance equation will not change if the structural heating terms are neglected ($\dot{\varepsilon}^e_{\alpha\beta}t\alpha\tilde{C}_{\alpha\beta\gamma\delta}\delta_{\gamma\delta}\simeq 0$ and $\dot{\chi}^e_{\alpha\beta}\frac{t^3}{12}\alpha\tilde{C}_{\alpha\beta\gamma\delta}\delta_{\gamma\delta}\simeq 0$) and the plastic dissipation. The former is done because the time variation of elastic deformations is very slow and the latter is done because we deal herein with brittle materials which have very small inelastic dissipation.

Moreover, with the free energy (7.30), the constitutive equations defining entropy with both average and gradient component are gained with:

$$\eta=\frac{\partial\Psi}{\partial\vartheta},\varsigma=\frac{\partial\Psi}{\partial\varphi} \tag{7.32}$$

Thermomechanical coupling still remains through boundary conditions. Namely, the classical format of radiative heat exchange is exploited herein, implying that within each cell the resultant flux over each surface should be equal to zero, which can be written as:

$$q^{\pm}_{n,i}\leftarrow q^{\pm}_{n,i}-\sigma\sum_{j=1}^{n_{\text{surf}}}\varepsilon_j A_j F_{ij}\left(\theta^{\pm}_j\right)^4 \tag{7.33}$$

where $q^{\pm}_{n,i}$ represents the outgoing flux, and the last term represents the contribution of incoming fluxes dependent upon the Stefan–Boltzmann constant σ, the surface emittance ε_j, the area of the surface A_j, the relative shape factor of each set of surfaces F_{ij} and the surface temperature to the power four.

It is evident that each such equation is nonlinear in temperature terms (or in terms of the mid-surface and temperature gradient); thus, it requires several iterations at the global level to converge. In addition, the modification of such an equation in (7.33) has to be done each time the fracture of the barriers changes the cell configuration. Please refer to Colliat, Ibrahimbegović and Davenne(2005) for a more detailed discussion of this issue. A detailed theoretical formulation is given in Section 7.6.

7.4 OPERATOR SPLIT SOLUTION PROCEDURE WITH VARIABLE TIME STEPS

The set of equations governing the semi-discretized problem of the thermomechanical coupling of shells consists of the nonlinear algebraic equations expressing mechanics equilibrium equations, along with differential equations describing the heat flow. These equations are accompanied by the evolution equations of internal variables (plastic strain and hardening variables), which are defined and solved at the local level, at each Gauss numerical integration point. The problem can thus formally be written as:

$$\underset{\sim}{r} := \begin{bmatrix} \underset{\sim}{r}^u\left(\underset{\sim}{d}^u, \underset{\sim}{d}^\theta, \varepsilon^p(\theta), \chi^p(\theta), \xi(\theta)\right) \\ M^{\theta\theta}\dot{\underset{\sim}{d}}^\theta - \underset{\sim}{r}^\theta\left(\underset{\sim}{d}^\theta, \varepsilon^p(\theta), \chi^p(\theta), \xi(\theta)\right) \end{bmatrix} = \underline{0} \tag{7.34}$$

$$\begin{pmatrix} \dot{\varepsilon}^p - \sum_i \dot{\gamma}^i \dfrac{\partial \Phi_i}{\partial n} \\ \dot{\chi}^p - \sum_i \dot{\gamma}^i \dfrac{\partial \Phi_i}{\partial m} \\ \dot{\xi} - \sum_i \dot{\gamma}^i \dfrac{\partial \Phi_i}{\partial q} \end{pmatrix} = \underset{\sim}{0}, \Phi_i \leq 0 \ \ \forall \text{GNP} \tag{7.35}$$

For the rate-independent constitutive response, the dependence of real-time is therefore present only in the heat transfer problem. However, since all the equations are, in general, tightly coupled, the same time parameter can be employed throughout. The system is thus first recast in a form where only algebraic equations will appear by integrating the evolution equations for internal variables as well as the heat transfer equation using the backward Euler scheme. The problem thus reduces to:

given: $\underset{\sim}{d}^u, \underset{\sim}{d}^\theta, \left(\varepsilon_n^p, \chi_n^p, \xi_n^p\right)\Big|_{\text{GNP}} \quad \forall \text{GNP}$

find: $\underset{\sim}{d}_{n+1}^u, \underset{\sim}{d}_{n+1}^\theta, \left(\varepsilon_{n+1}^p, \chi_{n+1}^p, \xi_{n+1}^p\right)\Big|_{\text{GNP}}$

such that:

$$\underset{\sim}{r}_{n+1} := \begin{bmatrix} \underset{\sim}{r}^u\left(\underset{\sim}{d}_{n+1}^u, \underset{\sim}{d}_{n+1}^\theta, \varepsilon_{n+1}^p, \chi_{n+1}^p, \xi_{n+1}\right) \\ M^{\theta\theta}\dfrac{\underset{\sim}{d}_{n+1}^\theta - \underset{\sim}{d}_n^\theta}{\Delta t} - \underset{\sim}{r}_{n+1}^\theta\left(\underset{\sim}{d}_{n+1}^\theta, \varepsilon_{n+1}^p, \chi_{n+1}^p, \xi_{n+1}\right) \end{bmatrix} = \underline{0} \tag{7.36}$$

$$\begin{pmatrix} \varepsilon_{n+1}^p - \varepsilon_n^p - \sum_i \lambda^i \dfrac{\partial \Phi_i}{\partial n} \\ \chi_{n+1}^p - \chi_n^p - \sum_i \lambda^i \dfrac{\partial \Phi_i}{\partial m} \\ \xi_{n+1} - \xi_n - \sum_i \lambda^i \dfrac{\partial \Phi_i}{\partial q} \end{pmatrix} = 0, \Phi_i \le 0 \ \forall \text{GNP} \tag{7.37}$$

It is interesting to note that the coupling effect in the discretized heat transfer equation is present between temperatures on one side and internal variables on the other side, not only through the plastic heating effect but also through the plastic deformation which induces modification of the cell assembly and determines the heat radiation conditions. Therefore, even when the plastic heating is neglected, which appears to be justified for brittle models of this kind where no large ductile deformations occur, the coupling still persists through a special remeshing procedure which one has to provide.

By reducing the coupling effect to a minimum, the linearized form of the system of algebraic equations to be solved can be written as:

$$\begin{bmatrix} \underset{\sim}{K}^{uu} & \underset{\sim}{K}^{u\theta} \\ 0 & \dfrac{M^{\theta\theta}}{\Delta t} + \underset{\sim}{K}^{\theta\theta} \end{bmatrix} \cdot \begin{pmatrix} \Delta \underset{\sim}{d}^u \\ \Delta \underset{\sim}{d}^\theta \end{pmatrix} = -\underset{\sim}{r}_{n+1}^{(k)} \tag{7.38}$$

These equations are solved for the converged values of the internal variables, which would ensure the plastic admissibility of the stress state $\Phi_i \le 0$.

7.5 NUMERICAL EXAMPLES

7.5.1 Example 1 — Circular Ring Heating

For this example, an analytical solution exists to verify the model. A circular ring with average radius R and initial temperature θ_0 is considered. With respect to the chosen reference frame and cylindrical coordinates (r, ϕ, z), temperature θ_1 is applied to all points which satisfy ϕ = 0/= 0 a h1 (see Figure 7.1).

For convenience, we recast this problem using dimensionless expressions and write:

$$\overline{\varphi} = \frac{\varphi}{2\pi}, \theta = \frac{\theta - \theta_1}{\theta_0 - \theta_1}, F_o = \frac{at}{R^2} \tag{7.39}$$

where:

- a is the thermal diffusivity [m²/s] and
- F_o the Fourier number.

The analytic solution of the heat equation for the middle fiber at $r = R$ can be written as:

$$\overline{\theta}(\overline{\varphi}, F_o) = \sum_{n=0}^{\infty} \frac{4}{(2n+1)\pi} \sin((2n+1)\pi\overline{\varphi}) e^{\left(-\frac{2n+1}{2} F_o\right)} \tag{7.40}$$

In particular, at the point $\overline{\varphi} = 1/2$, the dimensionless temperature evolves according to:

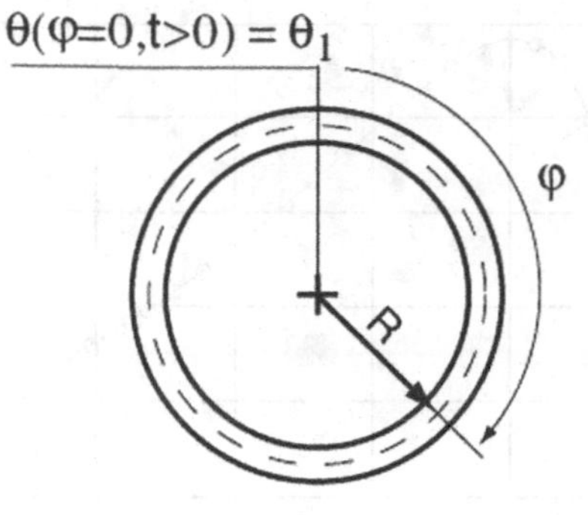

Figure 7.1 Circular ring transient problem.

$$\bar{\theta}(F_o) = \sum_{n=0}^{\infty} \frac{(-1)^n}{2n+1} e^{\left(-\frac{2n+1}{2}F_o\right)} \tag{7.41}$$

Coupled thermomechanical analysis was performed limited to the elastic regime by using the finite element model constructed with a presented shell element. Only half of the cylinder has been modeled, due to the symmetry, using a six element mesh (Figure 7.2).

Evolution of the bending moment at the opposite point $\bar{\varphi} = 1/2$ is shown in Figure 7.3. The peak value is reached during the transient heat transfer phase and followed by a decrease to zero limit in the subsequent steady-state phase. Comparison between the exact solution and the numerical results is illustrated in Figure 7.4 which pertains to the evolution of the dimensionless temperature for the same point. An excellent match is observed. However, this is not the case for other fibers placed at $r \neq R$. The reasoning for such is that in the transient phase the temperature distribution along the r coordinates could be described using Bessel's functions of the first kind. In the steady-state, the limit distribution is a logarithmic one. Thus, it is only as the thickness of the element decreases that such a limit expression

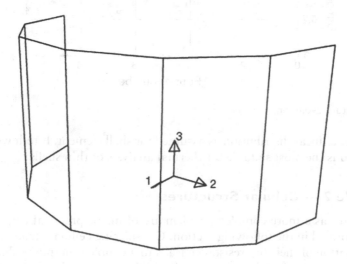

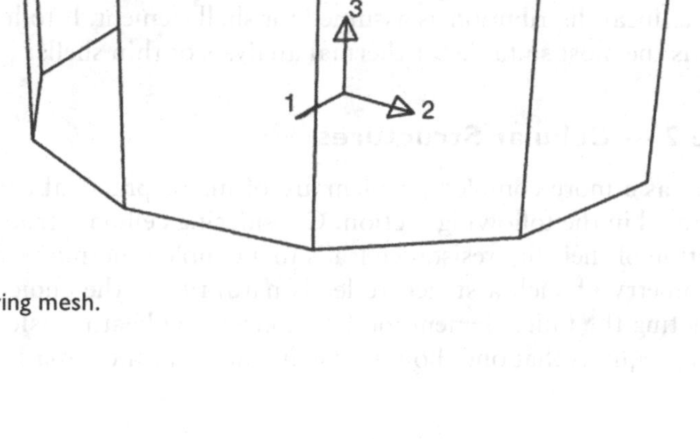

Figure 7.2 Circular ring mesh.

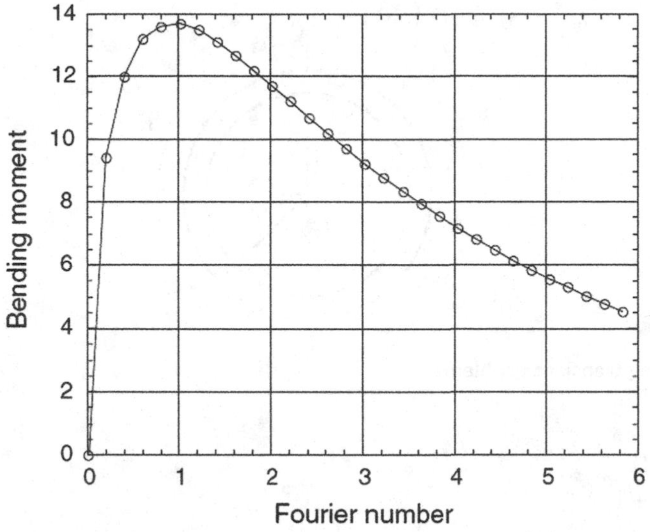

Figure 7.3 Bending moment.

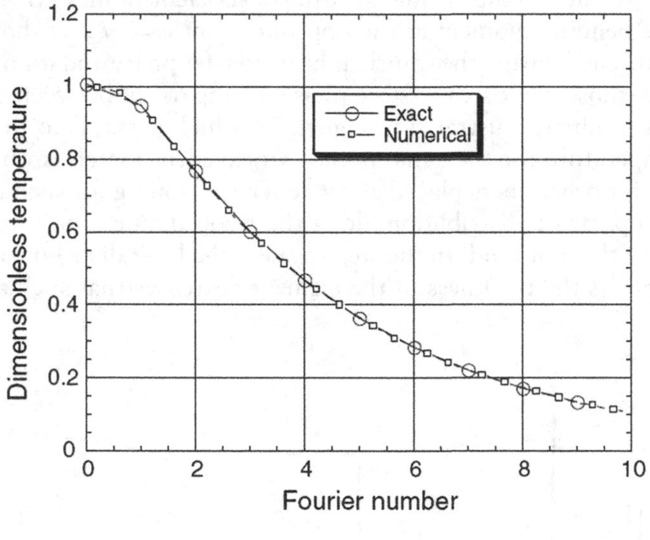

Figure 7.4 Temperature evolution.

becomes closer to a linear distribution as assumed for shell element. It follows that the strategy we developed is the most suitable for thermal analysis of thin shells.

7.5.2 Example 2 — Cellular Structures

Cellular structures as a more complex problem are of major practical engineering interest and will be elaborated in the following section. Considering cellular structures, like hollow bricks, the evaluation of their fire resistance leads to a coupled thermomechanical problem. Moreover, the geometry of such a structure leads naturally to the choice of flat shell elements for constructing the finite element model. Concerning heat transfer, the large variation of temperature requires that one should take into account the radiative heat exchanges.

Like clay bricks, the flue block consists of a hollow structure made of clay, which is vertically assembled in order to form a vertical stack, whose role is to evacuate smoke. The stack is submitted to an important temperature increase on the inner face, which may lead to the appearance of cracks and a loss of efficiency in the smoke evacuation. In order to study the cracking of the flue block, a thermomechanical analysis employing nonlinear mechanical model is to be employed. Only a quarter of the model is analyzed due to the symmetry of the structure, and mesh of 16 shell elements and five solid radiative elements is used (see Figure 7.5).

On the inner face, boundary conditions of the third kind are applied with a driven convection temperature which increases from 0° to 1000° in 10 min. Figure 7.6 shows the comparison of the corresponding average temperature evolution at four different positions, with decreasing values from inside to outside. It is noted that the temperature gradient between the inner face and the outer face is still very large even after 10 min. This gradient is at the origin of thermally induced stresses which can lead to flue cracking and eventually failure.

Since one of the interests is to eliminate the risk of cracking, it is important to have a look into the thermal strain field and induced stresses. The stress evolution in a horizontal direction for the outer face is presented in Figure 7.7. Since the inner face is heated, this part is generally in tension. Moreover, the supplementary global bending effect of the block is produced by the 2D nature of the problem and a non-uniform distribution of the temperature on the inner face. Both the peak stress level and the delay effect before this peak is reached, which are very clearly seen in Figure 7.6. The latter is a major factor in trying to estimate the block quality.

7.5.3 Example 3 — Hollow Brick Wall

A thermomechanical coupling is considered in the cellular units placed within a brick wall. It is assumed that the geometry and the loading allow for exploitation of the periodicity conditions. This implies that the analysis can be carried out on a single cellular unit, isolated

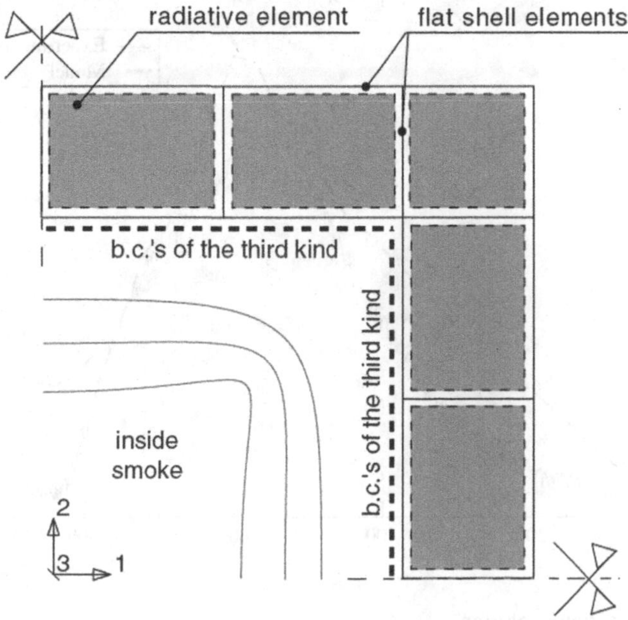

Figure 7.5 Flue block-mesh.

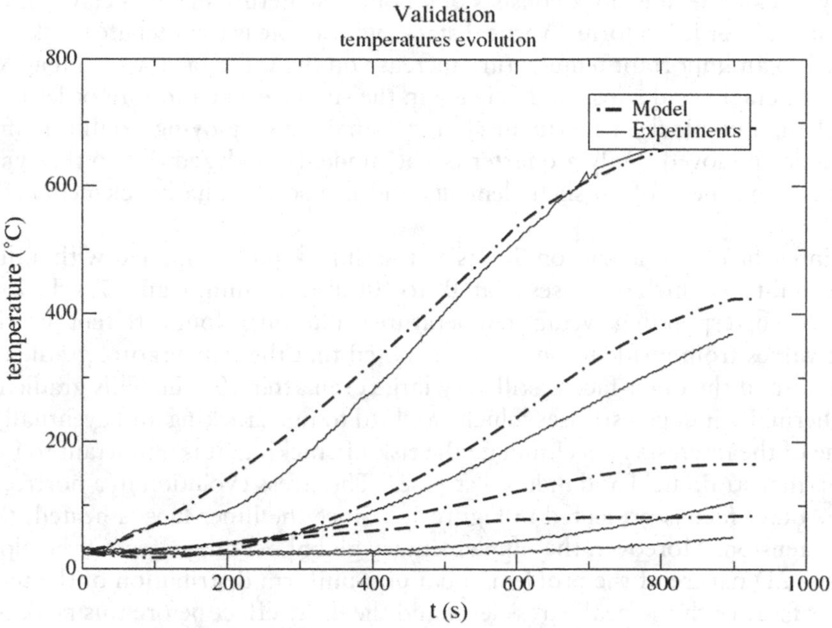

Figure 7.6 Flue block-temperature evolution from inside to outside.

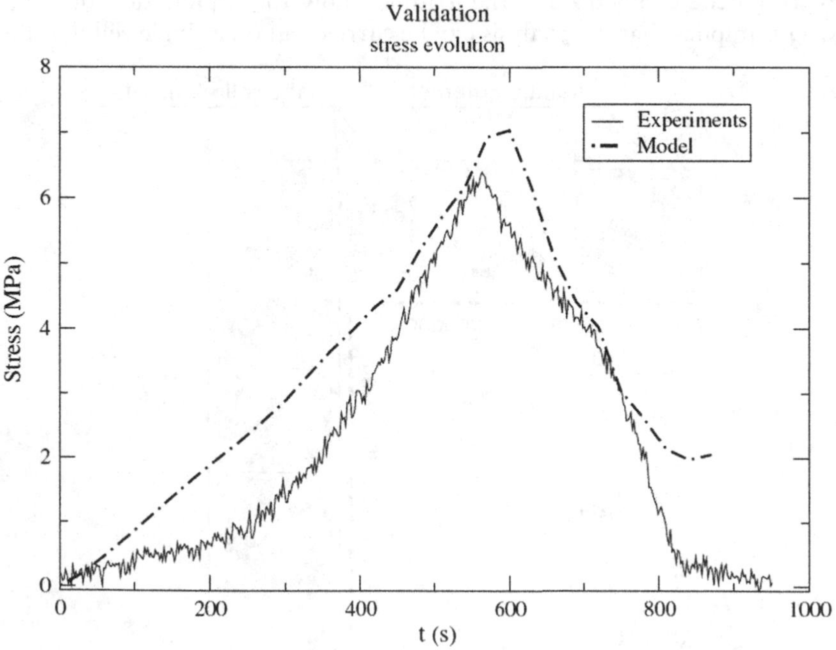

Figure 7.7 Flue block-reaction evolution.

from the whole structure at the level of interface with neighboring units, by applying the corresponding boundary conditions which assure periodicity. More precisely, for the typical unit assembly in a brick wall (see Figure 7.8) with only partial overlapping of successive layers, the same periodicity conditions are enforced only over half of the brick.

Therefore, the domain which is retained in the analysis corresponds to one typical unit of the size 570×200×200 mm^3. This domain also includes half of the vertical and horizontal joints with a thickness equal to 10 mm.

The chosen finite element mesh (see Figure 7.9) consists of three vertical layers of flat shell elements which brings the total number of this element to 384 for the entire brick. Another subtlety of the model is the choice which is made for representing the interface joints. This one is modeled using solid elastic elements covering the cells of the brick placed only at the top. However, the latter does not introduce any non-symmetry into the problem, considering the periodicity in the boundary conditions.

Both mechanical and thermal loading is applied in this case. The mechanical loading is supposed to represent the dead load on the brick chosen as a compressive loading of 1.3 MPa, which is introduced directly at the level of each element as the initial compressive loading in the bricks and remaining constant afterward. The thermal loading is then applied, in terms of the uniform temperature field applied only at the brick facet exposed to fire. The time evolution of this temperature field is given as:

$$\theta(t) = \theta_0 + 345\log(8t + 1) \tag{7.42}$$

where:

θ_0 is the initial temperature and
t the time in minutes.

The mechanical and thermal properties of the brick material and interface are given in Table 7.1.

Figure 7.10 shows the location of the cells.

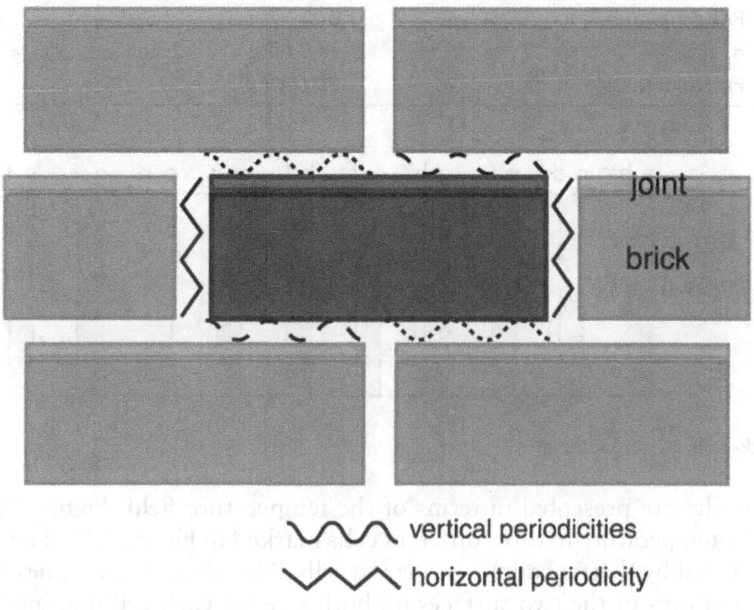

Figure 7.8 Unit assembly in a brick wall and periodic BC.

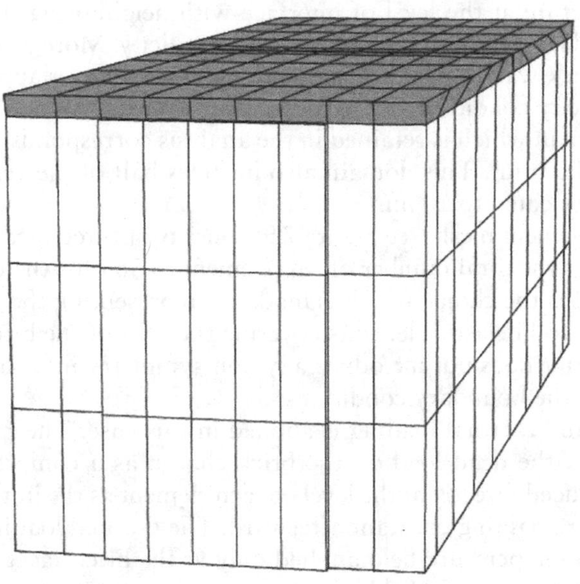

Figure 7.9 FE mesh for a cellular unit.

Table 7.1 Mechanical and Thermal Properties of the Brick and Interface

Mechanical and Thermal Properties	*Brick*	*Interface*
Density	1.870	2.100
Heat capacity	836 J kg^{-1} K^{-1}	950 J kg^{-1} K^{-1}
Conductivity (parallel to flakes)	0.55 Wm^{-1} K^{-1}	1.55 Wm^{-1} K^{-1}
Conductivity (perpendicular to flakes)	0.35 Wm^{-1} K^{-1}	
Thermal expansion coefficient at θ_{ref}	7.10^{-6} K^{-1}	1.10^{-6} K^{-1}
Young's modulus	12 GPa	15 GPa
Poisson's ratio	0.2	0.25
σ_y at θ_{ref}	14.5 MPa	
Fracture energy	80 J m^{-2}	

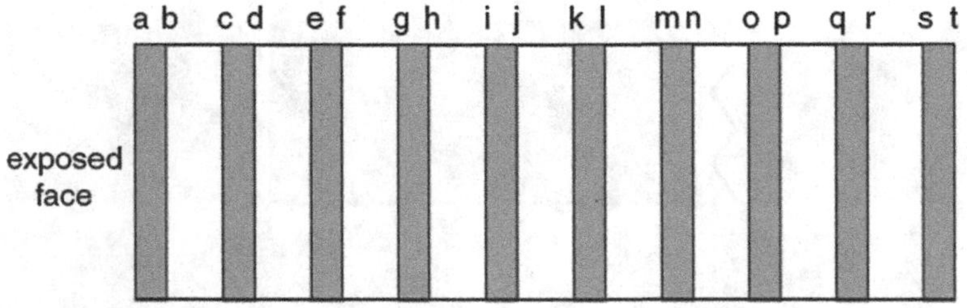

Figure 7.10 Brick-cells facets locating.

First, the results are presented in terms of the temperature field. Figure 7.11 shows the evolution of the temperature in three different cells marked in Figure 7.10. The experimental results are provided by thermocouples inside the cells. Therefore, these values are compared with the temperatures of the two surfaces on both sides of each cell obtained by the finite element analysis. The comparison shows that the model is able to capture the temperature

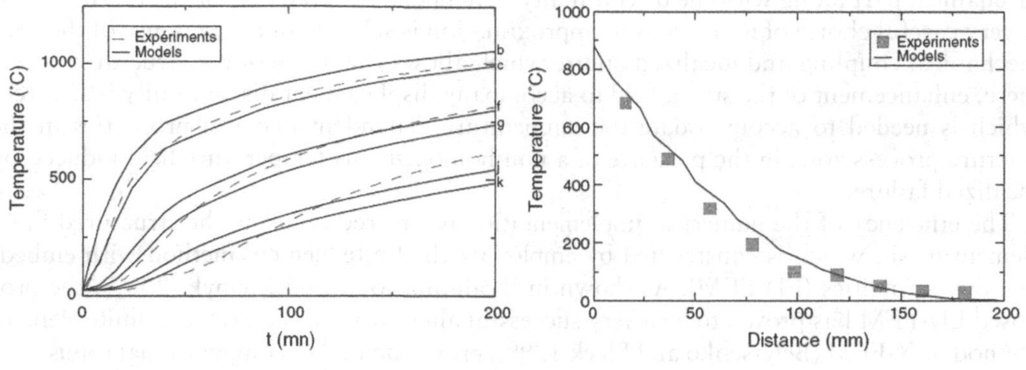

Figure 7.11 **Brick-temperature evolution and profile after 48 mm.**

evolutions even far away from the exposed face of the wall. This result is confirmed by Figure 7.11b which shows a temperature profile 48 min after the beginning of heating. The key point here is that in order to obtain such a good result the introduction of radiative exchanges into the heat transfer model is necessary.

From a mechanical point of view, Figure 7.12a shows the evolution of the sum of vertical reactions at selected nodes. Each curve corresponds to a line of nodes parallel to the exposed face of the wall and positive values are for compression (with a prestressed initial value due to constant mechanical loading). Figure 7.12b shows the comparison of the horizontal displacement of the wall built with ten rows of bricks. This bending is due to the temperature gradient through the wall. It is shown that the stiffness provided by the analysis is quite correct even if the displacement is slightly over-estimated by the absence of mechanical boundary conditions.

7.6 DETAILED THEORETICAL FORMULATION OF LOCALIZED THERMOMECHANICAL COUPLING PROBLEM

A detailed theoretical formulation for coupled thermomechanical failure problems that take into account both the fracture process zone and softening behavior at a localized failure zone is presented herein. A model is developed that describes the localized thermomechanical failure; introducing the displacement and deformation discontinuity for the

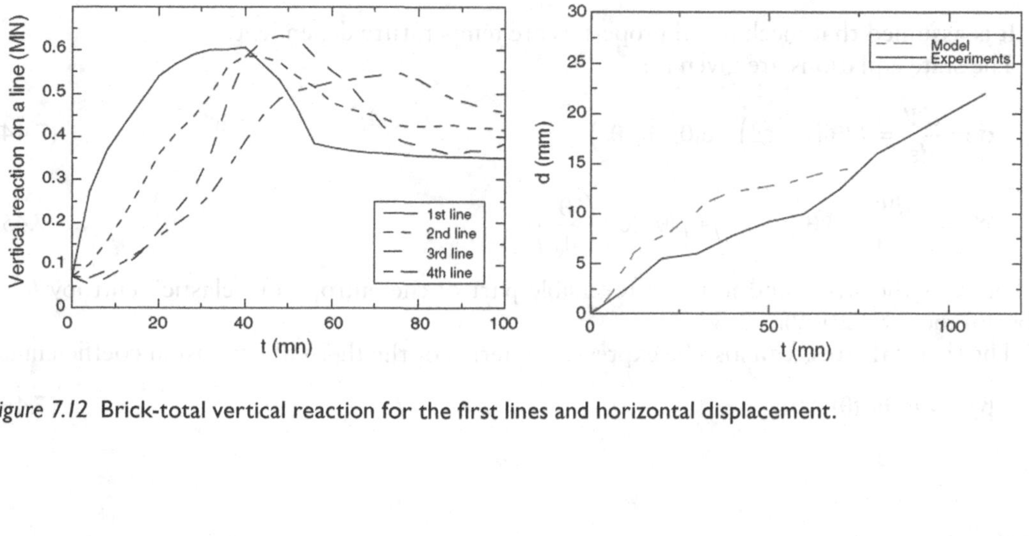

Figure 7.12 **Brick-total vertical reaction for the first lines and horizontal displacement.**

mechanical part along with the discontinuity in temperature gradient for the thermal part. A very careful choice of finite element approximation is selected in the presence of thermo-mechanical coupling and localized failure which allows the usage of the structured mesh. Here, enhancement of the strain field to accompany displacement discontinuity is selected, which is needed to accommodate the temperature dependent material properties in the fracture process zone in the presence of a non-homogeneous temperature field induced by localized failure.

The efficiency of the numerical implementation is ensured by using the structured finite element mesh, which is constructed by employing the finite element method with embedded discontinuities (ED-FEM). As shown in Ibrahimbegović and Melnyk (2007), the proposed ED-FEM has proven to be a very successful alternative to the extended finite element method or X-FEM (Belytschko and Black 1999), providing higher computational robustness with the discontinuities in displacement and in heat flux defined at the element level.

7.6.1 Continuum Thermo-Plastic Model and its Balance Equation

The free energy of the continuum thermo-plastic consists of three components: mechanical energy, thermal energy and thermomechanical energy:

$$\Psi(\varepsilon,\varepsilon^p,\xi,\theta) = \underbrace{\frac{1}{2}E\left(\varepsilon-\varepsilon^p\right)^2 - q\xi}_{\Psi_m} + \underbrace{\rho c\left[(\theta-\theta_0)-\theta\ln\left(\frac{\theta}{\theta_0}\right)\right]}_{\Psi_t} \underbrace{-\beta(\theta-\theta_0)\left(\varepsilon-\varepsilon^p\right)}_{\Psi_{tm}} \tag{7.43}$$

where:

E is Young's modulus,
ε is the total strain,
ε^p is the plastic strain,
q is the stress-like variable associated with hardening,
ξ is the hardening variable,
ρ is the mass density,
θ is the temperature,
θ_0 is the reference temperature,
c is the density heat capacity and
β is the thermal stress (for unit temperature change in the material).

It is assumed that mechanical properties are temperature dependent.

The state equations are given as:

$$\sigma := \frac{d\Psi}{d\varepsilon} = E(\theta)\left(\varepsilon-\varepsilon^p\right) - \beta(\theta)(\theta-\theta_0) \tag{7.44}$$

$$\eta^e := -\frac{d\Psi}{d\theta} = \beta(\theta)\left(\varepsilon-\varepsilon^p\right) + \rho(\theta)c(\theta)n\left(\frac{\theta}{\theta_0}\right) \tag{7.45}$$

where σ is the stress and η^e is the reversible part of the entropy or "elastic" entropy (e.g. Ibrahimbegović 2009).

The thermal stress can also be expressed in terms of the thermal expansion coefficient α:

$$\beta(\theta) = E(\theta)\alpha(\theta) \tag{7.46}$$

By taking the last result into account, (7.44) can be rewritten in an alternative form:

$$\sigma := \frac{d\Psi}{d\varepsilon} = E(\theta)\left[\left(\varepsilon-\varepsilon^p\right)-\underbrace{\alpha(\theta)(\theta-\theta_0)}_{\varepsilon_{th}}\right] = \sigma_m + \sigma_{th} \tag{7.47}$$

where ε_{th} denotes the thermal deformation, while σ_m denotes the mechanical part and σ_{th} the thermal part of the stress. Denoting with η^p the irreversible or "plastic" part of the "total" entropy η (with the additive split of entropy, $\eta = \eta^e + \eta^p$), the local form of internal dissipation rate can be expressed as follows:

$$0 \le \mathcal{D}_{\text{int}} := \theta\dot{\eta} + \sigma\dot{\varepsilon} - \frac{de}{dt} = \theta\dot{\eta} + \sigma\dot{\varepsilon} - \frac{d\left(\Psi + \eta^e\theta\right)}{dt} \tag{7.48}$$

where $e = \Psi + \eta^e\theta$ is the internal energy. The additive split of the dissipation rate into a mechanical and thermal part can be obtained with:

$$0 \le \mathcal{D}_{\text{int}} = \theta\left(\dot{\eta}^e + \dot{\eta}^p\right) + \sigma\dot{\varepsilon} - \underbrace{\frac{d\Psi}{d\theta}}_{-\eta^e}\dot{\theta} - \underbrace{\frac{d\Psi}{d\varepsilon}}_{\sigma}\dot{\varepsilon} - \underbrace{\frac{d\Psi}{d\varepsilon^p}}_{-\sigma}\dot{\varepsilon}^p - \underbrace{\frac{d\Psi}{d\xi}}_{-q}\dot{\xi} - \theta\dot{\eta}^e - \eta^e\dot{\theta} \tag{7.49}$$

$$0 \le \mathcal{D}_{\text{int}} = \underbrace{\theta\dot{\eta}^p}_{\mathcal{D}_{\text{ther}}} + \underbrace{\sigma\dot{\varepsilon}^p + q\dot{\xi}}_{\mathcal{D}_{\text{mech}}} \tag{7.50}$$

The temperature dependent yield criterion for the material in the fracture process zone is defined as:

$$\phi(\sigma, q, \theta) := |\sigma| - \left(\sigma_y(\theta) - q(\theta)\right) \le 0 \tag{7.51}$$

where:

$\sigma_y(\theta)$ is the initial yield stress of the material at temperature θ and
$q(\theta)$ is the stress-like hardening variable controlling the evolution of the yield threshold.

The temperature dependence of these two variables is expressed in the following equations:

$$\sigma_y(\theta) = \sigma_y\left[1 - \omega(\theta - \theta_0)\right] \tag{7.52}$$

$$q = -K(\theta)\xi; K(\theta) = K\left[1 - \omega(\theta - \theta_0)\right] \tag{7.53}$$

where σ_y and K are the values at the reference temperature θ_0.

The evolution laws of the state variables are established by the second law of thermodynamics, in which the internal dissipation reaches maximum value. In particular, the Kuhn–Tucker condition is used to find the maximum of internal dissipation $\mathcal{D}_{\text{int}}$ among the admissible stress values with $\phi(\sigma, q, \theta) \le 0$. This can be defined as the corresponding constrained minimization:

$$\underbrace{\max \mathcal{D}_{\text{int}}(\sigma, q, \theta)}_{\phi(\sigma,q,\theta)\le 0} \Leftrightarrow \min\max L(\sigma, q, \theta, \dot{\gamma});$$
$$L(\sigma, q, \theta, \dot{\gamma}) = -\mathcal{D}_{\text{int}}(\sigma, q, \theta) + \dot{\gamma}\phi(\sigma, q, \theta) \tag{7.54}$$

The corresponding optimality conditions can be written as follows:

$$0=\frac{\partial L}{\partial \sigma}\rightarrow \dot{\varepsilon}^p=\dot{\gamma}\frac{\partial \phi}{\partial \sigma}=\dot{\gamma}\text{sign}(\sigma) \tag{7.55}$$

$$0=\frac{\partial L}{\partial q}\rightarrow \dot{\xi}=\dot{\gamma}\frac{\partial \phi}{\partial q}=\dot{\gamma} \tag{7.56}$$

$$0=\frac{\partial L}{\partial \theta}\rightarrow \dot{\eta}^p=\dot{\gamma}\frac{\partial \phi}{\partial \theta}=\dot{\gamma}(\sigma_y+K\xi)\omega \tag{7.57}$$

where $\dot{\gamma}$ is the Lagrange multiplier.

The balance equations for this problem are obtained using the force equilibrium equation and the first law of thermodynamics. The force equilibrium equation can be written as:

$$-\rho\frac{d^2u}{dt^2}+\frac{d\sigma}{dx}+b=0 \tag{7.58}$$

where:

- ρ is the mass density,
- u is the displacement,
- σ is the stress and
- b is the distributed load.

The energy balance is then established by using the first principle:

$$\frac{d}{dt}\left[e+\frac{1}{2}\rho\left(\frac{du}{dt}\right)^2\right]=b\frac{du}{dt}+\frac{d}{dx}\left(\sigma\frac{du}{dt}\right)+R-\frac{dQ}{dx} \tag{7.59}$$

where:

- e is the internal energy density,
- R is distributed heat supply and
- Q is the heat flux.

The last equation can be rewritten explicitly as:

$$\frac{de}{dt}+\rho\frac{du}{dt}\left(\frac{d^2u}{dt}\right)=b\frac{du}{dt}+\frac{d\sigma}{dx}\frac{du}{dt}+\sigma\frac{\partial^2u}{\partial x\partial t}+R-\frac{dQ}{dx} \tag{7.60}$$

By combining this result with the force equilibrium equation, the reduced form of the first principle is obtained:

$$\frac{de}{dt}=\sigma\frac{d\varepsilon}{dt}+R-\frac{dQ}{dx} \tag{7.61}$$

By exploiting the Legrendre transformation, $e=\Psi+\eta^e\theta$, the free energy potential ψ can be further introduced:

$$\frac{de}{dt}=\frac{d\Psi}{dt}+\dot{\theta}\eta^e+\theta\dot{\eta}^e \tag{7.62}$$

$$\frac{de}{dt}=\underbrace{\frac{d\Psi}{d\theta}}_{-\eta^e}\dot{\theta}+\underbrace{\frac{d\Psi}{d\varepsilon}}_{\sigma}\dot{\varepsilon}+\underbrace{\frac{d\Psi}{d\varepsilon^p}}_{-\sigma}\dot{\varepsilon}^p+\underbrace{\frac{d\Psi}{d\xi}}_{-q}\dot{\xi}+\dot{\theta}\eta^e+\theta\dot{\eta}^e \tag{7.63}$$

Replacing this expression with (7.62), the final form of the balance equations is obtained:

$$\theta\dot{\eta}^e = -\frac{dQ}{dx} + \underbrace{\sigma\dot{\varepsilon}^p + q\dot{\xi}}_{\mathcal{D}_{\text{mech}}} + R \tag{7.64}$$

$$\theta\dot{\eta} = -\frac{dQ}{dx} + \mathcal{D}_{\text{int}} + R \tag{7.65}$$

It is noted that the definition of thermal dissipation in (7.48) allows for the final result in (7.62). By further considering only quasi-static loading applications, (7.56) and (7.62) can be recast as the final form of the balance equations:

$$\begin{aligned} 0 &= \frac{d\sigma}{dx} + b \\ \theta\dot{\eta} &= -\frac{dQ}{dx} + \mathcal{D}_{\text{int}} + R \end{aligned} \tag{7.66}$$

7.6.2 Thermodynamics Model for Localized Failure and Modified Balance Equation

7.6.2.1 Thermodynamics Model

When the localized failure happens, the free energy is decomposed into a regular part in the fracture process zone and an irregular part of the free energy at the localized failure point:

$$\Psi\left(\varepsilon, \varepsilon^p, \xi, \theta\right) = \bar{\Psi}\left(\bar{\varepsilon}, \bar{\varepsilon}^p, \theta, \bar{\xi}\right) + \delta_{\bar{X}}\bar{\bar{\Psi}}\left(\bar{\bar{\xi}}, \theta_1\right) \tag{7.67}$$

where:

- $\bar{*}$ denotes the regular part and
- $\bar{\bar{*}}$ represents the irregular part of the potential
- θ denotes the temperature in any position and
- θ_1 denotes the temperature at the localized failure point $\bar{x}$.

In (7.67) above, the irregular part of the energy is limited to the localized failure point by using $\delta_{\bar{x}}$, the Dirac-delta function:

$$\delta_{\bar{x}}(x) = \begin{cases} \infty; & x = \bar{x} \\ 0; & \text{otherwise} \end{cases} \tag{7.68}$$

The regular part of the free energy pertains to the fracture process zone, and it keeps the same form as written in (7.43). The localized free energy is assumed to be equal to:

$$\bar{\bar{\Psi}}\left(\bar{\bar{\xi}}, \theta_1\right) = \frac{1}{2}\bar{\bar{K}}(\theta_1)\bar{\bar{\xi}}^2 \tag{7.69}$$

where $\bar{\bar{\xi}}$ is the internal variable quantifying the softening behavior due to the localized failure. The chosen quadratic form of the softening potential in (7.69) further allows us to obtain the corresponding stress-like internal variable:

$$\bar{\bar{q}}\left(\theta, \bar{\bar{\xi}}\right) := \frac{d\bar{\bar{\Psi}}\left(\bar{\bar{\xi}}\right)}{d\bar{\bar{\xi}}} = -\bar{\bar{K}}(\theta_1)\bar{\bar{\xi}} \tag{7.70}$$

This variable drives the current ultimate stress value to zero, when the failure process is activated, as confirmed by the corresponding yield criterion:

$$\bar{\bar{\phi}}\left(t_{\bar{x}},\bar{\bar{q}}\right) := \left|t_{\bar{x}}\right| - \left(\sigma_u\left(\theta_1\right) - \bar{\bar{q}}\left(\theta_1,\bar{\bar{\xi}}\right)\right) \le 0 \tag{7.71}$$

where:

$\left|t_{\bar{x}}\right|$ is the traction at the localized failure point $\bar{x}$
$\sigma_u(\theta_1)$ is the initial value of the ultimate stress.

The mechanical properties at the localized failure are assumed to have the same dependence on temperature as the bulk part; hence, it can be written in the form:

$$\sigma_u\left(\theta\right) = \sigma_u\left[1 - \omega\left(\theta_1 - \theta_0\right)\right] \tag{7.72}$$

$$\bar{\bar{K}}\left(\theta\right) = \bar{\bar{K}}\left[1 - \omega\left(\theta_1 - \theta_0\right)\right] \tag{7.73}$$

where σ_u and $\bar{\bar{K}}$ are, respectively, the ultimate stress and softening modulus at the reference temperature.

Once the localized failure occurs, the crack-opening (further denoted as $\bar{\bar{u}}(t)$, see Figure 7.13) contributes to a "jump" or irregular part in the displacement field.

The total displacement field is thus the sum of the regular (smooth) part and irregular part:

$$u\left(x,t\right) = \bar{u}\left(x,t\right) + \bar{\bar{u}}\left(t\right)\left\{H_{\bar{x}}\left(x\right) - \varphi\left(x\right)\right\} \tag{7.74}$$

where $H_{\bar{x}}$ is the Heaviside function introducing the displacement jump:

$$H_{\bar{x}} = \begin{cases} 0, x \le \bar{x} \\ 1, x > \bar{x} \end{cases} \tag{7.75}$$

In (7.74) above, $\varphi(x)$ is a (smooth) function, introduced to limit the influence of the displacement jump within the "failure" domain. The usual choice for $\varphi(x)$ in the finite element implementation pertains to the shape functions of the selected interpolation. For example, for a 1D truss-bar with two nodes and element length l_e, it can be selected as:

$$\varphi\left(x\right) = N_2\left(x\right) = \frac{x}{l_e} \tag{7.76}$$

The corresponding illustrations for $H_{\bar{x}}(x)$ and $\varphi(x)$ in a two-node truss-bar element are given in Figure 7.14.

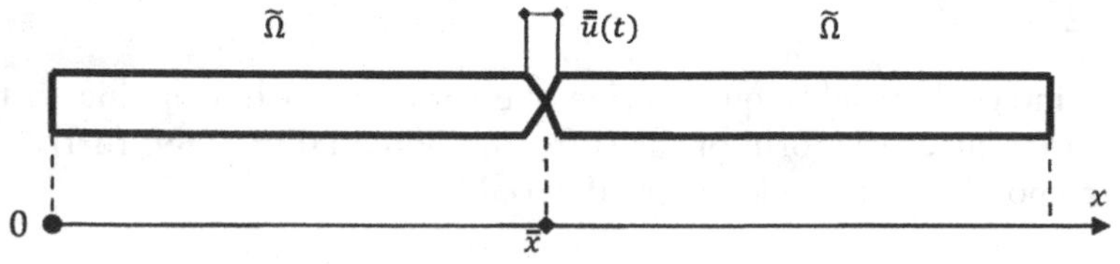

Figure 7.13 Displacement discontinuity at localized failure for the mechanical load.

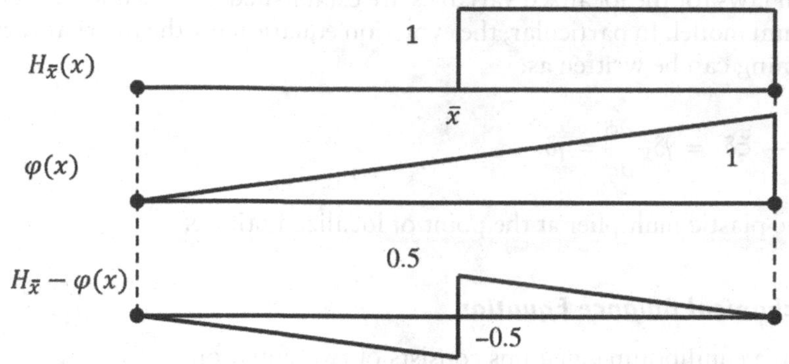

Figure 7.14 Displacement discontinuity for a two-node truss-bar element: Heaviside function $H_{\bar{x}}(x)$ and $\varphi(x)$.

Denoting with $\hat{u}(x,t)$ the continuous part of the displacement field, and with $\bar{\bar{u}}(t)$ the "jump" in displacement, an additive decomposition of displacement field can be written as:

$$u(x,t) = u(x,t) + \bar{\bar{u}}(t)H_{\bar{x}}(x) \tag{7.77}$$

The corresponding strain field can then be obtained by exploiting the kinematic equation:

$$\varepsilon(x,t) := \frac{du}{dx} = \frac{\partial}{\partial x}\left[\hat{u}(x,t) + \bar{\bar{u}}(t)H_{\bar{x}}(x)\right] = \hat{\varepsilon} + \delta_{\bar{x}}(x)\bar{\bar{u}}(t) \tag{7.78}$$

The rate of internal dissipation can then be written as:

$$\begin{aligned} 0 \le \mathcal{D}_{\text{int}} &= \theta\dot{\eta} + \sigma\dot{\varepsilon} - \frac{d}{dt}e\left(\bar{\varepsilon}^e, \bar{\xi}, \bar{\bar{\xi}}, \theta\right) = \\ &= \theta\dot{\eta} + \sigma\dot{\varepsilon} - \frac{d}{dt}\left[\Psi\left(\bar{\varepsilon}^e, \bar{\xi}, \bar{\bar{\xi}}, \theta\right) + \eta^e\theta\right] = \\ &= \theta\dot{\eta}^p + \dot{\bar{\varepsilon}}^e\left[\sigma - \frac{\partial}{\partial\varepsilon}\bar{\Psi}\left(\bar{\varepsilon}^e, \bar{\xi}, \theta\right)\right] + \sigma\dot{\bar{\varepsilon}}^p - \frac{\partial\Psi}{\partial\bar{\xi}}\dot{\bar{\xi}} - \delta_{\bar{x}}\frac{\partial\bar{\bar{\Psi}}\left(\bar{\bar{\xi}}, \theta_1\right)}{\partial\bar{\bar{\xi}}}\dot{\bar{\bar{\xi}}} + \\ &\quad + \delta_{\bar{x}}\sigma_{\bar{x}}\dot{\bar{\bar{u}}}(t) + \frac{d\varphi(x)}{dx}\dot{\bar{\bar{u}}}(t) \end{aligned} \tag{7.79}$$

For the elastic loading case where the rate of internal variables and the internal dissipation are equal to zero, the constitutive stress equation reads:

$$\sigma := \frac{d}{d\varepsilon}\bar{\Psi}\left(\bar{\varepsilon}^e, \bar{\xi}, \theta\right) = E(\theta)\left(\varepsilon - \varepsilon^p\right) - \beta(\theta)(\theta - \theta_0) \tag{7.80}$$

For the bulk material, this equation remains the same as presented in (7.44). Taking all this into account, the final expression for the internal dissipation in a plastic loading case can be obtained, where the correct interpretation is be given in terms of distribution (e.g. Ibrahimbegović and Brancherie 2003):

$$\mathcal{D}_{\text{int}} = \int_{\Omega^e} \mathcal{D}_{\text{int}} dx = \int_{\Omega^e} \left(\theta\dot{\eta}^p + \sigma\dot{\bar{\varepsilon}}^p + \bar{q}\dot{\bar{\xi}}\right)dx + \left.\left(\bar{\bar{q}}\dot{\bar{\bar{\xi}}}\right)\right|_{\bar{x}} \tag{7.81}$$

The evolution laws for the localized variables are established in the same way as for the classical continuum model. In particular, the evolution equation for the internal variables which control softening can be written as:

$$0 = \frac{d\bar{\bar{L}}^{\Omega}}{d\bar{\bar{q}}} \rightarrow \dot{\bar{\bar{\xi}}}\delta_{\bar{x}} = \dot{\bar{\bar{\gamma}}}\delta_{\bar{x}} \frac{\partial \bar{\bar{\phi}}}{\partial \bar{\bar{q}}} = \dot{\bar{\bar{\gamma}}}\delta_{\bar{x}} \tag{7.82}$$

where $\dot{\bar{\bar{\gamma}}}$ is the plastic multiplier at the point of localized failure.

7.6.2.2 Mechanical Balance Equation

The set of force equilibrium equations consists of two equations:

1. the local force equilibrium (established for the entire bulk domain)

$$0 = \frac{d\sigma}{dx} + b \tag{7.83}$$

2. the stress orthogonality condition for defining the traction at the localized failure point

$$0 = t_{\bar{x}} + \int_{\Omega^e} \frac{d\varphi(x)}{dx} \sigma_{\bar{x}} dx \tag{7.84}$$

7.6.2.3 Local Balance of Energy at the Localized Failure Point

For the regular part, the local energy balance is still described by a thermodynamic continuum model:

$$\theta\dot{\eta} = -\frac{dQ}{dx} + \mathcal{D}_{\text{int}} + R \tag{7.85}$$

The corresponding state equation reads:

$$\eta^e = -\frac{d\psi}{d\theta} = \beta\left(\varepsilon - \varepsilon^p\right) + \rho c \ln\left(\frac{\theta}{\theta_0}\right) \rightarrow \theta\dot{\eta}^e = \theta\beta\left(\dot{\varepsilon} - \dot{\varepsilon}^p\right) + \rho c\dot{\theta} \tag{7.86}$$

By considering that $\eta = \eta^e + \eta^p$, $\mathcal{D}_{\text{int}} = \mathcal{D}_{\text{mech}} + \mathcal{D}_{\text{ther}}$ and $\mathcal{D}_{\text{ther}} = \theta\dot{\eta}^p$, the local energy balance can finally be rewritten in the format equivalent to the heat transfer equation:

$$\rho c\dot{\theta} = -\frac{dQ}{dx} + \mathcal{D}_{\text{mech}} - \theta\beta\left(\dot{\varepsilon} - \dot{\varepsilon}^p\right) + R \tag{7.87}$$

where $\mathcal{D}_{\text{mech}}$ acts as an additional heat source. This equation holds at any point of the material in the regular part.

It is further considered that at the localized failure point, the material has no more ability to store heat, which implies setting the heat capacity to zero ($\rho c = 0$). It is also taken into account that at the localized failure point there is no heat source ($R = 0$) nor thermal stress ($b = 0$). Therefore, the mechanical dissipation at the localized failure can be balanced only against the change of heat flux. Moreover, the local energy balance equation at the localized failure point ought to be interpreted in the distribution sense, resulting in the corresponding jump in the heat flux:

$$0 = -\left[\frac{dQ}{dx}\right]_{\bar{x}} + \int_{\Omega} \delta_{\bar{x}} \bar{\bar{\mathcal{D}}}_{\text{mech}} \rightarrow [\![Q]\!] = \left(\bar{\bar{q}}\dot{\bar{\bar{\xi}}}\right)\Big|_{\bar{x}} \tag{7.88}$$

where the mechanical dissipation $\bar{\bar{\mathcal{D}}}_{\text{mech}}$ acts as the heat source at the failure point. As indicated in (7.87) above, this results in the corresponding "jump" of the heat flux through the localized failure section. The jump in the heat flux leads to a change of the temperature gradient at the localized point. In the finite element implementation, one needs additional shape functions for the describing not only displacement but also temperature field.

7.7 EMBEDDED DISCONTINUITY FINITE ELEMENT METHOD (ED-FEM) IMPLEMENTATION

7.7.1 Domain Definition

A 1D heterogeneous truss-bar subjected simultaneously to mechanical loading (including distributed load $b(x)$ and prescribed displacements at both ends) and heat transfer along the bar (Figure 7.15) is considered. The material heterogeneity is the direct result of temperature dependent material parameters under a heterogeneous temperature field. In particular, the bar is built of an elastoplastic material, occupying two different sub-domains separated by a localized failure point at $\bar{x}$:

$$\Omega_e = \Omega_e^1 \cap \Omega_e^2; \Omega_e = [0, l_e]; \Omega_e^1 = [0, \bar{x}]; \Omega_e^2 = [\bar{x}, l_e] \tag{7.89}$$

The localized mechanical failure is assumed to happen at the interface $\bar{x}$ (see Figure 7.16).

In the following, the indices "1" is used for all the thermodynamics variables related to sub-domain Ω_e^1, and the indices "2" to the second sub-domain Ω_e^2.

7.7.2 "Adiabatic" Operator Split Solution Procedure

Given positive experience (e.g. Kassiotis, Colliat and Matthies 2009), the operator split method is selected based upon adiabatic split to solve this problem. In the most general

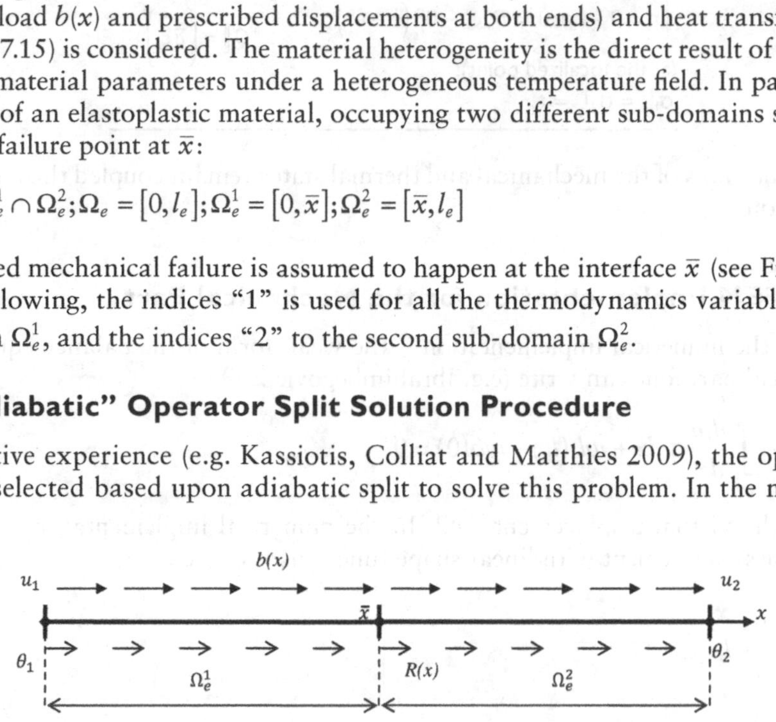

Figure 7.15 Heterogeneous two-phase material for a truss-bar, with phase-interface placed at $\bar{x}$.

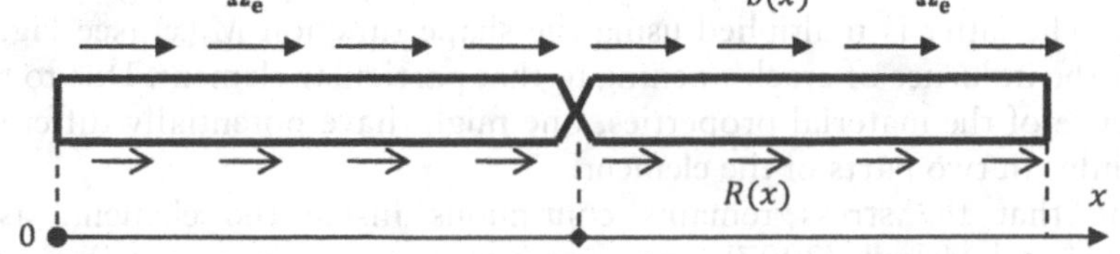

Figure 7.16 Two sub-domain Ω_e^1 and Ω_e^2 separated by a localized point at $\bar{x}$.

case with active localized failure, the coupled thermomechanical problem is described by a set of mechanical balance equations defined in (7.80) and (7.81), accompanied by the energy balance equations in (7.84) and (7.85). Solving all of these equations simultaneously is certainly not the most efficient option. In order to increase the solution efficiency, one can choose between two possible operator split implementations: isothermal and adiabatic (e.g. Ibrahimbegović 2009). The isothermal operator split is not capable of providing the stability of the computation (e.g. Kassiotis, Colliat and Matthies 2009). Therefore, one focuses only upon the adiabatic operator split method. In this method, the problem is divided into two phases, with each one contributing to the change of temperature:

Phase 1—the mechanical part with "adiabatic" condition	Phase 2—the thermal part
$\begin{cases} 0 = \dfrac{d\sigma}{dx} + b \\ \dot{\eta} = 0 \rightarrow \rho c\dot{\theta} = \mathcal{D}_{\text{mech}} - \theta\beta\left(\dot{\varepsilon} - \dot{\varepsilon}^{p}\right) \end{cases}$	$\rho c\dot{\theta} = -\dfrac{dQ}{dx} + R$ $[\![Q]\!] = \left(\bar{\bar{q}}\dot{\bar{\bar{\xi}}}\right)\Big\vert_{\bar{x}}$
(at the localized point): $\sigma_1\big\vert_{\bar{x}} = \sigma_2\big\vert_{\bar{x}} = t_{\bar{x}}$	

The computations of the mechanical and thermal states remain coupled through the adiabatic condition.

7.7.3 ED-FEM Implementation for the Mechanical Part

The basis of the numerical implementation is the weak form of the balance equations. For the mechanical part, one can write (e.g. Ibrahimbegović 2009):

$$\int_{\Omega^e} wb\,dx - \int_{\Omega^e} \frac{dw}{dx}\sigma\,dx + wb(l_e) - wb(0) = 0 \tag{7.90}$$

where w is the virtual displacement field. In the numerical implementation, the simplest two-node truss-bar element with linear shape functions is chosen:

$$N_1(x) = 1 - \frac{x}{l_e} \tag{7.91}$$

$$N_2(x) = \frac{x}{l_e} \tag{7.92}$$

where l_e is the element length. When the localized failure occurs, a displacement discontinuity at the failure point is introduced, with parameter $\alpha_1^m(t)$ representing the crack-opening displacement. The latter is multiplied using the shape function $M_1(x)$ (see Figure 7.17), in order to limit the influence of crack-opening to that particular element. Due to the temperature dependence of the material properties, one might have potentially different values of Young's modulus in two parts of the element.

Considering that the stress remains continuous inside the element, as shown in Ibrahimbegović and Melnyk (2007), one must introduce the corresponding strain discontinuity at the localized failure point. This is carried out by using the shape function $M_2(x)$ shown in Figure 7.18 with the corresponding parameter $\alpha_2^m(t)$. It is noted that both $M_1(x)$

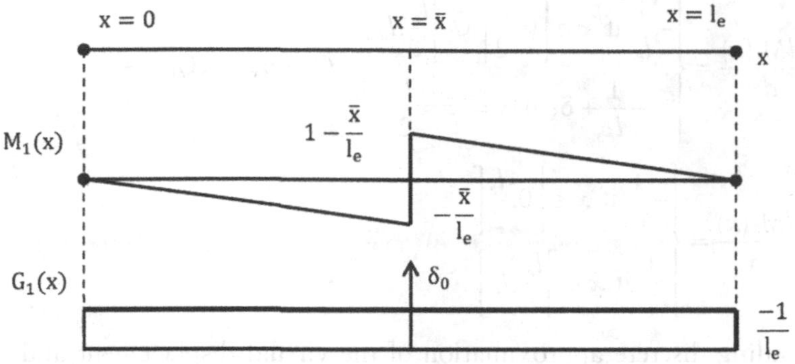

Figure 7.17 Displacement discontinuity shape function two sub-domain $M_1(x)$.

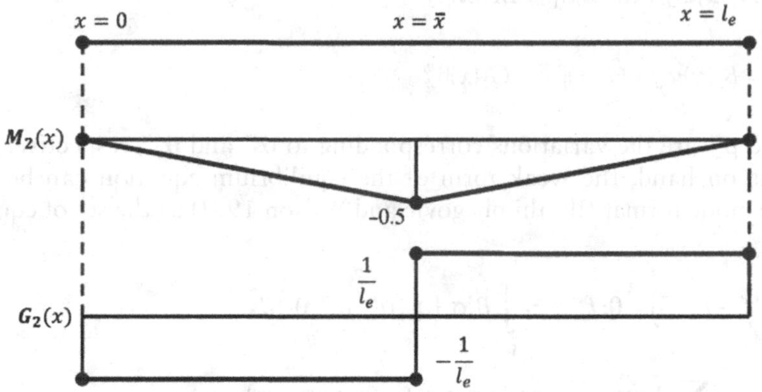

Figure 7.18 Strain discontinuity shape function $M_2(x)$ and its derivatives.

and $M_2(x)$ are chosen with respect to the localized failure that occurs in the middle of the element, so that $\bar{x} = l_e / 2$. Thus, the displacement field interpolation can be written as:

$$u(x,t) = N_a(x)d_a(t) + M_1(x)\alpha_1^m(t) + M_2(x)\alpha_2^m(t) \tag{7.93}$$

with

$$M_1(x) = H_{\bar{x}} - N_2(x) = \begin{cases} -\dfrac{x}{l_e} \text{ if } x \in \left[0, \dfrac{l_e}{2}\right[\\ 1 - \dfrac{x}{l_e} \text{ if } x \in \left]\dfrac{l_e}{2}, l_e\right] \end{cases} \tag{7.94}$$

$$M_2(x) = \begin{cases} -\dfrac{x}{l_e} \text{ if } x \in \left[0, \dfrac{l_e}{2}\right[\\ \dfrac{x}{l_e} - 1 \text{ if } x \in \left]\dfrac{l_e}{2}, l_e\right] \end{cases} \tag{7.95}$$

The corresponding strain interpolation can then be written as:

$$\varepsilon(x,t) = \frac{du(x,t)}{dx} = B_a(x)d_a(t) + G_1(x)\alpha_1^m(t) + G_2(x)\alpha_2^m(t) \tag{7.96}$$

$$G_1(x) = \frac{dM_1(x)}{dx} = \begin{cases} -\dfrac{1}{l_e} \text{ if } x \in \left[0, \dfrac{l_e}{2}\right[\cup \left]\dfrac{l_e}{2}, l_e\right] \\ -\dfrac{1}{l_e} + \delta_{\bar{x}} \text{ if } x = \bar{x} = \dfrac{l_e}{2} \end{cases} = \tilde{G}_1 + \delta_{\bar{x}=\frac{l_e}{2}}; \tilde{G}_1 = -\frac{1}{l_e} \tag{7.97}$$

$$G_2(x) = \frac{dM_2(x)}{dx} = \begin{cases} -\dfrac{1}{l_e} \text{ if } x \in \left[0, \dfrac{l_e}{2}\right[\\ \dfrac{1}{l_e} \text{ if } x \in \left]\dfrac{l_e}{2}, l_e\right] \end{cases} \tag{7.98}$$

The corresponding discrete approximation of the virtual displacement and strain can be written in an equivalent form:

$$w(x,t) = N_a(x)w_a + M_1(x)\beta_1^m + M_2(x)\beta_2^m \tag{7.99}$$

$$\frac{dw(x,t)}{dx} = B_a(x)w_a + \tilde{G}_1(x)\beta_1^m + G_2(x)\beta_2^m \tag{7.100}$$

where β_1^m and β_2^m are the variations corresponding to α_1^m and α_2^m, respectively. With these interpolations on hand, the weak form of the equilibrium equation can be recast in an incompatible mode format (Ibrahimbegović and Wilson 1991) as the set of equations:

$$\begin{cases} A_{e=1}^{\text{nel}}\left(f^{\text{int},e} - f^{\text{ext},e}\right) = 0; f^{\text{int},e} = \displaystyle\int_{\Omega_i^e} B_a^e \sigma_i\left(d_a, \alpha_1^m, \alpha_2^m, \theta_i\right) dx \\ h_1^e = 0; h_1^e = \displaystyle\sum \int_{\Omega_i^e} \tilde{G}_1^e \sigma_i\left(d_a, \alpha_1^m, \alpha_2^m, \theta_i\right) dx + t_{\bar{x}}\left(\alpha_1^m, \alpha_2^m\right) \\ h_2^e = 0; h_2^e = \displaystyle\sum \int_{\Omega_i^e} G_2^e \sigma_i\left(d_a, \alpha_1^m, \alpha_2^m, \theta_i\right) dx \end{cases} \tag{7.101}$$

Given highly the nonlinear material behavior, this set of equations ought to be solved using an iterative scheme. If Newton's method is used, systematic use of the consistent linearization (e.g. Ibrahimbegović 2009) is used, where the corresponding incremental stress–strain relation has to be obtained. It is noted that the chosen isoparametric elements provide continuum consistent interpolation, and furthermore that the continuum and discrete tangent moduli remain the same in a 1D setting (e.g. Ibrahimbegović 2009). Thus, one starts with the consistent linearization of the continuum problem to obtain the stress rate constitutive equation, one in each sub-domain "i":

$$\dot{\sigma}_i = E_i\left(\dot{\bar{\varepsilon}}_i - \dot{\bar{\varepsilon}}_i^p\right) - \beta_i\dot{\theta}_i \tag{7.102}$$

The time derivative of temperature can be computed by imposing the adiabatic step:

$$\theta_i\dot{\eta}_i^e = \theta_i\beta_i\left(\dot{\bar{\varepsilon}}_i - \dot{\bar{\varepsilon}}_i^p\right) + \rho_i c_i\dot{\theta}_i = 0 \rightarrow \dot{\theta}_i = -\frac{\beta_i\theta_i}{\rho_i c_i}\left(\dot{\bar{\varepsilon}}_i - \dot{\bar{\varepsilon}}_i^p\right) \tag{7.103}$$

Combining the last two results, finally it is obtained with:

$$\dot{\sigma}_i = C_i^{e,\text{ad}}\left(\dot{\bar{\varepsilon}}_i - \dot{\bar{\varepsilon}}_i^p\right); C_i^{e,\text{ad}} = E_i + \frac{\beta_i^2\theta_i}{\rho_i c_i} \tag{7.104}$$

where $C_i^{e,\text{ad}}$ denotes the adiabatic tangent modulus. For sub-domain *i*, undergoing elastic loading, with $\dot{\bar{\varepsilon}}_i^p = 0$, the constitutive equation can simplified as:

$$\dot{\sigma}_i = C_i^{e,\text{ad}}\dot{\bar{\varepsilon}}_i \tag{7.105}$$

On the other hand, if sub-domain *i* undergoes plastic loading, the consistency condition requires:

$$\dot{\bar{\phi}}_i(\sigma_i, q_i, \theta_i) = \frac{d\phi_i}{d\sigma_i}\dot{\sigma}_i + \frac{d\phi_i}{d\xi_i}\dot{\bar{\xi}}_i + \frac{d\phi_i}{d\theta_i}\dot{\theta}_i = 0 \tag{7.106}$$

With the expression $\bar{\phi}_i$ chosen herein, (7.101) can further be simplified to:

$$\text{sign}(\sigma_i)\dot{\sigma}_i - K_i\dot{\bar{\bar{\xi}}}_i + \left(\sigma_y^i + K_i\bar{\xi}_i\right)\omega_i\dot{\theta}_i\dot{\bar{\phi}}_i = 0 \tag{7.107}$$

By using (7.98), the constitutive equation in rate form is obtained:

$$\begin{aligned}\dot{\sigma}_i &= \text{sign}(\sigma_i)\left(\sigma_y^i + K_i\bar{\xi}_i\right)\omega_i\frac{\beta_i\theta_i}{\rho_i c_i}\dot{\bar{\varepsilon}}_i \\ &+\left[K_i - \text{sign}(\sigma_i)\left(\sigma_y^i + K_i\bar{\xi}_i\right)\omega_i\frac{\beta_i\theta_i}{\rho_i c_i}\right]\dot{\bar{\varepsilon}}_i^p\end{aligned} \tag{7.108}$$

From (7.102), one has:

$$\dot{\bar{\varepsilon}}_i^p = \dot{\bar{\varepsilon}}_i - \frac{\dot{\sigma}_i}{C_i^{e,\text{ad}}} \tag{7.109}$$

Combining (7.103) and (7.104), the constitutive equation for a plastic domain "*i*" can be formulated:

$$\begin{aligned}\dot{\sigma}_i &= \text{sign}(\sigma_i)\left(\sigma_y^i + K_i\bar{\xi}_i\right)\omega_i\frac{\beta_i\theta_i}{\rho_i c_i}\dot{\bar{\varepsilon}}_i \\ &+\left[K_i - \text{sign}(\sigma_i)\left(\sigma_y^i + K_i\bar{\xi}_i\right)\omega_i\frac{\beta_i\theta_i}{\rho_i c_i}\right]\left(\dot{\bar{\varepsilon}}_i - \frac{\dot{\sigma}_i}{C_i^{e,\text{ad}}}\right)\end{aligned} \tag{7.110}$$

$$\dot{\sigma}_i = \underbrace{\frac{K_i C_i^{e,\text{ad}}}{C_i^{e,\text{ad}} + K_i - \text{sign}(\sigma_i)\left(\sigma_y^i + K_i\bar{\xi}_i\right)\omega_i\frac{\beta_i\theta_i}{\rho_i c_i}}}_{C_i^{p,\text{ad}}}$$

In conclusion, the following constitutive equation can be employed:

$$\dot{\sigma}_i = C_i^{\text{ad}}\dot{\bar{\varepsilon}}_i; C_i^{\text{ad}} = \begin{cases} C_i^{e,\text{ad}}; & \text{if } \bar{\phi}_i < 0 \text{ elastic case} \\ C_i^{p,\text{ad}}; & \text{if } \bar{\phi}_i = 0 \text{ plastic case}\end{cases} \tag{7.111}$$

where $C_i^{e,\text{ad}}$ and $C_i^{p,\text{ad}}$ are defined in (7.99) and (7.105), respectively. To solve the problem, a two operator split is employed (e.g. Ibrahimbegović 2009) with "local" and "global" phases of computation. The former provides the internal variables, while the latter gives the nodal values of displacement.

7.7.4 ED-FEM Implementation for the Thermal Part

The local form of the energy balance equation can be written as:

$$\theta\dot{\eta}_i^e = -\frac{\partial Q}{\partial x} + \sigma_i\dot{\bar{\varepsilon}}_i^p + \bar{q}_i + \delta_{\bar{x}}\left(\bar{\bar{q}}\dot{\bar{\bar{\xi}}}\right) + R_i \tag{7.112}$$

In each of the two sub-domains, the heat transfer obeys the Fourier heat conduction law:

$$Q_i = -k_i\frac{\partial\theta_i}{\partial x} \tag{7.113}$$

By using the evolution equation for "elastic" entropy, the local energy balance can be rewritten in an equivalent form to the heat equation:

$$\rho c\dot{\theta}_i = k_i\frac{d^2\theta_i}{dx^2} + R_0^i + \bar{\bar{\mathcal{D}}}_{\text{mech}}\delta_{\bar{x}}(x); \bar{\bar{\mathcal{D}}}_{\text{mech}} = \bar{\bar{q}}\dot{\bar{\bar{\xi}}} \tag{7.114}$$

where:

$$R_0^i = \sigma_i\dot{\bar{\varepsilon}}_i^p + \bar{q}_i\dot{\bar{\xi}}_i - \theta_i\beta_i\left(\dot{\bar{\varepsilon}}_i - \dot{\bar{\varepsilon}}_i^p\right) + R_i \tag{7.115}$$

The strong form (7.114) is further transferred into the weak form by introducing an arbitrary temperature field, denoted as ϑ, and by applying the virtual work laws:

$$\int_0^l \vartheta\left(\rho c\dot{\theta}_i - k_i\frac{d^2\theta_i}{dx^2} - R_0^i - \bar{\bar{\mathcal{D}}}_{\text{mech}}\delta_{\bar{x}}\right)dx = 0 \tag{7.116}$$

After integration by part, we can finally obtain the following weak form:

$$\int_0^l \vartheta\rho_i c_i\dot{\theta}_i dx + \int_0^l \frac{d\vartheta}{dx}k_i\frac{d\theta_i}{dx}dx = \int_0^l \vartheta\left(R_0^i + \delta_{\bar{x}}\bar{\bar{\mathcal{D}}}_{\text{mech}}\right)dx \tag{7.117}$$

A two-node truss-bar element is considered. The nodal values of temperature and the weighting temperature at node i are denoted as $d_{\theta i}$ and $d_{\vartheta i}$, respectively. Thus, $\mathbf{d}_\theta$ and $\mathbf{d}_\vartheta$ denote the real and the arbitrary nodal temperature vector, respectively. For a two-node element, one has:

$$\mathbf{d}_\theta = \begin{Bmatrix} d_{\theta 1} \\ d_{\theta 2} \end{Bmatrix}; \mathbf{d}_\vartheta = \begin{Bmatrix} d_{\vartheta 1} \\ d_{\vartheta 2} \end{Bmatrix}$$

The real and weighting temperature fields along the element are constructed with interpolation shape functions. Furthermore, the jump of temperature gradient at the localized failure point, is represented by an additional shape function, the same one for the real and weighting temperature fields:

$$\theta_i(x) = N_a(x)d_\theta^a(t) + M_2(x)\alpha_2^t(t) \tag{7.118}$$

$$\vartheta(x) = N_a(x)d_\vartheta^a + M_2(x)\beta_2^t \tag{7.119}$$

where $N_a(x)$ and $M_2(x)$ are defined in (7.91) and illustrated in Figure 7.18 for a two-node truss-bar element, whereas at $\alpha_2^t(t)$ is the variable controlling the "jump" in temperature gradient. It is noted that: $\theta_1(t) = 1/2\left(d_\theta^1(t) + d_\theta^2(t)\right) + \alpha_2(t)$, where θ_1 is the temperature at the interface (at the middle of the element).

By considering these interpolations, the weak form (7.112) is finally reduced to:

$$\int_{\Omega_i^e}\Big[N_a(x)\rho_i c_i N_a(x)\dot{d}_\theta^a + N_a(x)\rho_i c_i M_2(x)\alpha_2(t) + B_a(x)k_i B_a(x)d_\theta^a \tag{7.120}$$

$$+B_a(x)k_i G_2(x)\alpha_2(t)\Big]dx = Q_\theta^a + N_a(\bar{x})\bar{\bar{\mathcal{D}}}_{\text{mech}}$$

$$\int_{\Omega_i^e}\Big[M_2(x)\rho_i c_i N_a(x)\dot{d}_\theta^a + M_2(x)\rho_i c_i M_2(x)\dot{\alpha}_2(t) + G_2(x)k_i B_a(x)d_\theta^a \tag{7.121}$$

$$+G_2(x)k_i G_2(x)\alpha_2(t)\Big]dx = M_2(\bar{x})\bar{\bar{\mathcal{D}}}_{\text{mech}}$$

Finally, the finite element equations to be solved for the "thermal" phase are given with:

$$\begin{cases}\mathbf{M}^e\dot{\mathbf{d}}_\theta^e + P^e\dot{\alpha}_2^e + \mathbf{K}^e\mathbf{d}_\theta^e + \mathbf{F}^e\alpha_2^e = \mathbf{Q}_\theta^1\\ \mathbf{P}^{e^T}\dot{\mathbf{d}}_\theta^e + \mathbf{L}^e\dot{\alpha}_2^e + \mathbf{F}^{e^T}\mathbf{d}_\theta^e + \mathbf{H}^e\alpha_2^e = \mathbf{Q}_\theta^2\end{cases} \tag{7.122}$$

where:

$$\mathbf{M}_{2x2}^e = \sum\int_{\Omega_i^e} N_a(x)\rho_i c_i N_a(x)dx$$

$$\mathbf{M}_{2x2}^e = \frac{l_e}{24}\begin{bmatrix}7\rho_1c_1+\rho_2c_2 & 2(\rho_1c_1+\rho_2c_2)\\ 2(\rho_1c_1+\rho_2c_2) & \rho_1c_1+7\rho_2c_2\end{bmatrix} \tag{7.123}$$

$$\mathbf{P}_{1x2}^e = \sum\int_{\Omega_i^e} N_a(x)\rho_i c_i M_2(x)dx$$

$$\mathbf{P}_{1x2}^e = -\frac{l_e}{24}\begin{Bmatrix}2\rho_1c_1+\rho_2c_2\\ \rho_1c_1+2\rho_2c_2\end{Bmatrix} \tag{7.124}$$

$$\mathbf{K}_{2x2}^e = \sum\int_{\Omega_i^e} B_a(x)k_i B_a(x)dx$$

$$\mathbf{K}_{2x2}^e = \frac{(k_1+k_2)}{2l_e}\begin{bmatrix}1 & -1\\ -1 & 1\end{bmatrix} \tag{7.125}$$

$$\mathbf{F}_{1x2}^e = \sum\int_{\Omega_i^e} B_a(x)k_i G_2(x)dx$$

$$\mathbf{F}_{1x2}^e = \frac{k_1-k_2}{2l_e}\begin{Bmatrix}1\\ -1\end{Bmatrix} \tag{7.126}$$

$$\mathbf{L}_{1x1}^e = \sum\int_{\Omega_i^e} M_2(x)\rho_i c_i M_2(x)dx$$

$$\mathbf{L}_{1x1}^e = \frac{l_e}{24}(\rho_1c_1+\rho_2c_2) \tag{7.127}$$

$$\mathbf{H}^{e}_{1x1} = \sum \int_{\Omega_i^e} G_2(x)k_iG_2(x)dx$$

$$\mathbf{H}^{e}_{1x1} = \frac{(k_1 + k_2)}{2l_e} \tag{7.128}$$

$$\mathbf{Q}^{1}_{\boldsymbol{\theta}1x2} = Q^{a}_{\theta} + N_a(\bar{x})\bar{\bar{\mathcal{D}}}_{\text{mech}};$$

$$\mathbf{Q}^{1}_{\boldsymbol{\theta}1x2} = \frac{l_e}{8}\begin{Bmatrix} 3R_0^1 + R_0^2 \\ R_0^1 + 3R_0^2 \end{Bmatrix} + \frac{1}{2}\bar{\bar{\mathcal{D}}}_{\text{mech}}\begin{Bmatrix} 1 \\ 1 \end{Bmatrix} \tag{7.129}$$

$$\mathbf{Q}^{2}_{\boldsymbol{\theta}1x1} = M_2(\bar{x})\bar{\bar{\mathcal{D}}}_{\text{mech}} = -0.5\bar{\bar{\mathcal{D}}}_{\text{mech}} \tag{7.130}$$

There are many methods capable of solving the time-dependent equation (7.117) (e.g. Ibrahimbegović 2009). Here the Newmark integration scheme is chosen. Assuming that the heat transfer problem lasts for a duration $[0,T]$, This duration can be divided into n increments: $[t_0 = 0, t_1, \ldots, t_i, \ldots, t_{n-1}, t_n = T]$ with the time step $h = t_{i+1} - t_i$.

Given the state variables at time i:

$$d_i = d_\theta(t_i),\ \dot{d}_i = \upsilon_i = \dot{d}_\theta(t_i)$$

$$\alpha_2^i = \alpha_2^i(t_i),\ \dot{\alpha}_2^i = {}_2^i = \dot{\alpha}_2(t_i)$$

Find the state variables at the time $i+1$:

$$d_{i+1} = d_\theta(t_{i+1})$$

$$\upsilon_{i+1} = \dot{d}_{i+1} = \dot{d}_\theta(t_{i+1})$$

$$\alpha_2^{i+1} = \alpha_2(t_{i+1})$$

$$\dot{\alpha}_2^{i+1} = \mho_2^{i+1} = \dot{\alpha}_2^i(t_{i+1})$$

By considering the equation of Newmark: $\Delta\upsilon_i = \frac{\gamma}{\beta\Delta t}\Delta d$ and $\Delta\mho = \frac{\gamma}{\beta\Delta t}\Delta\alpha_2$ and by linearization, (7.75) becomes:

$$\begin{cases} \left(\frac{\gamma}{\beta\Delta t}\mathbf{M}^e + \mathbf{K}^e\right)\Delta d + \left(\frac{\gamma}{\beta\Delta t}\mathbf{P}^e + \mathbf{F}^e\right)\Delta\alpha_2 = \boldsymbol{R}_1^i \\ \left(\frac{\gamma}{\beta\Delta t}\mathbf{P}^{e^T} + \mathbf{F}^{e^T}\right)\Delta d + \left(\frac{\gamma}{\beta\Delta t}L^e + H^e\right)\Delta\alpha_2 = R_2^i \end{cases} \tag{7.131}$$

where the residuals are computed using the following equation

$$\begin{cases} R_1^i = Q_\theta^1 - M^e v_i - P^e\mho_2^i - K^e d_i - F^e\alpha_2^i \\ R_2^i = Q_\theta^2 - P^{e^T} v_i - L^e\mho_2^i - F^{e^T} d_i - H^e\alpha_2^i \end{cases} \tag{7.132}$$

From $(7.84)_2$, the expression of $\Delta\alpha_2$ from $\Delta\boldsymbol{d}$ is established:

$$\Delta\alpha_2 = \left(\frac{\gamma}{\beta\Delta t}L^e + H^e\right)^{-1} R_2^i - \left(\frac{\gamma}{\beta\Delta t}L^e + H^e\right)^{-1}\left(\frac{\gamma}{\beta\Delta t}\boldsymbol{P}^{e^T} + \boldsymbol{F}^{e^T}\right)\Delta\boldsymbol{d} \tag{7.133}$$

Combining (7.133) and $(7.131)_1$, Δd is obtained from the current nodal temperature and the current value of heat flux "jump" as the solution of:

$$\left[\frac{\gamma}{\beta\Delta t}\mathbf{M}^e+\mathbf{K}^e-\left(\frac{\gamma}{\beta\Delta t}\mathbf{P}^e+\mathbf{F}^e\right)\left(\frac{\gamma}{\beta\Delta t}L^e+H^e\right)^{-1}\left(\frac{\gamma}{\beta\Delta t}P^{e^T}+F^{e^T}\right)\right]\Delta d= \tag{7.134}$$
$$R_1^i-\left(\frac{\gamma}{\beta\Delta t}\mathbf{P}^e+\mathbf{F}^e\right)\left(\frac{\gamma}{\beta\Delta t}L^e+H^e\right)^{-1}R_2^i$$

or:

$$\hat{\mathrm{K}}\Delta d=\hat{F} \tag{7.135}$$

Once Δd is known, $\Delta\alpha_2$ can be computed. The nodal temperature and the "jump" in heat flow through the localized mechanical failure at the next time step can be updated using the formula:

$$d_{i+1}=d_i+\Delta d,\alpha_2^{i+1}=\alpha_2^i+\Delta\alpha_2$$

7.8 NUMERICAL EXAMPLES

7.8.1 Example 1 — Simple Tension Imposed Temperature Example with a Fixed Mesh

In this example, a steel bar 5 mm long is elaborated. The bar is built-in at the left end and subjected to an imposed displacement at the right end. The imposed displacement increases 1.6×10^{-4} mm in each step. Simultaneously, the right end of the bar is heated, and its temperature is raised from 0°C to 1000°C, with a 10°C increase in each step. The temperature at the left end is kept equal to 0°C. The loading increases until the localized failure of the bar. The problem geometric data and loading program are described in Figures 7.19 and 7.20, respectively.

The problem is subsequently considered for three different variations of the material properties: (i) the material properties are independent of temperature, (ii) the material properties are linearly dependent on temperature and (iii) the material properties are non-linearly dependent on temperature (following a suggestion given by regulation Eurocode (CEN-Eurocode1 2003)).

7.8.1.1 Material Properties Independent on Temperature

In this case, the material properties of the bar are assumed to be constant regardless of the temperature change. The chosen values for the material parameters are given in Table 7.2.

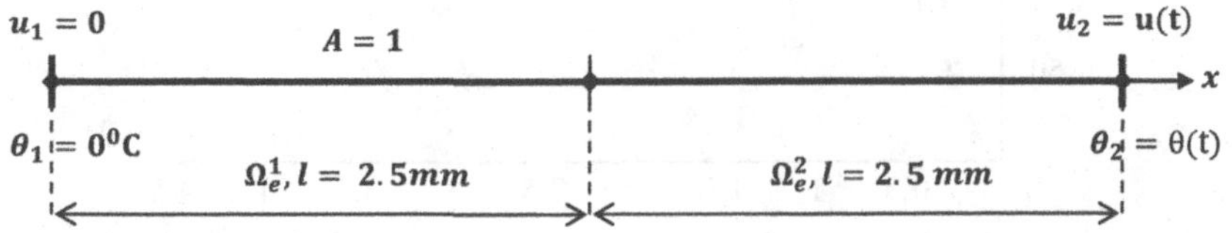

Figure 7.19 Bar subjected to imposed displacement and temperature applied simultaneously.

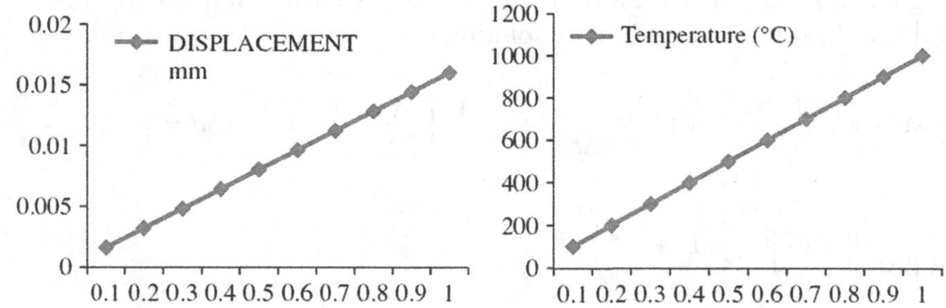

Figure 7.20 Time variation of imposed displacement and temperature.

Table 7.2 Material Properties of Steel Bar

Material Properties	*Value*	*Dimension*
Young's modulus (E)	205 000	MPa
Initial yield stress (σ_y)	250	MPa
Ultimate stress (σ_u)	300	MPa
Plastic hardening modulus (K_p)	20 000	MPa
Localized softening modulus $\left(\bar{\bar{K}}\right)$	−30 000	MPa m^{-1}
Mass density (ρ)	$7\,865 \times 10^{-9}$	$Ns^{-2}\,mm^{-4}$
Thermal conductivity (k)	45	$Ns^{-1}\,K^{-1}$
Heat specific (c)	0.46×10^{9}	$mm^2\,s^{-2}\,K^{-1}$
Thermal elongation (α)	0.00001	

The computed results for stress–strain curves in two sub-domains are presented in Figure 7.21, while the force–displacement curve of the bar is given in Figure 7.22.

In Table 7.3 and Figure 7.23, the resulting time evolution of temperature and its distribution along the bar is presented. For the case with material properties independent on temperature, it can be concluded that there is no difference in the strain values between the two sub-domains. The "jump" in temperature gradient (αth), which appears at the localized failure point, also remains very small. The computed dissipation due to plasticity in the

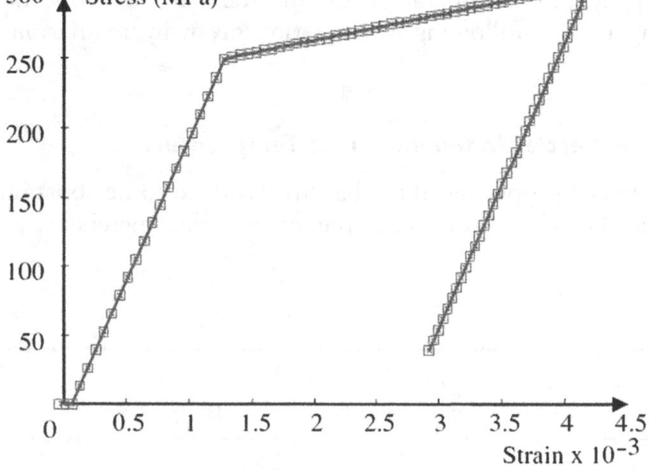

Figure 7.21 Stress–strain curves in two sub-domains (line for 1st sub-domain, squares for 2nd).

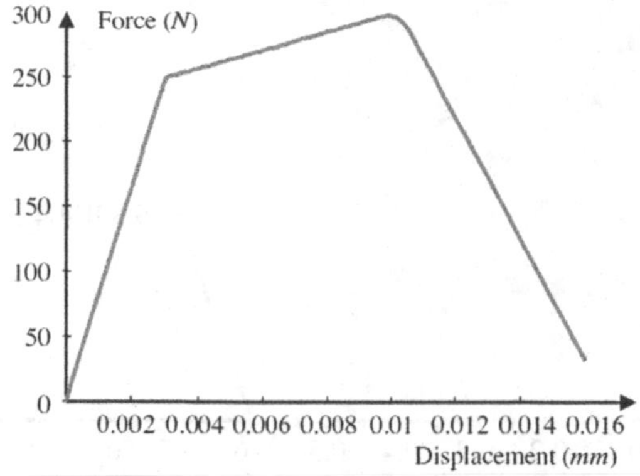

Figure 7.22 Force–displacement curve of the bar.

Table 7.3 Time Evolution of Temperature along the Bar

Time	θ *at* $x=0$	θ *at* $x=0.25l_e$	θ *at* $x=0.50l_e$	θ *at* $x=0.75l_e$	θ *at* $x=l_e$	$\Delta\theta$
0.0	0.0000	0.0000	0.0000	0.0000	0.0000	0.0000
0.1	0.0000	25.0000	50.0000	75.0000	100.0000	0.0000
0.2	0.0000	50.0000	100.0000	150.0000	200.0000	0.0000
0.3	0.0000	75.0000	150.0000	225.0000	300.0000	0.0000
0.4	0.0000	100.0000	200.0000	300.0000	400.0000	0.0000
0.5	0.0000	125.0000	250.0000	375.0000	500.0000	0.0000
0.6	0.0000	150.0000	300.0000	450.0000	600.0000	0.0000
0.7	0.0000	175.0005	350.0000	525.0005	700.0000	0.0000
0.8	0.0000	200.0007	400.0014	600.0007	800.0000	0.0014
0.9	0.0000	225.0008	450.0015	675.0008	900.0000	0.0015
1.0	0.0000	250.0008	500.0016	750.0008	1000.0000	0.0016

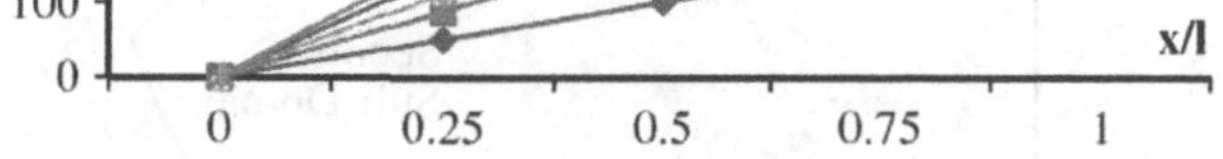

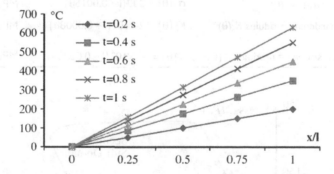

Figure 7.23 Distribution of temperature (°C) along the bar at chosen values of time.

fracture process zone is 36.63 Nmm, while the dissipation due to localized failure is 29.44 Nmm. In summary, the total mechanical dissipation in the bar is equal to 66.07 Nmm.

7.8.1.2 Material Properties are Linearly Dependent on Temperature

In this example, the same mechanical material properties of the steel bar are chosen as in the first example (see Table 7.2), and they hold only at a reference temperature (equal to 0°C) (see Figure 7.24).

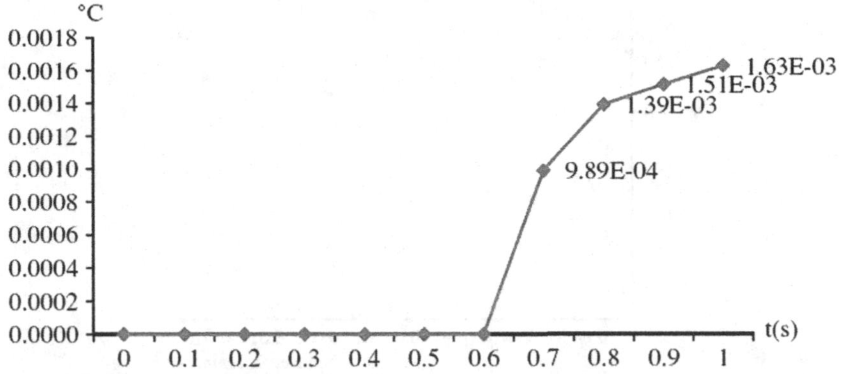

Figure 7.24 Evolution of $\Delta\theta$ vs. time (in K).

For other temperature values, they vary linearly according to the expression given in Table 7.4.

The thermal material properties are independent on temperature and equal to those in the first example. The resulting stress–strain curves in the two sub-domains and the resulting force–displacement diagram are presented in Figures 7.25 and 7.26, respectively.

The distribution of temperature along the bar is shown in Table 7.5 and Figure 7.27.

In this example, the total plastic dissipation and the total localized dissipation are 14.08 Nmm and 13.82 Nmm, respectively. Thus, the total mechanical dissipation is equal to 27.90 Nmm.

Table 7.4 Material Properties of Steel Bar Linearly Dependent

Material Properties	*Expression*	*Dimension*
Young's modulus $E(\theta)$	$E(\theta) = 2.05\times10^5(1-0.0008\theta)$	MPa
Initial yield stress $\sigma_y(\theta)$	$\sigma_y(\theta) = 250(1-0.001\theta)$	MPa
Ultimate stress $\sigma_u(\theta)$	$\sigma_u(\theta) = 300(1-0.0015\theta)$	MPa
Plastic hardening modulus $K_p(\theta)$	$K_p(\theta) = 2\times10^4(1-0.0008\theta)$	MPa
Localized softening modulus $\bar{\bar{K}}(\theta)$	$\bar{\bar{K}}(\theta) = -3\times10^4(1-0.0008\theta)$	MPa m^{-1}

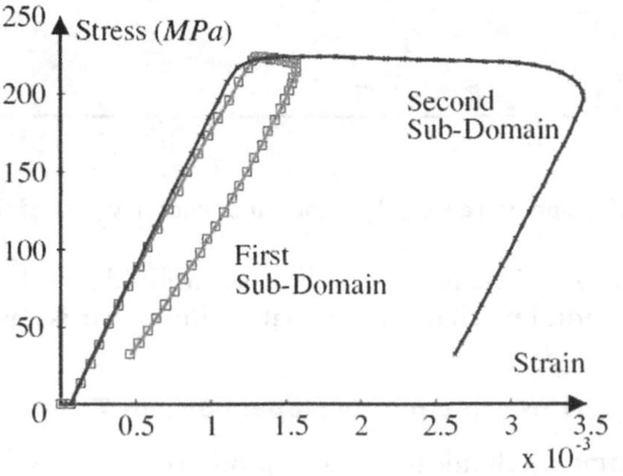

Figure 7.25 Stress–strain curves in two sub-domains (bold line for 1st sub-domain, squared line for 2nd).

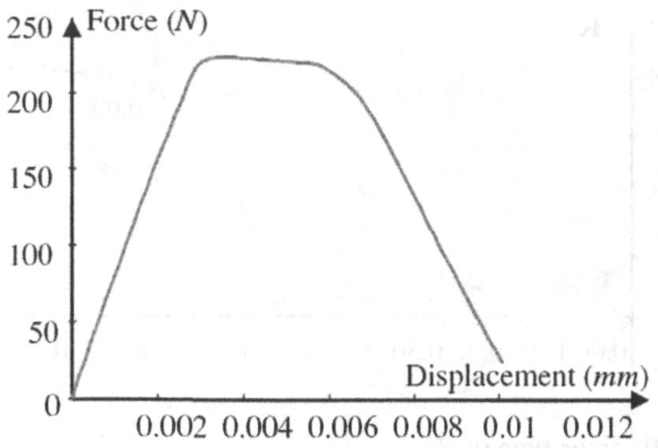

Figure 7.26 Force–displacement curve.

Table 7.5 Time Evolution of Temperature along the Bar

Time	θ *at x=0*	θ *at x=0.25l*$_e$	θ *at x=0.50l*$_e$	θ *at x=0.75l*$_e$	θ *at x=l*$_e$	Δθ
0.000	0.0000	0.0000	0.0000	0.0000	0.0000	0.0000
0.1000	0.0000	25.0000	50.0000	75.0000	100.0000	0.0000
0.2000	0.0000	50.0000	100.0000	150.0000	200.0000	0.0000
0.3000	0.0000	75.0000	150.0000	225.0000	300.0000	0.0000
0.3500	0.0000	87.5000	175.0000	262.5000	350.0000	0.0000
0.4000	0.0000	100.0000	199.9999	300.0000	400.0000	−0.0001
0.4500	0.0000	112.5002	225.0005	337.5002	450.0000	0.0005
0.5000	0.0000	125.0004	250.0007	375.0004	500.0000	0.0007
0.5500	0.0000	137.5004	275.0008	412.5004	550.0000	0.0008
0.6000	0.0000	150.0004	300.0008	450.0004	600.0000	0.0008
0.6300	0.0000	157.5004	315.0008	472.0004	630.0000	0.0008

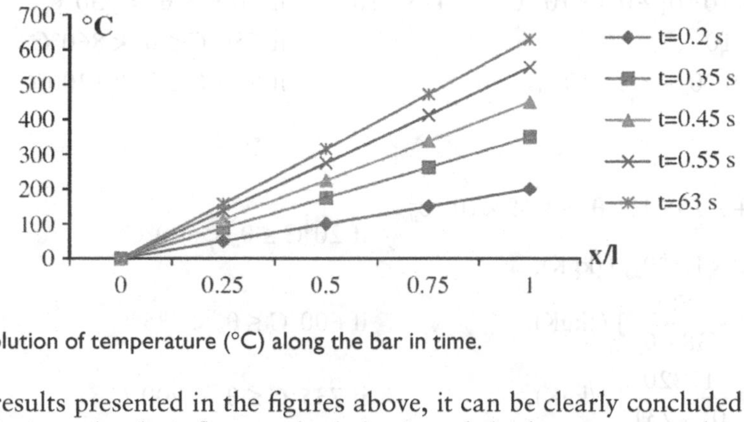

Figure 7.27 Evolution of temperature (°C) along the bar in time.

From the results presented in the figures above, it can be clearly concluded that the temperature variations deeply influence the behavior of the bar. In particular, the displacement at the end of the bar when failure occurs reduces from 0.016 mm to 0.011 mm, the initial yield stress falls down to approximately 225 MPa from 250 MPa and so the ultimate strength reduces from 300 MPa to about 220 MPa. The total dissipation in this example is also reduced, from 66.07 Nmm to 27.90 Nmm. Figure 7.25 indicates that the variation of the temperature field leads to a significant difference in the material behavior and computed stress–strain curves in two parts of the bar. As seen from Figure 7.28, the "jump" in temperature gradient accompanying localized failure remains relatively small.

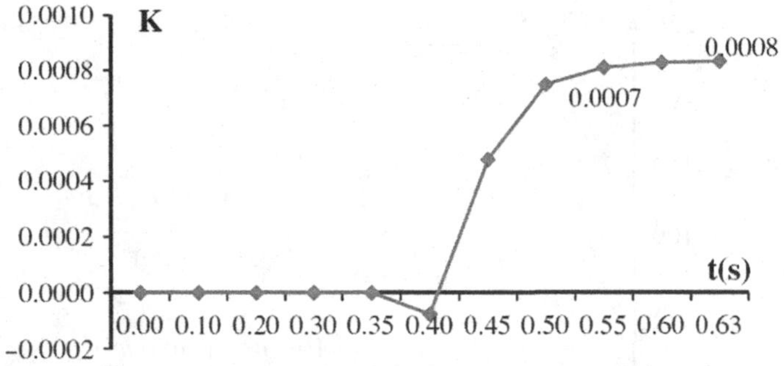

Figure 7.28 Evolution of Δθ versus time (in K).

7.8.1.3 Material Properties Non-Linearly Dependent on Temperature (Eurocode 1993-1-2)

In Eurocodes (CEN-Eurocode1 2003), the material properties of the steel bar subjected to thermal loading are not constant but dependent on temperature as multi-linear functions. Based on those regulations, the evolution of mechanical properties as functions of temperature can be established as shown in Table 7.6.

where ω_* are the temperature dependent coefficients. The values of the temperature dependent coefficients for yield stress, ultimate strength and Young's modulus are taken from Eurocode 1993-1-2. The corresponding values of the coefficients for the plastic hardening modulus and localized softening modulus are taken as the same as the one for Young's modulus. All the values used for these coefficients are presented in Table 7.7 and Figure 7.29.

The evolution of thermal properties is also taken from Eurocodes (CEN-Eurocode1 2003).

Thermal Elongation

$$\alpha = \begin{cases} 1.2\times10^{-5}\theta_a + 0.4\times10^{-8}\theta_a^2 - 2.416\times10^{-4}, & \text{if } 20°\text{C} \le \theta_a < 750°\text{C} \\ 1.1\times10^{-2}, & \text{if } 750°\text{C} \le \theta_a < 860°\text{C} \\ 2\times10^{-5}\theta_a - 6.2\times10^{-3}, & \text{if } 860°\text{C} \le \theta_a < 1200°\text{C} \end{cases}$$

Specific Heat

$$c = \begin{cases} 425 + 7.73\times10^{-1}\theta_a - 1.69\times10^{-3}\theta_a^2 + 2.22\times10^{-6}\theta_a^3 \text{J}/(\text{kgK}), & \text{if } 20°\text{C} \le \theta_a < 600°\text{C} \\ 666 + \dfrac{13002}{738 - \theta_a}\text{J}/(\text{kgK}), & \text{if } 600°\text{C} \le \theta_a < 735°\text{C} \\ 545 + \dfrac{17820}{\theta_a - 731}\text{J}/(\text{kgK}), & \text{if } 735°\text{C} \le \theta_a < 900°\text{C} \\ 650\text{J}/(\text{kgK}), & \text{if } \theta_a \ge 900°\text{C} \end{cases}$$

Thermal Conductivity

$$k = \begin{cases} 54 - 3.33\times10^{-2}\theta_a \text{W}/(\text{mK}), & \text{if } 20°\text{C} \le \theta_a < 800°\text{C} \\ 27.3\,\text{W}/(\text{mK}), & \text{if } 800°\text{C} \le \theta_a \le 1200°\text{C} \end{cases}$$

Table 7.6 Material Properties of a Linearly Dependent Steel Bar

Material Properties	*Expression*	*Dimension*
Young's modulus $E(\theta)$	$E(\theta) = 2.05\times10^5\left(1-\omega_E(\theta-20)\right)$	MPa
Initial yield stress $\sigma_y(\theta)$	$\sigma_y(\theta) = 250\left(1-\omega_{\sigma_y}(\theta-20)\right)$	MPa
Ultimate stress $\sigma_u(\theta)$	$\sigma_u(\theta) = 300\left(1-\omega_{\sigma_u}(\theta-20)\right)$	MPa
Plastic hardening modulus $K_p(\theta)$	$K_p(\theta) = 2\times10^4\left(1-\omega_{K_p}(\theta-20)\right)$	MPa
Localized softening modulus $\bar{\bar{K}}(\theta)$	$\bar{\bar{K}}(\theta) = -3\times10^4\left(1-\omega_{\bar{\bar{K}}}(\theta-20)\right)$	MPa m^{-1}

Table 7.7 Temperature Dependent Coefficients

$\theta(°C)$	ω_{σ_u}	ω_{σ_y}	ω_E	ω_{K_p}	$\omega_{\bar{\bar{K}}}$
0.000	0.00000	0.00000	0.00000	0.00000	0.00000
20	0.00000	0.00000	0.00000	0.00000	0.00000
100	0.00000	0.00000	0.00000	0.00000	0.00000
200	0.00000	0.00107	0.00056	0.00056	0.00056
300	0.00000	0.00138	0.00071	0.00071	0.00071
400	0.00000	0.00153	0.00079	0.00079	0.00079
500	0.00046	0.00133	0.00083	0.00083	0.00083
600	0.00091	0.00141	0.00119	0.00119	0.00119
700	0.00113	0.00136	0.00128	0.00128	0.00128
800	0.00114	0.00122	0.00117	0.00117	0.00117
900	0.00107	0.00109	0.00106	0.00106	0.00106
1000	0.00098	0.00099	0.00097	0.00097	0.00097
1100	0.00091	0.00091	0.00091	0.00091	0.00091
1200	0.00085	0.00085	0.00085	0.00085	0.00085

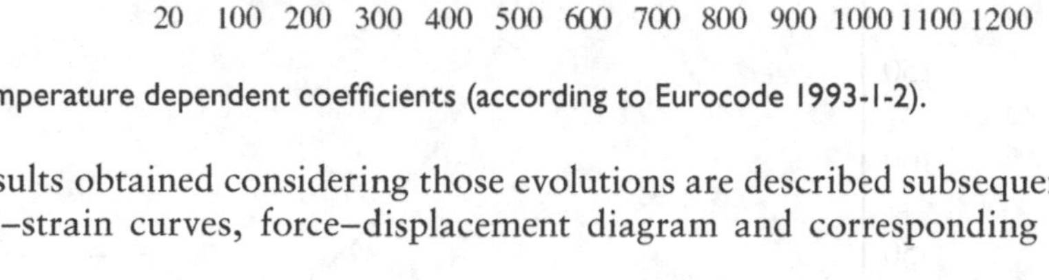

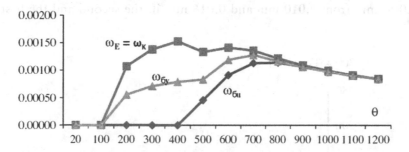

Figure 7.29 Temperature dependent coefficients (according to Eurocode 1993-1-2).

The main results obtained considering those evolutions are described subsequently in terms of the stress–strain curves, force–displacement diagram and corresponding temperature variations.

Figure 7.30 clearly shows the large difference in strain between the two sub-domains, both before and after the initiation of a localized failure. Mathematically, this difference is due to different values of α_1^m and α_2^m (see (7.93)). Before the initiation of localized failure, the difference in temperature will lead to a difference in tangent modulus between the two sub-domains, which results in the appearance of α_2^m which represents the difference in strain

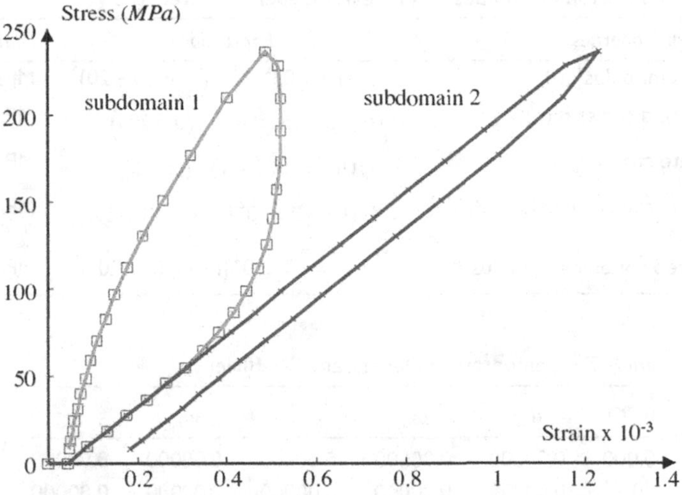

Figure 7.30 Stress–strain curves in two sub-domains (squared line for subdomain 1, bold line for subdomain 2).

between the two sub-domains. After a localized failure occurs, α_1^m increases and contributes to the different behaviors in two parts of the bar (see Figure 7.31).

From Table 7.8 and Figure 7.32, it can be seen that the temperature distribution is nonlinear. Its gradient changes in the middle of the bar. This change can be computed through α_2 (see (7.71)). It is noted that the magnitude of α_2 increases and then decreases with time (see Table 7.8 and Figure 7.33). However, Figure 7.21 also shows that the change in temperature gradient is relatively small in comparison to the temperature at the localized failure point (the maximum ratio of $\Delta\theta / \theta_l$ (θ_l is the temperature of the localized point) is approximately 0.0136%), and therefore does not significantly contribute to the final results.

In this example, once again, a reduction in the strength of the bar is observed (see Figure 7.31): the maximum displacement that can be applied to the bar now reduces to roughly 0.006 mm from 0.010 mm and 0.016 mm in the second and the first example, respectively.

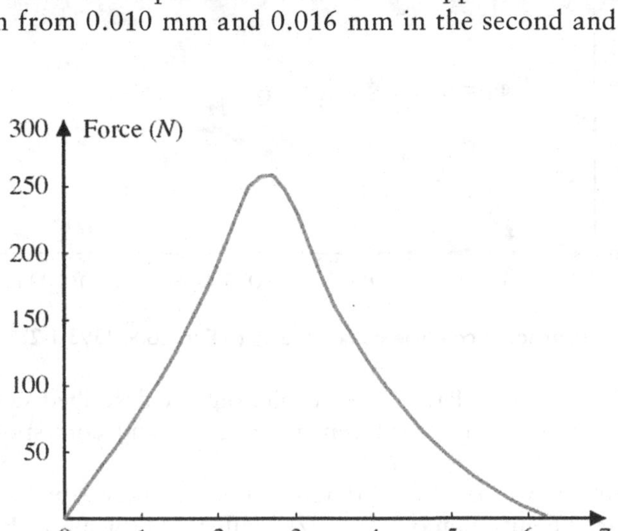

Figure 7.31 Force–displacement diagram for the bar.

Table 7.8 Distribution of Temperature Along the Bar

Time	θ *at x = 0*	θ *at x = 0.25l*$_e$	θ *at x = 0.50l*$_e$	θ *at x = 0.75l*$_e$	θ *at x = l*$_e$	Δθ
0.000	0.0000	0.0000	0.0000	0.0000	0.0000	0.0000
0.0500	0.0000	12.5000	25.0000	37.5000	50.0000	0.0000
0.1000	0.0000	25.0000	50.0000	75.0000	100.0000	0.0000
0.1500	0.0000	37.5000	75.0000	112.5000	150.0000	0.0000
0.2000	0.0000	49.9998	99.9996	149.9998	200.0000	−0.0004
0.2500	0.0000	62.5020	125.0041	187.5020	250.0000	0.0041
0.3000	0.0000	75.0053	150.0106	225.0053	300.0000	0.0106
0.3500	0.0000	87.5094	175.0188	262.5094	350.0000	0.0188
0.3900	0.0000	97.5133	195.0265	292.5133	390.0000	0.0265

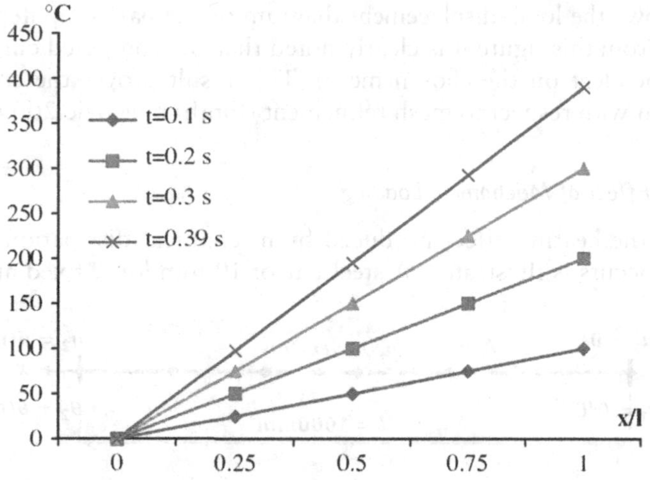

Figure 7.32 Distribution of temperature (°C) along the bar due to time.

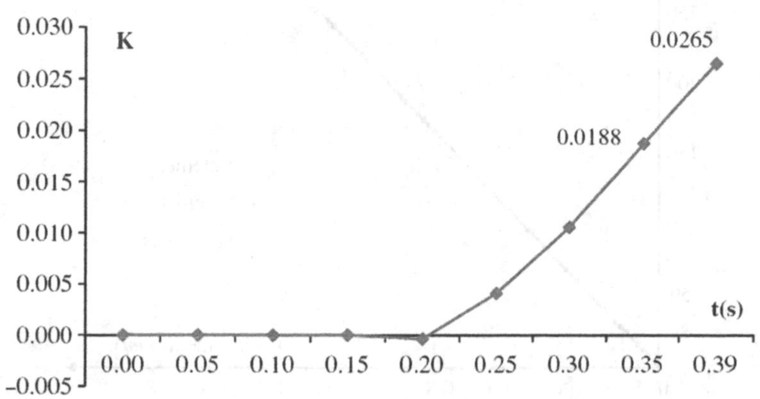

Figure 7.33 Evolution of Δθ vs. time (in K) where we computed $\Delta\theta = \theta_{x=0.5l_e} - 0.5(\theta_{x=0} + \theta_{x=l_e})$.

The total mechanical dissipation along the bar is significantly smaller than the second and the first example (15.01 Nmm in comparison to 27.90 Nmm and 66.07 Nmm). The majority contribution comes from the localized dissipation: 10.55 Nm in comparison with the total plastic dissipation: 4.47 Nm.

7.8.1.3.1 Mesh Refinement, Convergence and Mesh Objectivity

The influence of the chosen number of elements upon the computed final results is presented further. The geometry description is given in Figure 7.34.

A steel bar is considered built-in at the left end and subjected to an imposed displacement at the right end (increasing linearly to 2 mm). Simultaneously, the right end of the bar is heated, and its temperature is raised from 0°C to 100°C. The temperature of the left end is kept constant and equal to 0°C. The material properties of the bar are considered as temperature independent and are the same as in Table 7.2, except localized softening failure which amounts to $(\bar{\bar{K}}) = 45\,\text{MPa m}^{-1}$.

Figure 7.35 shows the load–displacement diagram of the bar computed by using 3, 5, 7 and 9 elements. From this figure it is clearly noted that the computed curve after localized failure is not dependent on the chosen mesh). This result proves the convergence of the numerical solution with respect to mesh refinement (Ibrahimbegović 2009).

7.8.1.3.2 Heating Effect of Mechanical Loading

In this example, the heating effect produced by mechanical dissipation in a bar when a localized failure occurs is illustrated. A steel bar of 10 mm long, fixed at the left end and

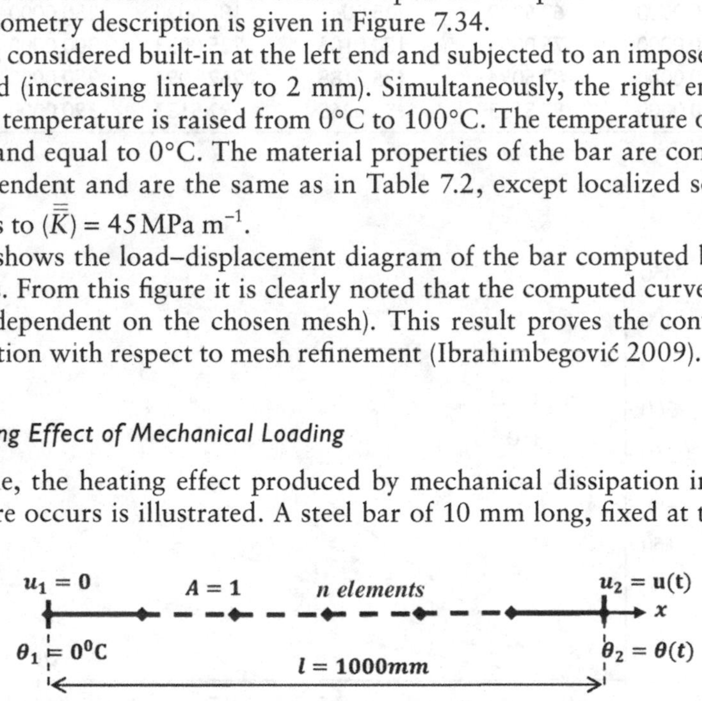

Figure 7.34 Bar subjected to imposed loading and imposed temperature.

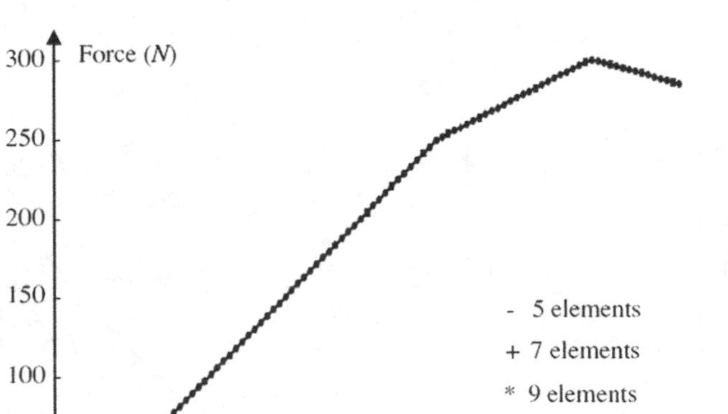

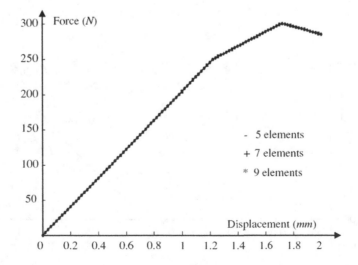

Figure 7.35 Load–displacement diagram with a different number of elements.

subjected to an increasing displacement (0.045 mm/s) at the right end until collapse is elaborated. The initial temperature is constant along the bar and equal to 0°C. Material properties of the bar are given in Table 7.2. Due to a problem in manufacturing, the ultimate stress at the middle point reduces to 299 MPa instead of 300 MPa in other parts (see Figure 7.36).

This problem is solved with two different meshes: five elements and nine elements. In these two meshes, the middle element represents the zone with a smaller ultimate stress (σ_u = 299 MPa). The localized failure will, therefore, occur in this element. The computed load–displacement diagram of the bar is given in Figure 7.37, while the evolution of temperature in the bar is shown in Figures 7.38 and 7.39.

The computed results clearly demonstrate the heating effect produced by the mechanical dissipation. Namely, the plastic dissipation equals the heat supply leading to the temperature increase. Initially, the dissipation in FPZ is equally distributed along the bar, so that the temperature at every part of the bar remains the same. However, with the start of localized failure, additional dissipation at the failure point acts as a concentrated heat supply. This further leads to a heat transfer process in the bar and results in the evolution of temperature, as shown in Figures 7.38 and 7.39.

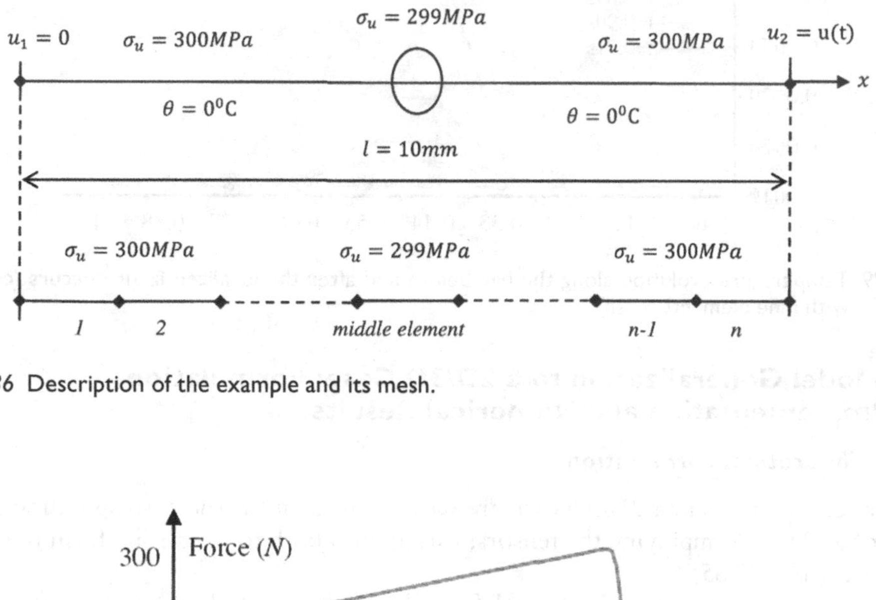

Figure 7.36 Description of the example and its mesh.

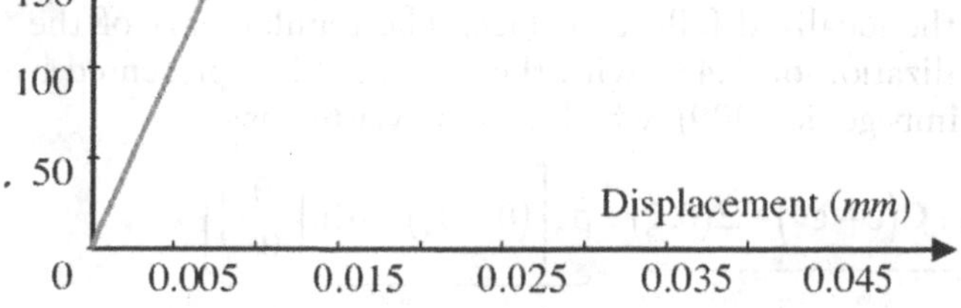

Figure 7.37 Load–displacement curve.

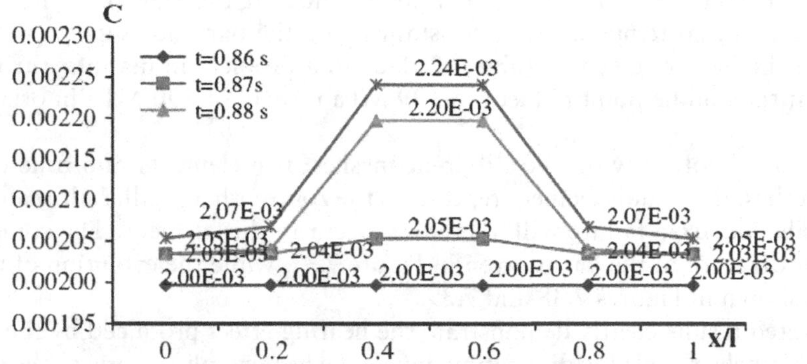

Figure 7.38 Temperature evolution along the bar before and after the localized failure occurs (computed with five elements mesh).

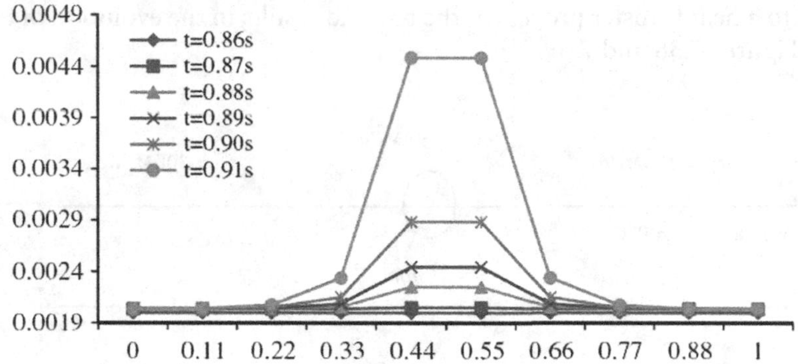

Figure 7.39 Temperature evolution along the bar before and after the localized failure occurs (computed with nine elements mesh).

7.8.2 Model Generalization to a 2D/3D Case: Formulation, Implementation and Numerical Results

7.8.2.1 Theoretical Formulation

The balance equations in a 2D/3D case are written the same as the corresponding generalization of a 1D case employing the tensors; this applies both to the strong form in (7.63) or the weak form in (7.85).

The first ingredient of the theoretical formulation in a 2D/3D case is the free energy, which can be written as an additive decomposition of the regular and irregular part:

$$\psi = \bar{\psi}\left(\varepsilon,\varepsilon^{p},\xi,\theta\right)+\delta_{\Gamma_s}(x)\bar{\bar{\psi}}\left(\bar{\bar{\xi}}\right) \tag{7.136}$$

where Γ_s denotes the localized failure surface. The regular part of the free energy is the appropriate generalization of (7.43), with the strain fields represented by the second-order tensors (e.g. Ibrahimbegović 2009), which can be written as:

$$\bar{\Psi} = \underbrace{\frac{1}{2}\left(\varepsilon-\varepsilon^{p}\right):\mathrm{C}\left(\varepsilon-\varepsilon^{p}\right)+\Xi(\theta,\xi)}_{\Psi_m}+\underbrace{\rho c\left[\left(\theta-\theta_0\right)-\theta\ln\left(\frac{\theta}{\theta_0}\right)\right]}_{\Psi_t} \\ \times\underbrace{-\left(\theta-\theta_0\right)\beta:\left(\varepsilon-\varepsilon^{p}\right)}_{\Psi_{th}} \tag{7.137}$$

The constitutive equations for the stress tensor are the tensor generalization of the 1D result in (7.44):

$$\sigma := \frac{d\Psi}{d\varepsilon} = \mathrm{C}\left(\varepsilon - \varepsilon^p\right) - \left(\theta - \theta_0\right)\beta \quad (7.138)$$

where the second-order stress tensor σ is defined in terms of the fourth-order elasticity tensor C and the second-order strain tensors ε, ε^p, along with the second-order tensor of thermal stress β. It is noted that in a 2D/3D case, the temperature is still a scalar field, and so is the elastic entropy:

$$\eta^e := -\frac{d\Psi}{d\theta} = \beta : \left(\varepsilon - \varepsilon^p\right) + \rho c \ln\left(\frac{\theta}{\theta_0}\right) \quad (7.139)$$

The second main ingredient of the model is the yield criterion, indicating the creation of the FPZ, which can be written in terms of the chosen norm of the stress tensor and hardening variable q:

$$\phi(\sigma, q, \theta) = \|\sigma\| - \left(\sigma_y - q\right) \le 0; q := -\frac{\partial \Xi}{\partial \xi} = -\mathrm{K}\xi \quad (7.140)$$

The stress orthogonality condition in (7.81) can be generalized to 2D/3D to define the stress traction vector:

$$\mathrm{t}_{\Gamma_s} = \int_{\Gamma_s} \tilde{G} \sigma_{\Gamma_s} d\Gamma_s \quad (7.141)$$

The latter is then used to define the corresponding yield criteria indicating the localized failure:

$$\bar{\bar{\phi}}_1(\sigma, q, \theta) = \mathrm{t}_{\Gamma_s} \cdot \mathrm{n} - \left(\bar{\bar{\sigma}}_f - \bar{\bar{q}}\right) \le 0 \quad (7.142)$$

where $\bar{\bar{\sigma}}_f$ is the ultimate stress at the start of localized failure, and $\bar{\bar{q}}$ is the softening variable controlling the evolution of plasticity threshold; the latter can be computed by assuming the simplest quadratic form of the softening potential:

$$\bar{\bar{\psi}}\left(\bar{\bar{\xi}}, \theta_1\right) = \frac{1}{2}\bar{\bar{K}}\left(\theta_1\right)\bar{\bar{\xi}}^2 \rightarrow \bar{\bar{q}}(\theta) := -\frac{d\bar{\bar{\psi}}\left(\bar{\bar{\xi}}\right)}{d\bar{\bar{\xi}}} = \bar{\bar{K}}\bar{\bar{\xi}} \quad (7.143)$$

The second law of thermodynamics can be used to obtain the total dissipation written as the additive decomposition of the regular part and irregular part in terms of distribution:

$$\mathcal{D}_{\text{int}} = \underbrace{\theta \dot{\eta}^p + \sigma : \dot{\varepsilon}^p + q\dot{\xi}}_{\bar{\mathcal{D}}_{\text{int}}} + \underbrace{\left(\bar{\bar{q}}\dot{\bar{\bar{\xi}}}\right)}_{\bar{\bar{\mathcal{D}}}_{\text{int}}} \delta_{\Gamma_s} \quad (7.144)$$

7.8.2.1.1 Finite Element Approximation

Once the localized failure occurs over surface Γ_s, the displacement field can be split into a regular part and an irregular part, which controls the strong discontinuity at the localized failure surface:

$$u(x,t) = \bar{u}(x,t) + \bar{\bar{u}}(x,t)\left(\mathcal{H}_{\Gamma_s}(x) - \varphi(x)\right) \quad (7.145)$$

The complete domain Ω is thus decomposed into two sub-domains Ω^+ and Ω^-, and the corresponding choice of $\varphi(x)$ is made to limit the influence of the particular discontinuity.

The discrete approximation is constructed by using the structured finite element mesh so that one again needs the embedded discontinuities to represent the localized failure surface. In term of the finite element approximation, the "jump" in the displacement field due to the localized failure is presented using a method of incompatible modes (e.g. Baker and Borst 2005; Bathe 1996). In particular, for a 2D case, and the simplest choice of discrete approximation based upon the three node CST element (e.g. see Bathe 1996), one can write the corresponding FE interpolation of displacement field:

$$\mathbf{u}(x,y) = \sum_{a=1}^{3} N_a(x,y)\mathbf{d}_a + M_1(x,y)\bar{\bar{\mathbf{u}}} + M_2(x,y)\alpha_2^a \tag{7.146}$$

where:

N_a are standard linear polynomial shape functions;
$M_1(x)$ is the approximation of displacement jump, and
$M_2(x)$ is the jump in the derivative.

It is noted that the choice of these incompatible modes is set with respect to the localized failure surface (see Figure 7.40).

A similar enhancement is chosen for constructing the discrete approximation of the temperature field within the element that is crossed by the localized surface failure. More precisely, it can be written:

$$\theta(x,y) = \sum_{a=1}^{3} N_a(x,y)d_a^{\theta} + M_2(x,y)\alpha_2^a = \mathbf{N}_a\mathbf{d}_a^{\theta} + M_2(x,y)\alpha_2^a \tag{7.147}$$

The final product of the finite element discretization is a set of nonlinear algebraic equations, with extra unknowns for each element crossed by the localized failure surface. The system is fully equivalent to the system in (7.96). Given this, one can again exploit the adiabatic operator split solution procedure by closely following the steps of development already presented in detail for a 1D case.

7.8.2.1.2 Numerical Results

Evaluating the fire resistance of RC structures in current practice relies upon the standard fire test, specified by ISO 834. However, the standard fire resistance tests should not be the only ones relied upon in order to determine the survival time of concrete elements under realistic fire scenarios, since they can often define less severe heating environments than

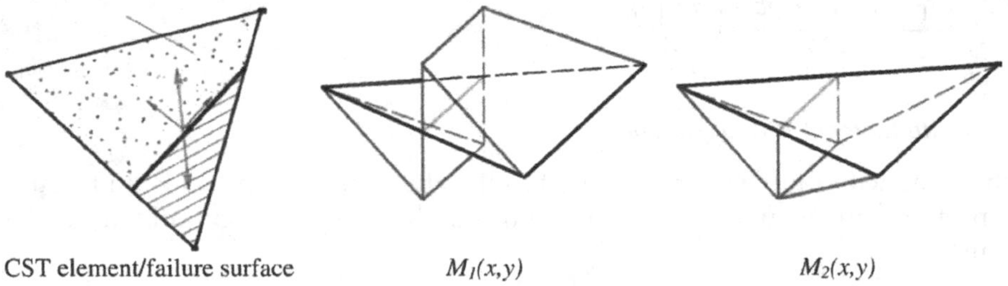

Figure 7.40 Additional shape functions for representing displacement and temperature fields.

real fires. One example from the nuclear industry with a more demanding fire scenario, as illustrated in Figure 7.41, considers more rapid time-evolutions of the temperature and heat flux than the standard tests.

The first example deals with plain concrete. More precisely, a concrete slab is considered (with no reinforcement) with dimensions of 2350 mm×2350mm×300 m. The behavior of the slab is modeled using the thermoplasticity model proposed in Ibrahimbegović, Colliat and Davenne (2005), which considers that the yield criterion is set in terms of the principal values of stress. The aim is to check if the proposed thermoplasticity model for localized failure can reproduce the corresponding fire-induced damage conditions in a slab submitted to a temperature gradient through the thickness. For that reason, the slab is heated on one side using the critical fire scenario for nuclear installation shown in Figure 7.41. This produces the bending effect in the slab and the typical damage pattern of tensile failure, as shown in Figure 7.42. It can be seen that the localized failure surface (presented by a corresponding red lines approximation in each element) is typical of slab failure in bending.

One can provide a more thorough analysis of the computed result. Namely, the chosen scenario characterizes fires where very high temperatures are reached in a few seconds. Due to this quick thermal load increase, leaving no time for diffusion, a high thermal gradient is located in a thin layer at the bottom of the slab. This kind of loading generates compressive stresses at the

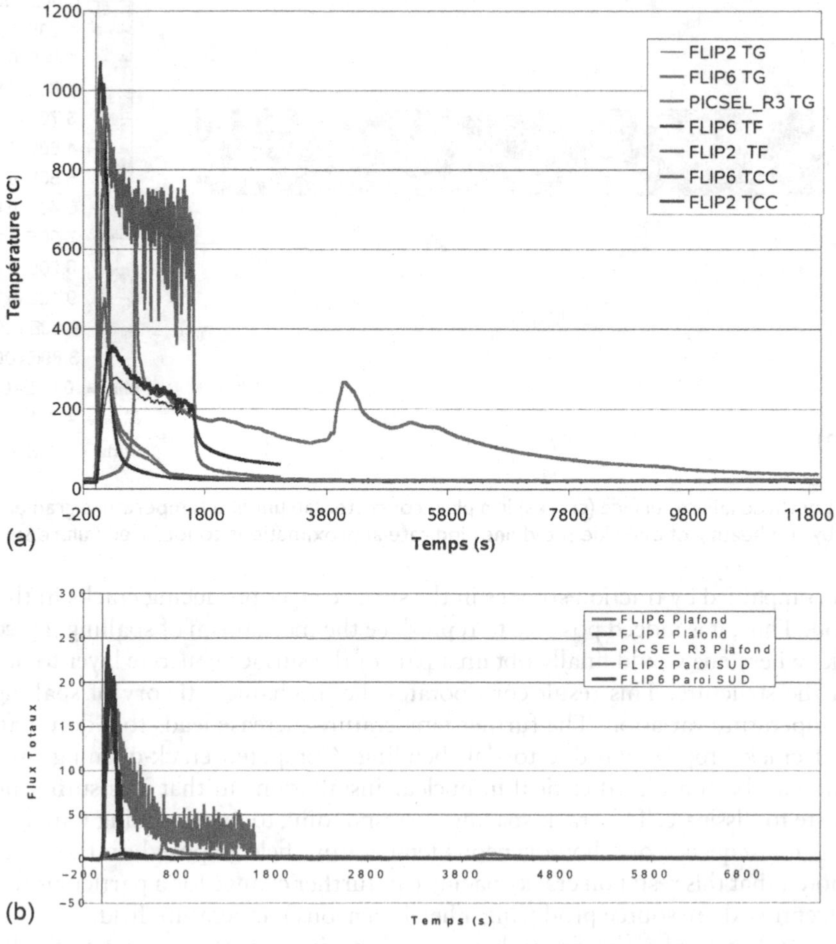

Figure 7.41 Fire scenario for nuclear installation in terms of temperature and flux variations.

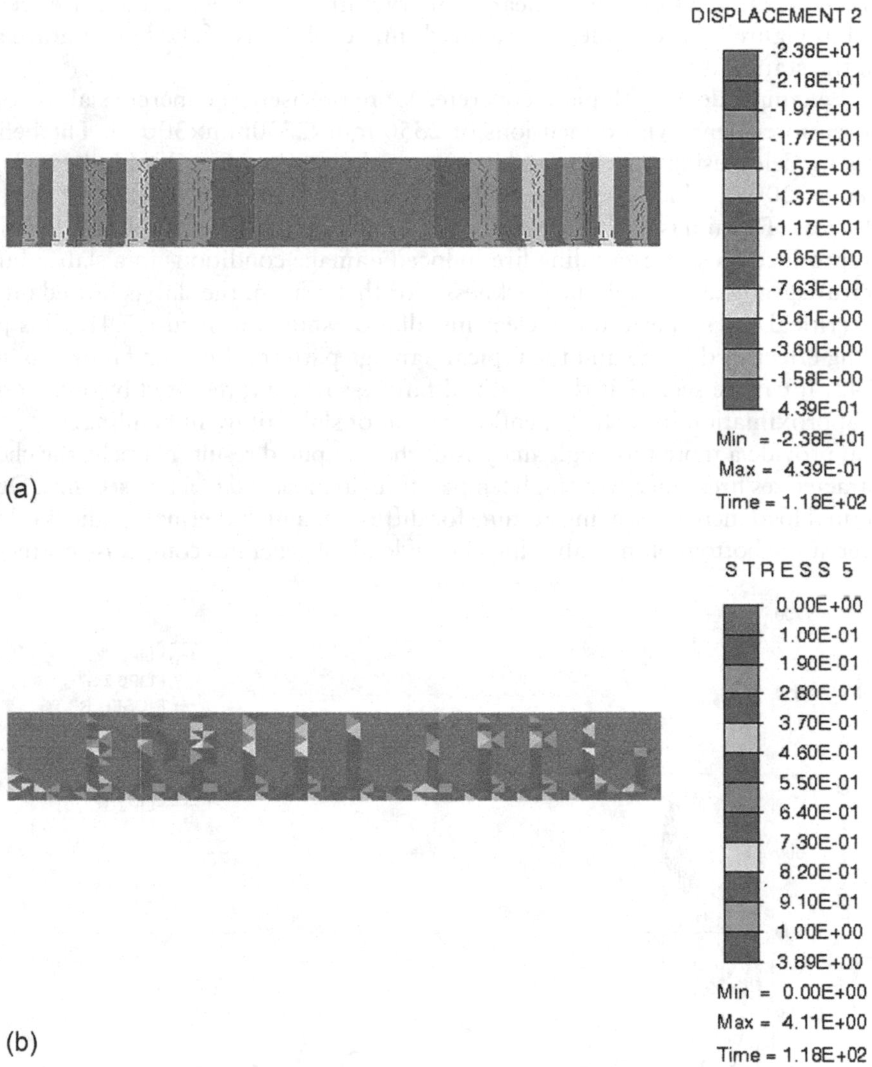

Figure 7.42 Localized failure surface (cracks) in a plain concrete slab under a temperature gradient produced by the heating of one side (bold lines indicate approximations to localized failure surface).

bottom, accompanied by traction stresses in the surface layer producing cracks in the horizontal direction. Thus, it is indeed possible to reproduce the mechanism of spalling, by connecting these cracks where one would finally obtain a part of the surface concrete layer to detach from the rest of the structure. This result corroborates the mechanical theory of spalling induced by steep temperature variation. The further temperature increase leads to a dominant vertical direction of crack propagation due to slab bending. Computed crack-opening, in the range of 1–4 mm, can be considered critical in nuclear installations in that it is sufficient to allow nuclear waste to dissipate. The crack-spacing corresponding to pure bending remains uniform, which is the consequence of a homogeneous temperature field imposed on the bottom of the slab. It is noted that this result on crack-spacing can further change for a particular fire scenario with a concentrated fire source producing a heterogeneous temperature field.

In the second example, the fire-induced cracking in a reinforced concrete slab is considered, which illustrates the positive influence of reinforcement in preventing or limiting

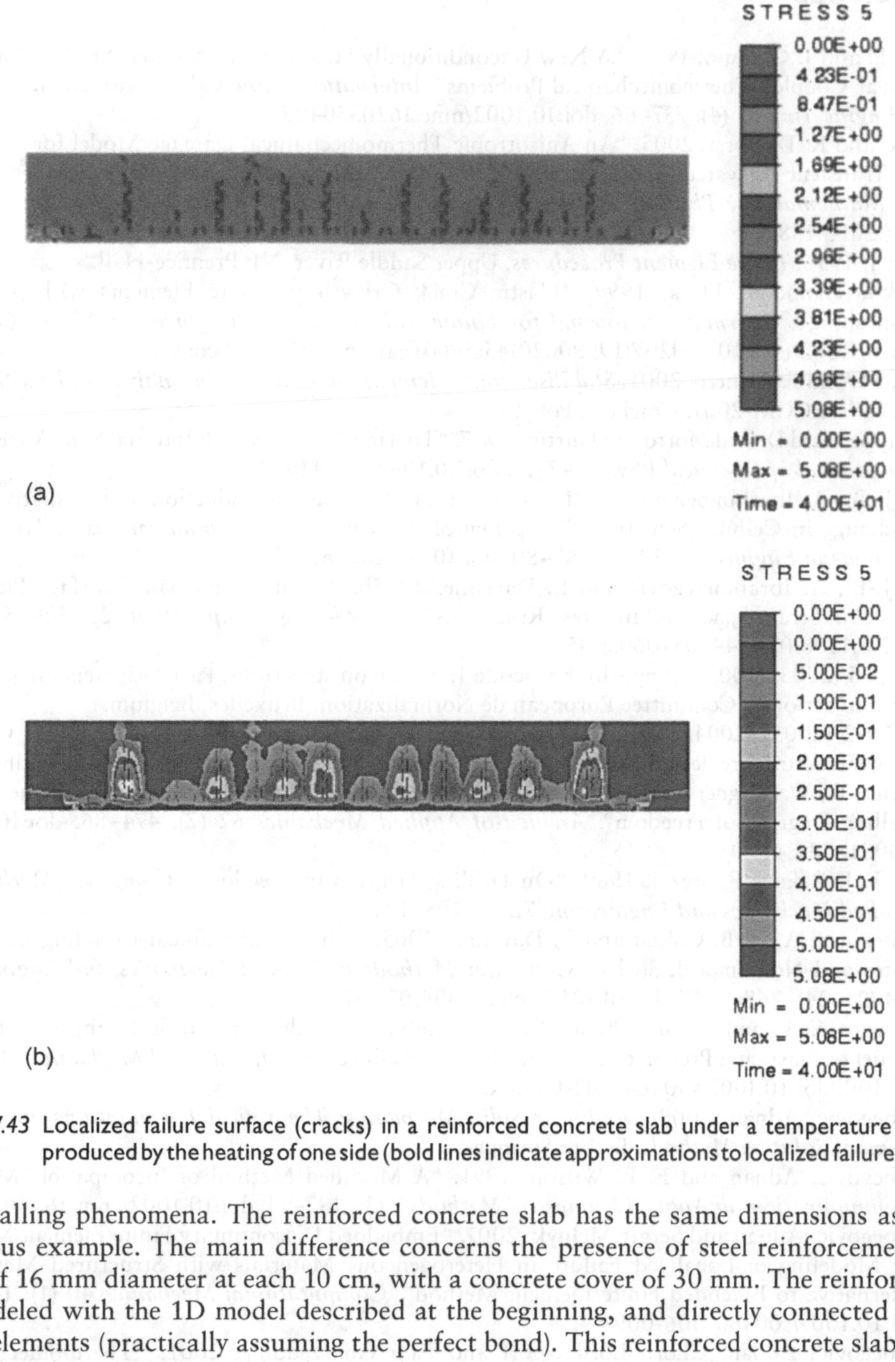

Figure 7.43 **Localized failure surface (cracks) in a reinforced concrete slab under a temperature gradient produced by the heating of one side (bold lines indicate approximations to localized failure surface).**

the spalling phenomena. The reinforced concrete slab has the same dimensions as in the previous example. The main difference concerns the presence of steel reinforcement with bars of 16 mm diameter at each 10 cm, with a concrete cover of 30 mm. The reinforcement is modeled with the 1D model described at the beginning, and directly connected to concrete elements (practically assuming the perfect bond). This reinforced concrete slab is then loaded with critical thermal loading as shown in Figure 7.41, and the computed results are presented in Figure 7.43.

In comparison with the first example, with a plain concrete slab, the crack pattern for a reinforced concrete slab is much less pronounced, which confirms that the reinforcement plays an important role. First, the vertical cracks are fewer and of smaller width. Second, there are no horizontal cracks which connect, and thus no mechanical spalling would occur in a case when reinforcement is used to strengthen the concrete slab.

REFERENCES

Armero, F. and J. C. Simo. 1992. "A New Unconditionally Stable Fractional Step Method for Non-Linear Coupled Thermomechanical Problems". *International Journal for Numerical Methods in Engineering* 35 (4): 737–66. doi:10.1002/nme.1620350408.

Baker, G. and R. D. Borst. 2005. "An Anisotropic Thermomechanical Damage Model for Concrete at Transient Elevated Temperatures". *Philosophical Transactions of the Royal Society A: Mathematical, Physical and Engineering Sciences* 363 (1836): 2603–28. doi:10.1098/rsta.2005.1589.

Bathe, K. J. 1996. *Finite Element Procedures*. Upper Saddle River, NJ: Prentice-Hall.

Belytschko, T. and T. Black. 1999. "Elastic Crack Growth in Finite Elements with Minimal Remeshing". *International Journal for Numerical Methods in Engineering* 45 (5): 601–20. doi:10.1002/(sici)1097-0207(19990620)45:5<601::aid-nme598>3.0.co;2.

Bischoff, M. and F. Armero. 2001. *Stabilised finite element methods for the analysis of heat transfer in shells*. ECCM-2001. Crackow, Poland.

Coleman, Bernard D. and Morton E. Gurtin. 1967. "Thermodynamics with Internal State Variables". *The Journal of Chemical Physics* 47 (2). doi:10.1063/1.1711937.

Colliat, J. B., A. Ibrahimbegović and L. Davenne. 2005a. "Heat Conduction and Radiative Heat Exchange in Cellular Structures Using Flat Shell Elements". *Communications in Numerical Methods in Engineering* 22 (3): 167–80. doi:10.1002/cnm.784.

Colliat, J.-B., A. Ibrahimbegović and L. Davenne. 2005b. "Saint-Venant Multi-surface Plasticity Model in Strain Space and in Stress Resultants". *Engineering Computations* 22 (5/9): 536–57. doi:10.1108/02644400510603005.

EN 1991-1-5 (2003). 2003. "(English): Eurocode 1: Action on structures. Part 1–5: General actions – thermal actions". Committee European de Normalization: Bruxelles, Belgium.

EN 1993-1-2 (2005). 2004. "(English): Eurocode 3: Design of steel structures. Part 1–2: General rules–Structure fire design". Committee European de Normalization: Bruxelles, Belgium.

Gruttmann, F., W. Wagner and P. Wriggers. 1992. "A Nonlinear Quadrilateral Shell Element with Drilling Degrees of Freedom". *Archive of Applied Mechanics* 62 (7): 474–86. doi:10.1007/bf00810238.

Hughes, T. J. R. and F. Brezzi. 1989. "On Drilling Degrees of Freedom". *Computer Methods in Applied Mechanics and Engineering* 72 (1): 105–121.

Ibrahimbegović, A., J. B. Colliat and L. Davenne. 2005. "Thermomechanical Coupling in Folded Plates and Non-Smooth Shells". *Computer Methods in Applied Mechanics and Engineering* 194 (21–24): 2686–707. doi:10.1016/j.cma.2004.07.052.

Ibrahimbegović, A. and D. Brancherie. 2003. "Combined Hardening and Softening Constitutive Model of Plasticity: Precursor to Shear Slip Line Failure". *Computational Mechanics* 31 (1–2): 88–100. doi:10.1007/s00466-002-0396-x.

Ibrahimbegović, Adnan. 2009. *Nonlinear Solid Mechanics: Theoretical Formulations and Finite Element Solution Methods*. Berlin: Springer.

Ibrahimbegović, Adnan and E. L. Wilson. 1991. "A Modified Method of Incompatible Modes". *Communications in Applied Numerical Methods* 7 (3): 187–94. doi:10.1002/cnm.1630070303.

Ibrahimbegović, Adnan and Sergiy Melnyk. 2007. "Embedded Discontinuity Finite Element Method for Modeling of Localized Failure in Heterogeneous Materials with Structured Mesh: An Alternative to Extended Finite Element Method". *Computational Mechanics* 40 (1): 149–55. doi:10.1007/s00466-006-0091-4.

Ibrahimbegović, Adnan, Chorfi Lotfi Chorf and Fadi Gharzeddine. 2001. "Thermomechanical Coupling at Finite Elastic Strain: Covariant Formulation and Numerical Implementation". *Communications in Numerical Methods in Engineering* 17 (4): 275–89. doi:10.1002/cnm.405.

Ibrahimbegović, Adnan, Robert L. Taylor and Edward L. Wilson. 1990. "A Robust Quadrilateral Membrane Finite Element with Drilling Degrees of Freedom". *International Journal for Numerical Methods in Engineering* 30 (3): 445–57. doi:10.1002/nme.1620300305.

Ibrahimbegović, Adnan and François Frey. 1993. "An Efficient Implementation of Stress Resultant Plasticity in Analysis of Reissner–Mindlin Plates". *International Journal for Numerical Methods in Engineering* 36 (2): 303–20. doi:10.1002/nme.1620360209.

Ibrahimbegović, Adnan and François Frey. 1994. "Stress Resultant Geometrically Nonlinear Shell Theory with Drilling Rotations—Part II. Computational Aspects". *Computer Methods in Applied Mechanics and Engineering* 118 (3–4): 285–308. doi:10.1016/0045-7825(94)90004-3.

Ibrahimbegović, Adnan, Fadi Gharzeddine and Lotfi Chorfi. 1998. "Classical Plasticity and Viscoplasticity Models Reformulated: Theoretical Basis and Numerical Implementation". *International Journal for Numerical Methods in Engineering* 42 (8): 1499–535. doi:10.1002/(sici)1097-0207(19980830)42:8<1499::aid-nme443>3.3.co;2-o.

Kassiotis, C., J.-B. Colliat, A. Ibrahimbegovic and H. G. Matthies. 2009. "Multiscale in Time and Stability Analysis of Operator Split Solution Procedures Applied to Thermomechanical Problems". Edited by Adnan Ibrahimbegović. *Engineering Computations* 26 (1/2): 205–23. doi:10.1108/02644400910924870.

Lewis, R. W., K. Morgan, R. H. Thomas and K. N. Seetharamu. 1996. *The Finite Element Method in Heat Transfer Analysis.* New York: John Wiley.

Libai, Avinoam and James G. Simmonds. 2005. *The Nonlinear Theory of Elastic Shells*. 2. Cambridge: Cambridge University Press.

Naghdi, P. M. 1972. *Theory of Shells, Handbuch der Physik*. Vol. VIa/2, New York: SpringerVerlag.

Ngo, Van-Minh, Adnan Ibrahimbegović and Delphine Brancherie. 2013. "Model for Localized Failure with Thermo-Plastic Coupling: Theoretical Formulation and ED-FEM Implementation". *Computers & Structures* 127: 2–18. doi:10.1016/j.compstruc.2012.12.0.

Simmonds, J. G. 2001. "A Simplified, Nonlinear Thermodynamic Theory of Beamshells". *Quarterly of Applied Mathematics* 59 (3): 401–12. doi:10.1090/qam/1848525.

Simo, J. C. and C. Miehe. 1992. "Associative Coupled Thermoplasticity at Finite Strains: Formulation, Numerical Analysis and Implementation". *Computer Methods in Applied Mechanics and Engineering* 98 (1): 41–104. doi:10.1016/0045-7825(92)90170-o.

Simo, J. C. and J. G. Kennedy. 1992. "On a Stress Resultant Geometrically Exact Shell Model. Part V. Nonlinear Plasticity: Formulation and Integration Algorithms". *Computer Methods in Applied Mechanics and Engineering* 96 (2): 133–71. doi:10.1016/0045-7825(92)90129-8.

Simo, J. C., D. D. Fox and M. S. Rifai. 1989. "On a Stress Resultant Geometrically Exact Shell Model. Part II: The Linear Theory; Computational Aspects". *Computer Methods in Applied Mechanics and Engineering* 73 (1): 53–59. doi:10.1016/0045-7825.

Simo, J. C., J. G. Kennedy and S. Govindjee. 1998. "Non-Smooth Multisurface Plasticity and Viscoplasticity. Loading/unloading Conditions and Numerical Algorithms". *International Journal for Numerical Methods in Engineering* 26 (10): 2161–85. doi:10.1002/nme.1620261003.

Stabler, J. and G. Baker. 2000. "On the Form of Free Energy and Specific Heat in Coupled Thermo-Elasticity with Isotropic Damage". *International Journal of Solids and Structures* 37 (34): 4691–713. doi:10.1016/s0020-7683(99)00292-9.

Chapter 8

Fluid-induced extreme loads

Among the multi-physics problems that are currently entering the mainstream of scientific research in computational mechanics (e.g. see Ibrahimbegović and Brank 2005; Oden et al. 2003), perhaps the most frequently studied are the problems of fluid–structure interaction, some dealing with problems in medicine, (e.g. Barcelos, Bavestrello and Maute 2006; Bathe and Zhang 2009; Bazilevs et al. 2008; Bazilevs, Calo and Zhang et al. 2006; Bazilevs et al. 2009a,b,c; Causin, Gerbeau and Nobile 2005; Degroote, Bathe and Vieren 2009; Deparis et al. 2006; Dettmer and Perić 2007), among others. The fluid–structure interaction is already an interesting problem in its own right with a vast number of important applications. The goal is to conduct the direct coupling of different codes developed for a particular sub-problem (i.e. either solid or fluid mechanics) into a single code.

A novel partitioned approach for a nonlinear fluid–structure interaction will be presented here (Kassiotis, Ibrahimbegovic and Matthies 2010; Kassiotis et al. 2011a,b). A combination of the finite volume method for fluid and finite element method for the structure will be utilized. This partitioned approach has been preferred in the earlier attempts to solve this problem (Belytschko, Yen and Mullen 1979; Felippa and Park 2004; Hughes and Liu 1978). The instabilities in the previous procedures are eliminated by the application of implicit schemes for each sub-problem, but also for the partitioned coupling. The resulting algorithm provides the computational robustness, which is of great interest not only for fluid–structure interaction but also for any other multi-physics problems of current interest, where one would like to re-use the available codes for any particular sub-problem in a more general framework.

The main advantage of the code-coupling approach for fluid–structure interaction concerns the fact that the coupling is limited only to the fluid–structure interface. Therefore, the main difficulty is reduced to enforcing the interface matching in respect of the two different discretization schemes, the finite element versus the finite volume, as well as the two different time-integration schemes and different time steps and their matching at the interface.

The strategy of re-using the stand-alone software for particular sub-problems is likely to become the most efficient way for the development of software products for multi-physics applications.

8.1 STRUCTURE AND FLUID FORMULATIONS

8.1.1 Structure Equation of Motion

Structure motion is based on the Lagrangian description. Namely structure domain Ω_s is considered with imposed displacements $\bar{u}$ on the Dirichlet boundary $\partial\Omega_{s,D}$ and moving under the loading of traction forces $\bar{t}$ on the Neumann boundary $\partial\Omega_{s,N}$and body force

$\bar{\boldsymbol{b}}$ applied in the whole domain Ω_s. This dynamic motion has to be computed in the time interval [0, T]. The Cauchy governing equation for the structure describes the momentum conservation. The strong form of this equation can be written with respect to the deformed configuration as follows; given $\bar{\boldsymbol{u}}$ on $\partial\Omega_{s,D}\times[0,T]$, $\bar{\boldsymbol{t}}$ and $\bar{\boldsymbol{b}}\times[0,T]$ in, find: $\boldsymbol{u}\in\Omega_s\times[0,T]$ u ∈ _s× [0, T] so that:

$$\nabla\cdot\underbrace{J\boldsymbol{\sigma}\mathbf{F}^{-T}}_{\mathbf{P}}+\rho_s\left(\bar{\boldsymbol{b}}-\partial_t^2\boldsymbol{u}\right)=0 \text{ in } \Omega_s\times[0,T] \tag{8.1}$$

where ρ_s denotes the material density of the solid domain, $\boldsymbol{u}$ its displacement field and $\partial_t^2\boldsymbol{u}$ the accelerations. The Cauchy stress tensor $\boldsymbol{\sigma}$ in the deformed configuration can be linked to the first Piola–Kirchhoff stress tensor **P**, formulated in the initial configuration through the gradient F of the deformation and its Jacobian J (e.g. Ibrahimbegović 2009).

To close this partial differential equations system, the displacements (or rather its derivatives) are linked with the stresses through constitutive law. For instance, an elastic material model based on the Saint-Venant–Kirchhoff constitutive equation is assumed that links the Cauchy stress tensor $\boldsymbol{\sigma}$ and the Green-Lagrange strain tensor E through:

$$\mathbf{F}^{-1}J\boldsymbol{\sigma}\mathbf{F}^{-T}=\mathscr{C}:\mathbf{E},\mathbf{E}=\frac{1}{2}\left(\mathbf{F}^T\mathbf{F}-I\right) \text{ and } \mathbf{F}=\mathbf{I}+\nabla\boldsymbol{u} \tag{8.2}$$

where $\mathscr{C}$ denotes the constitutive fourth-order elasticity tensor. Thus, due to geometric nonlinearities it is *a priori* impossible to directly find an exact solution to the problem defined above. The goal is to find the best approximation of the solution in a finite-dimensional space where the solution can be found numerically. The FEM approximation (Belytschko, Liu and Moran 2000; Ibrahimbegović 2009; Zienkiewics and Taylor 2005) derives from weak forms of the equilibrium in (8.1) and can be written for this problem as given $\bar{\boldsymbol{t}}$ on $\partial\Omega_{s,N}\times[0,T]$ and $\bar{\boldsymbol{b}}$ in $\Omega_s\times[0,T]$, finds $\boldsymbol{u}\in\mathscr{U}$ such that, for all$\delta\boldsymbol{u}\in\mathscr{U}_0$:

$$\mathscr{G}_s(\boldsymbol{u};\delta\boldsymbol{u}):=\int_{\Omega_s}\rho_s\partial_t^2\boldsymbol{u}\cdot\delta\boldsymbol{u}+\int_{\Omega_s}\boldsymbol{\sigma}:\nabla\delta\boldsymbol{u}-\int_{\Omega_s}\bar{\boldsymbol{b}}\cdot\delta\boldsymbol{u}-\int_{\partial\Omega_s}\bar{\boldsymbol{t}}\cdot\delta\boldsymbol{u}=0$$

where $\mathscr{U}$ and $\mathscr{U}_0$ are functional spaces for the solution and its variation.

The solid domain Ω_s is then discretized in a finite number of sub-domains or elements $\mathscr{T}_h=(\kappa_e)_{e=1,\dots,n_{el}}$ so that the whole space is covered with the finite elements that do not intersect. The solution space associated with this approximation solution is restrained to the space of continuous element-wise polynomial functions, which is denoted:

$$\mathscr{U}^h=\mathscr{U}\cap\left\{\boldsymbol{u}\in\mathscr{C}^0(\Omega_s)\big|\,\boldsymbol{u}\big|_\kappa\in\mathscr{P}^p(\kappa),\forall\kappa\in\mathscr{T}_h\right\} \tag{8.3}$$

where $\mathscr{P}^p(\kappa)$ is the space of polynomials of order p on κ. The same restriction $\mathscr{U}_0^h$ holds on the associated vector space. The semi-discretized FE problem is defined as: given $\bar{\boldsymbol{t}}$ on $\partial\Omega_{s,N}$ and $\bar{\boldsymbol{b}}$ in Ω_s, find $\boldsymbol{u}\in\mathscr{U}^h$ such that, for all $\delta\boldsymbol{u}\in\mathscr{U}_0^h$:

$$\mathscr{G}_s(\mathbf{u};\delta\mathbf{u})=0 \tag{8.4}$$

This semi-discrete problem can also be written in matrix notation by using the real valued vectors $\mathbf{u}\in\mathbb{R}^{nd\text{-}o\text{-}f}$:

$$\mathscr{R}_s(\mathbf{u}_s;\lambda):=\mathbf{M}_s\ddot{\mathbf{u}}_s+\mathbf{f}_s^{\text{int}}(\mathbf{u}_s)-\mathbf{f}_s^{\text{ext}}(\lambda)=0 \tag{8.5}$$

where:

$\mathbf{M}_s$ is the mass matrix

$\mathbf{f}_s^{\text{int}}$ is the internal force which is highly nonlinear if large deformation or complex material behavior is used and

$\mathbf{f}_s^{\text{ext}}$ is the external force vector.

Here the λ represents the boundary forces computed from the fluid flow problem and imposed on the fluid–structure interface. Each matrix and vector of this semi-discrete equation is properly defined by assembling locally computed array in each element with the polynomial basis N_e of $\mathscr{P}(\kappa_e)$:

$$\mathbf{M}_{s,e} = \int_{\kappa_e} \rho_s N_e^T N_e$$

$$\mathbf{f}_{s,e}^{\text{int}}(\mathbf{u}_e) = \int_{\kappa_e} \nabla N : \mathbf{P}(\mathbf{u}_e N_e) \tag{8.6}$$

$$\mathbf{f}_{s,e}^{\text{ext}} = \int_{\kappa_e} N_e^T \bar{b}_e + \int_{\partial\kappa_e} N_e^T \bar{t}_e$$

In order to complete the discretization process, the time-integration of the structural problem can be carried out by using standard time-stepping schemes (Belytschko 1983; Hughes, Pister and Taylor 1979). In particular, the generalized HHT-α method (Chung and Hulbert 1994) is used herein. The time interval $[0,T]$ is discretized into a finite number of time steps t_N such as t_0=0 and $t_{N_{\max}} = T$. In a typical time step $\Delta t = t_{N+1} - t_N$, the time derivatives of nodal displacement are approximated with:

$$\begin{aligned}
\mathbf{u}_{N+1} &= \mathbf{u}_N + \Delta t \dot{\mathbf{u}}_N + \Delta t^2 \left[\left(\frac{1}{2} - \beta\right)\ddot{\mathbf{u}}_N + \beta \ddot{\mathbf{u}}_{N+1}\right] \\
\dot{\mathbf{u}}_{N+1} &= \dot{\mathbf{u}}_N + \Delta t\left[(1-\gamma)\ddot{\mathbf{u}}_N + \gamma \ddot{\mathbf{u}}_{N+1}\right] \\
\mathbf{u}_{N+\alpha_f} &= (1-\alpha_f)\mathbf{u}_N + \alpha_f \mathbf{u}_{N+1} \\
\ddot{\mathbf{u}}_{N+\alpha_m} &= (1-\alpha_m)\ddot{\mathbf{u}}_N + \alpha_m \ddot{\mathbf{u}}_{N+1}
\end{aligned} \tag{8.7}$$

In the semi-discrete form of the solid equation of motion in (8.5), the acceleration $\ddot{\mathbf{u}}$ and the displacement $\mathbf{u}$ are evaluated at $t_{N+\alpha_f}$ and $t_{N+\alpha_m}$. For the elastic linear case, it is shown (Chung and Hulbert 1994) that there are optimum values for the parameters $\beta, \gamma, \alpha_f, \alpha_m$ for a given spectral radius $\rho_\infty \in [0,1]$ρ∞ ∈ [0, 1].

$$\begin{aligned}
\beta &= \frac{(1+\alpha_m-\alpha_f)^2}{4}, \gamma = \frac{1}{2} + \alpha_m - \alpha_f \\
\alpha_f &= \frac{1}{1+\rho_\infty}, \alpha_m = \frac{2-\rho_\infty}{1+\rho_\infty}
\end{aligned} \tag{8.8}$$

The spectral radius controls the numerical damping of the time-integration scheme. The damping decreases with smaller values of ρ_∞ which is maximum for ρ_∞= 0. For ρ_∞= 1 the method is the classic trapezoidal rule. Other time-integration schemes can be easily derived from this general formulation (Hilber, Hughes and Taylor 1977).

8.1.2 Free-Surface Flow

An Arbitrary Lagrangian–Eulerian description of a two-phase flow (taking into account both water and the surrounding air) is utilized when discretized by finite volume (e.g. Ferziger and Perić 2002). In this volume of fluid (V.O.F.) method a volume fraction is used to represent the interface; the main remaining issue is how to convert the interface without diffusing, dispersing or wrinkling it. V.O.F. methods use precise convection schemes that reconstruct the interface from the volume fraction distribution before advection.

For complex flows (with jets, cavitation and aeration in the sloshing wave), it is natural to consider the Navier– Stokes equations for two immiscible and incompressible flows (water and air, for instance) occupying transient domains $\Omega_i(t)$ so that the whole fluid domain is considered as $\Omega_f(t) = \Omega_1(t) \cup \Omega_2(t)$. The interface between both domains Ω_1 and Ω_2 is denoted as Γ. In space–time domain $\Omega_f(t) \times [0,T]$, the Navier–Stokes equations formulated in an ALE framework apply. For an arbitrary motion of the total fluid domain Ω_f described by a displacement field $\boldsymbol{u}_m$, it can be written as:

$$\begin{aligned} &\rho\partial_t \upsilon + \rho(\boldsymbol{\upsilon} - \dot{\boldsymbol{u}}_m)\nabla \cdot \boldsymbol{\upsilon} - \nabla \cdot 2\mu D(\boldsymbol{\upsilon}) = \\ &= -\nabla p + f_\Gamma + \rho g \text{ in } \Omega_f(t) \times [0,T] \\ &\nabla \cdot \boldsymbol{\upsilon} = 0 \qquad \text{in } \Omega_f(t) \times [0,T] \end{aligned} \tag{8.9}$$

where p denotes the pressure field and $\boldsymbol{\upsilon}$ the velocity. Symbol $\boldsymbol{g}$ is introduced which depicts the gravity field, with f_Γ showing the surface tension forces. It can be expressed as $f_\Gamma = \sigma\kappa\delta_\Gamma$, where σ is the surface tension, κ the curvature of the free-surface (i.e. interface between Ω_1 and Ω_2) and δ_Γ the mass distribution concentrated at the surface (equivalent to a Dirac distribution). Fluid material properties are the dynamic viscosity μ and the density ρ. To write a unique formulation in the whole domain $\Omega_f(t)$, we express the local material property values as the function of ι:

$$\rho = \iota\rho_1 + (1-\iota)\rho_2 \text{ and } \mu = \iota\mu_1 + (1-\iota)\mu_2 \text{ in } \Omega_f(t) \times [0,T] \tag{8.10}$$

where the characteristic function or fluid volume fraction ι is defined as:

$$\iota(\boldsymbol{x},t) = \begin{cases} 1, & \text{for } \boldsymbol{x} \in \Omega_1(t) \\ 0, & \text{for } \boldsymbol{x} \in \Omega_2(t) \end{cases} \tag{8.11}$$

It is observed that when the whole domain is filled with one fluid $(\iota = 1 \in \Omega_f)$, the classical Navier–Stokes equations in the ALE framework appear. The fluid volume fraction ι and the mass distribution δ_Γ are linked by the following relation:

$$\nabla\iota = \delta_\Gamma \boldsymbol{n} \tag{8.12}$$

To close the set of equations the conservation of ι has to be written. When no reaction between phases occurs, the fluid volume fraction evolves only by advection:

$$\partial_t\iota + (\boldsymbol{\upsilon} - \dot{\mathbf{u}}_m)\nabla\iota = 0 \tag{8.13}$$

The conservation equation system on the whole domain Ω_f in d dimensions can be written as a function of the $2d+2$ unknowns: $\boldsymbol{u}_m, \upsilon, p$ and ι are the solutions for equations (8.9) and (8.10) are verified. Using a fixed (Eulerian) grid creates difficulties in the in fluid–structure interaction problems because of moving domains at the fluid interface which follows the deformations of the structure. Application of Arbitrary Lagrangian–Eulerian (ALE) method where the whole grid is moved inside the fluid domain leads to the difficulty of pertaining the quality and the validity of the fluid inner mesh for different new shapes of the boundary. This is solved by building a suitable map for the domain motion given its interface displacement. The fluid displacement $\boldsymbol{u}_m$ is arbitrary inside domain Ω_f, but on the boundary it has to fulfill the condition:

$$\boldsymbol{u}_m = \bar{\boldsymbol{u}}_m \text{ on } \partial\Omega_f(t)\times[0,T] \tag{8.14}$$

Inside the fluid domain Ω_f, the fluid displacement is an arbitrary extension of $\bar{\boldsymbol{u}}_m|_{\partial\Omega_f}$:

$$\boldsymbol{u}_m = \text{Ext}\left(\bar{\boldsymbol{u}}_m|_{\partial\Omega_f}\right) \tag{8.15}$$

The latter is possible to construct by solving the Laplace smoothing equation:

$$\nabla\cdot(\gamma\nabla\boldsymbol{u}_m) = 0 \text{ in } \Omega_f \tag{8.16}$$

This kind of Laplace smoothing equation is known to have some limitations when the deformation of the fluid domain is governed by large rotations. In this case, it shows sufficient quality performance when the diffusion coefficient is made spatially dependent upon the distance to the fluid–structure interface.

The second major difference from solids concerns the most popular discretization technique for fluid in terms of the finite volume method. The latter will transform the weak form of the continuous equations in (8.9) to (8.16) into a set of algebraic equations that can be solved numerically.

The FV formulation can be written directly using an integrated form of the conservation equation written in (8.9). Another possibility, which is utilized here, is to consider the restriction of the solution space of weak form problems. The weak form of the Navier–Stokes equation can be written as defined in (Glowinski 2003): find $(\boldsymbol{u}_m, \boldsymbol{\upsilon}, p \text{ and } \iota) \in \mathscr{U}\times\mathscr{I}\times\mathscr{V}\times\mathscr{P}$ such that, for all $(\delta\boldsymbol{u}_m, \delta\boldsymbol{\upsilon}, \delta p \text{ and } \delta\iota) \in \mathscr{U}_0\times\mathscr{I}_0\times\mathscr{V}_0\times\mathscr{P}_0 :\in \boldsymbol{U}\times\boldsymbol{I}\times\boldsymbol{V}\times\boldsymbol{P}$:

$$\begin{aligned}\mathscr{G}_f := &\int_{\Omega_f}\nabla\cdot(\gamma\nabla\boldsymbol{u}_m)\delta\boldsymbol{u}_m + \int_{\Omega_f}\left(\partial_t\iota + (\boldsymbol{\upsilon}-\boldsymbol{u}_m)\nabla\iota\right)\delta\iota \times \int_{\Omega_f}\rho\partial_t\boldsymbol{\upsilon}\cdot\delta\boldsymbol{\upsilon} + \\ &+\int_{\Omega_f}\rho\nabla(\boldsymbol{\upsilon}-\dot{\boldsymbol{u}}_m)\otimes\boldsymbol{\upsilon}\cdot\delta\boldsymbol{\upsilon} - \int_{\Omega_f}\nabla\cdot\mu_f\mathbf{D}(\boldsymbol{\upsilon})\cdot\delta\boldsymbol{\upsilon} + \int_{\Omega_f}p\nabla\cdot\delta\boldsymbol{\upsilon} + \\ &+\int_{\Omega_f}\nabla\cdot\boldsymbol{\upsilon}\delta p + [\text{B.C. in a weak form}] = 0\end{aligned} \tag{8.17}$$

where $\mathscr{U},\mathscr{I},\mathscr{V}$, and $\mathscr{P}$ are suitable functional spaces for the solution fields.

For this method, the whole of volume Ω_f is divided into a set of discrete elements, here called discrete volumes $(\kappa_{f,e})_{e=1,n_{el}}$ covering the whole domain ($\Omega_f = \bigcup_{e=1}^{n_{el}}\kappa_{f,e}$) without overlapping $\bigcap_{e=1}^{n_{el}}\kappa_{f,e} = \emptyset$). For finite element discretization, the solution spaces are restricted to suitable spaces of piecewise polynomial functions over the set of discrete elements. For the finite volume discretization used herein, the test functions were chosen in the space of the characteristic discrete volume functions. For instance, the velocity can be approximated as:

$$\mathscr{V}^h = \mathscr{V}\cap\left\{\boldsymbol{\upsilon}|\boldsymbol{\upsilon}|_\kappa \in \text{span}(\iota_\kappa), \forall\kappa\mathscr{I}^h(\Omega_f)\right\} \tag{8.18}$$

where ι_κ is the characteristic function of the element defined as:

$$\iota_\kappa : \Omega_f \to \mathbb{R}$$
$$x \to \begin{cases} 1 & \text{if } x \in \kappa \\ 0 & \text{if } x \in \Omega / \kappa \end{cases} \tag{8.19}$$

The same kind of restriction holds for the solution spaces of (8.17). Therefore, the functions are piecewise constant by elements, and with the restriction of the weak formulation it gives: $\left(u_m^h, \iota^h, v^h, p^h\right) \in \mathscr{U}^h \times \mathscr{I}^h \times \mathscr{V}^h \times \mathscr{V}^h$, such that, for all: $\left(\delta u_m^h, \delta\iota^h, \delta v^h, \delta p^h\right) \in \mathscr{U}_0^h \times \mathscr{I}_0^h \times \mathscr{V}_0^h \times \mathscr{V}_0^h : \mathscr{G}_f\left(\left(u_m^h, \iota^h, v^h, p^h\right); \delta u_m^h, \delta\iota^h, \delta v^h, \delta p^h\right) = 0$. The divergence terms in (8.17) can be written in terms of flux at the boundary of the volume controls using the Gauss theorem.

Hence, the weak formulation can be written as:

$$0 = \sum_\kappa \left\{ \oint_{\partial\kappa} d\boldsymbol{\Gamma} \cdot (\gamma \nabla u_m) \right\} - [B.C.]$$

$$0 = \sum_\kappa \left\{ \int_\kappa \partial_t \iota + \oint_{\partial\kappa} d\boldsymbol{\Gamma} \cdot (v - \dot{u}_m)\iota \right\} - [B.C.]$$

$$0 = \sum_\kappa \left\{ \int_\kappa \rho \partial_t v + \oint_{\partial\kappa} \rho d\boldsymbol{\Gamma} \cdot (v - \dot{u}_m) \otimes v - \oint_{\partial\kappa} d\boldsymbol{\Gamma} \cdot \mu_f \mathbf{D}(v) + \oint_{\partial\kappa} p d\boldsymbol{\Gamma} \right\} - [B.C.]$$

$$0 = \sum_\kappa \left\{ \oint_{\partial\kappa} d\boldsymbol{\Gamma} \cdot v \right\} - [B.C.]$$

where $d\boldsymbol{\Gamma}$ is the elementary surface vector. It is interesting to see that there is no continuity requirement for the solution (contrary to classical FE), and therefore the approximate solutions need not be defined at the interface. Their flux can be computed without being imposed by the restriction of the solution space to the FV space. The difficulty that arises is to build an accurate representation of the fluxes at the boundaries from a piecewise constant field. On each control volume, three levels of numerical approximations are applied to build the boundary fluxes: *interpolation* to express variable values at the control volume surface in terms of nodal values (depending on where the variable is stored); *differentiation* to build convective and diffusive fluxes the value of the gradient of the quantity of interest—or at least its approximation—is required; *integration* to approximate surface and volume integral using quadrature formulas.

The semi-discrete form of the discretized fluid problem can be written in a matrix form as follows. The fluid mesh motion considers that $\mathbf{u}_m$ is imposed by the motion of interface $\mathbf{u}$:

$$\mathscr{R}_m(\mathbf{u}_m, \mathbf{u}) := \mathbf{K}_m \mathbf{u}_m - \mathbf{D}_m \mathbf{u} = \mathbf{0} \tag{8.20}$$

where $\mathbf{D}_m$ is a projection/restriction operator and $\mathbf{K}_m$ governs the extension of the boundary displacement (see (8.16)). The volume fraction ι, the d components of velocity $\mathbf{v}$ and pressure p are coupled through a set of nonlinear equations. Written in a matrix form, it gives the following semi-discrete problem:

$$\mathscr{R}_f\left(\iota, \mathbf{v}_f, \mathbf{p}_f, \mathbf{u}_m\right) := \begin{bmatrix} \mathbf{M}_\iota \dot{\iota} + \mathbf{N}_\iota\left(\mathbf{v}_f - \dot{\mathbf{u}}_m\right)\iota \\ \mathbf{M}_f(\iota)\dot{\mathbf{v}}_f + \mathbf{N}_f\left(\iota, \mathbf{v}_f - \dot{\mathbf{u}}_m\right)\mathbf{v} + \mathbf{K}_f(\iota)\mathbf{v}_f + \mathbf{B}_f p_f - \mathbf{f}_f(\iota) \\ \mathbf{B}_f^T \mathbf{v}_f \end{bmatrix} = \mathbf{0} \tag{8.21}$$

where $\mathbf{M}_\iota$ and $\mathbf{N}_\iota$ are the matrices associated with the advection problem of the fluid volume fraction, $\mathbf{M}_f$ is a positive definite mass matrix, $\mathbf{N}_f$ is an unsymmetrical advection matrix, $\mathbf{K}_f$ is the conduction matrix describing the diffusion terms and $\mathbf{B}_f$ is for the gradient matrix, whereas $\mathbf{f}_f$ is the discretized nodal loads on the flow. This matrix form also takes into account the boundary conditions; special care has to be taken concerning the discretization of boundary conditions—and especially normal flux—when using the finite volume method (Ghidaglia and Pascal 2005).

For a given motion of the fluid domain, the coupling between the mesh deformation and the Navier–Stokes equation is weak in the sense that no variable, such as velocity v or pressure p, influences the fluid domain deformation under imposed boundary displacements. The coupling between the mesh motion problem and the fluid momentum equation can, therefore, be ensured explicitly.

The next that needs to be solved is the choice of velocity variation in the time step $\Delta t = t_{N+1} - t_N$. Due to the fact that the mesh motion is arbitrary and does not rely on any physical phenomenon, it is *a priori* possible to take any velocity evolution on the window $[T_N, T_{N+1}]$ enabling the initial mesh deformation to be equal to $\boldsymbol{u}_{m,N}$, and the final mesh deformation is equal to $\boldsymbol{u}_{m,N+1}$.

The geometric conservation law demands a numerical scheme to reproduce exactly and independently from the mesh motion a constant solution. This condition can be found in the literature for ALE formulation discretized either by the finite volume (Demirdžić and Perić 1988) or the stabilized finite element methods (Förster, Wall and Ramm 2006). It is proved (Farhat, Geuzaine and Grandmont 2001) that the velocity of the dynamic mesh needs to be computed for all first- and second-order time-integration schemes (like implicit Euler or Crank–Nicholson):

$$\dot{\mathbf{u}}_m = \frac{\mathbf{u}_{m,N+1} - \mathbf{u}_{m,N}}{\Delta t} \tag{8.22}$$

The volume fraction function is supposed to be sharp at the interface between water and gas, and, therefore, standard FV discretization that can be strongly diffusive cannot be applied, as they would smear the interface. One way to guarantee a sharp and bounded solution is by using a numerical scheme designed for the multidimensional advection equation (LeVeque 1996). In this case, explicit treatment referred to as MULES (multidimensional universal limiter with explicit solution) is used (e.g. Behzadi, Issa and Rusche 2004; Ubbink and Issa 1999).

8.2 FLUID–STRUCTURE INTERACTION PROBLEM COMPUTATIONS

8.2.1 Theoretical Formulation Of Fluid–Structure Interaction Problem

The interaction problem of motion of the fluid (denoted further with *f*) and structure (denoted with *s*) is elaborated. For the sake of generality, it is supposed that these problems are nonlinear and time-dependent. This interaction is handled by the classical direct force-motion transfer (DFMT, e.g. Felippa and Park 2004; Ross et al. 2009) that can formally be expressed in terms of corresponding Steklov–Poincaré operator (e.g. Deparis et al. 2006):

$$\begin{aligned} \mathscr{S}_i : \mathscr{H}^{1/2}(\Gamma) &\rightarrow \mathscr{H}^{1/2}(\Gamma) \\ \boldsymbol{u}_i &\rightarrow \boldsymbol{\lambda}_i \end{aligned} \tag{8.23}$$

The Steklov–Poincaré operator gives the evolution of dual field $\boldsymbol{\lambda}_i$ for an imposed primal field $\boldsymbol{u}_i$ on the interface space and time domain $\Gamma \times [0,T]$. In fact, this operator requires the computation of the fluid and structure problem on the complete space–time domain, governed by Navier–Stokes equations for incompressible flow and nonlinear dynamics equations for structural motion. For the problems of this kind, the Steklov–Poincaré operator is not available analytically, but rather as the results of numerical approximations. These results not only depend on the chosen model, material properties and boundary conditions, but also on discretization techniques, time-integration algorithms, discrete equations solvers etc.

The interface matching considered in the following is based on two classical mechanics principles:

Continuity of primal quantities or the perfect matching condition over the interface:

$$\boldsymbol{u}_s = \boldsymbol{u}_f = \boldsymbol{u} \ on \ \Gamma \times [0,T] \tag{8.24}$$

where $\boldsymbol{u}$ denotes the value of the primal variable at the interface. In the continuum setting, the time derivatives of this condition give the equivalent equations for velocity $\boldsymbol{v} = \dot{\boldsymbol{u}}$and acceleration $\boldsymbol{a} = \dot{\boldsymbol{v}}$:

$$\boldsymbol{v}_s = \boldsymbol{v}_f = \boldsymbol{v};\ \boldsymbol{a}_s = \boldsymbol{a}_f = \boldsymbol{a} \text{ on } \Gamma \times [0,T] \tag{8.25}$$

However, as the result of time discretization, these conditions are no longer equivalent.

Equilibrium of dual quantities or action–reaction principle, which implies:

$$\boldsymbol{\lambda}_f + \boldsymbol{\lambda}_s = 0 \text{ on } \Gamma \times [0,T] \tag{8.26}$$

The action–reaction principle in (8.26) can be reformulated using the Steklov–Poincaré operator defined in (8.23), resulting with the so-called Steklov–Poincaré formulation (e.g. Deparis et al. 2006):

$$\text{Find: } \boldsymbol{u} \text{ on } \Gamma \times [0,T], \text{so that: } S_f(\boldsymbol{u}) + S_s(\boldsymbol{u}) = 0 \tag{8.27}$$

Using the inverse of the first Steklov–Poincaré operators allows to rewrite the equilibrium of dual quantities as the following fixed-point equation that concerns only the unknown at the interface:

$$\text{Find: } \boldsymbol{u} \text{ on } \Gamma \times [0,T], \text{so that}: \boldsymbol{u} = S_s^{-1}\left(-S_f(\boldsymbol{u})\right) \tag{8.28}$$

The fixed-point equation can be reformulated in order to get a root equation:

$$\text{Find: } \boldsymbol{u} \text{ on } \Gamma \times [0,T], \text{so that}: S_s^{-1}\left(-S_f(\boldsymbol{u})\right) - \boldsymbol{u} = 0 \tag{8.29}$$

where all the requested quantities are defined at the interface.

It is recalled that these Steklov–Poincaré operators are associated with the semi-discrete form of the continuum equations. Namely, the fluid problem is defined by an ALE formulation (e.g. Demirdžić and Perić 1988; Hughes, Liu and Zimmermann 1981) of the Navier–Stokes equations. The latter considers the fluid mesh motion that $\mathbf{u}_m$ is imposed by the motion of the interface u and can be written as:

$$R_m(\mathbf{u}_m, \mathbf{u}) := \mathbf{K}_m \mathbf{u}_m - \mathbf{D}_m \mathbf{u} = 0 \tag{8.30}$$

where $\mathbf{D}_m$ is a projection/restriction operator and $\mathbf{K}_m$ governs the extension of the boundary displacement either by a diffusion process or a pseudo-solid equation (e.g. Ferziger and Perić 2002). The fluid flow in this moving domain is described by the semi-discrete Navier–Stokes equations:

$$R_f(\mathbf{v}_f, p_f; \mathbf{u}_f) := \begin{bmatrix} \mathbf{M}_f \dot{\mathbf{v}}_f + \mathbf{N}_f(\mathbf{v}_f - \dot{\mathbf{u}}_m)\mathbf{v} + \mathbf{K}_f \mathbf{v}_f + \mathbf{B}_f \mathbf{p}_f - \mathbf{f}_f \\ \mathbf{B}_f^T \mathbf{v}_f \end{bmatrix} = 0 \tag{8.31}$$

where $\mathbf{v}_f, \dot{\mathbf{v}}_f$ and $\mathbf{p}_f$ are fluid velocity, its derivative and fluid pressure $\mathbf{M}_f$ is a (positive definite) mass matrix, $\mathbf{N}_f$ is a (non-symmetric) advection matrix, $\mathbf{K}_f$ is the matrix with diffusion terms, $\mathbf{B}_f$ stands for the gradient matrix, and $\mathbf{f}_f$ is the driving force on the flow. The discretization process leading to fluid equations of motion and incompressibility constraint in (8.31) is carried out by the finite volume method (Ferziger and Perić 2002).

The finite element method (Zienkiewics and Taylor 2005) is used for the structure subproblem, resulting with structural semi-discrete equations of motion as set in (8.27)

With this notation on hand, the explicit form of the Steklov–Poincaré operator for fluid and for structure, which can be expressed in terms of algorithms 1 and 2, is delivered Tables 8.1 and 8.2.

Table 8.1 Algorithm 1—Steklov-Poincaré Operator for Fluid $\mathbf{S}_f$

Require: Fluid state variable at the time T_N
1: Impose displacement of mesh at fluid–structure interface:
$u_m = u$ on $\Gamma \times [T_N, T_{N+1}]$
2: Solve mesh intern nodes displacement:
$R_m(u_m, u) = 0$ on $\Omega_f \times [T_N, T_{N+1}]$
3: Solve fluid problem in ALE formulation:
$R_f(\upsilon_f, p_f; u_f) = 0$ on $\Omega_f \times [T_N, T_{N+1}]$
4: Get boundary traction force at the interface:
$\lambda = -\sigma_f n$ on $\Gamma \times [T_N, T_{N+1}]$

Table 8.2 Algorithm 2—Poincaré-Steklov Operator for Fluid S_s^{-1}

Require: Solid state variable at the time T_N
1: Impose boundary traction force at fluid–structure interface:
$\sigma_s n = \lambda$ on $\Gamma \times [T_N, T_{N+1}]$
2: Solve structure problem:
$R_s(u_s, \lambda) = 0$ on $\Omega_s \times [T_N, T_{N+1}]$
3: Get boundary displacement at the interface:
$u = u_s$ on $\Gamma \times [T_N, T_{N+1}]$

8.3 NUMERICAL EXAMPLES

8.3.1 Example 1a — Flexible Appendix in a Flow

An example which is elaborated in this part was first proposed in Wall and Ramm (1988), and subsequently also studied in Dettmer and Perić (2007), Hübner, Walhorn and Dinkler (2004), Matthies, Niekamp and Steindorf (2006); and has been used as a benchmark for fluid–structure interaction.

A fixed square bluff body, with a flexible appendix attached to it, immersed within an incompressible flow (see Figure 8.1) filling the whole domain is considered. At a sufficiently long distance from this body, the flow is uniform with an imposed velocity $\bar{\mathfrak{v}}$; the corresponding value of the Reynolds number with respect to the characteristic size of the obstacle is $\mathscr{R} = 330$. For this Reynolds number, the flow exhibits a transient behavior with vortices separating from the corner of the square. These vortices induce an alternative drop and increase in the pressure field behind the rigid bluff body at a frequency that depends on the Reynolds number and the shape of the bluff body (Roshko 1952). The vortex shedding induces oscillations of the flexible appendix.

The thickness of the appendix as well as the material properties is chosen so that its first eigenfrequency is close to the frequency of the vortex shedding (Table 8.3).

The first eigenfrequency of the problem f = 3.03 1/s obtained considering a linear material is closed to the natural frequency of vortex shedding behind a square bluff body at Reynolds $\mathscr{R} = 330$.

The chosen fluid discretization contains 5,080 FV cells (i.e. around 20 × 103 dof), which is quite sufficient to get an accurate representation of the flow, and also of the fluid loading on the structure. The model accuracy is comparable to the one used in Dettmer and Perić (2007), built with 4,300 finite elements. The PISO algorithm and the Euler implicit scheme with a time step $\Delta t = 0.004s$ are used. The PCG solver is used for pressure correction step and mesh motion equations and PBiCG for the momentum predictor.

The flexible appendix is discretized with 20 nine-node elements, with quadratic polynomes description of large structural displacement. Neo-Hookean and Saint-Venant–Kirschoff

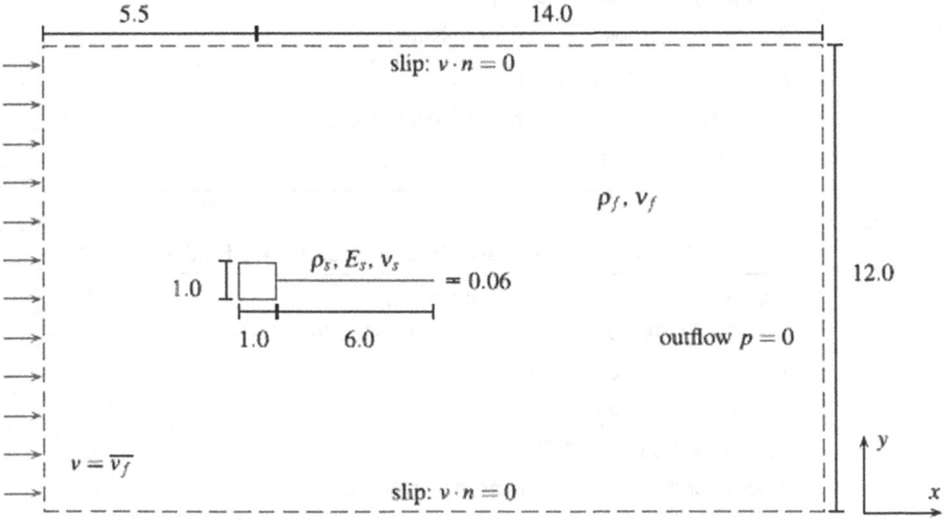

Figure 8.1 The benchmark used for fluid–structure interaction problems.

Table 8.3 Material Properties f Steel Bar

FLUID	
Density ρ_f	1.18×10^{-3} kg/m^3
ν_f	1.54×10^{-1} m^2/s air 20°C
Imposed velocity at the left-hand side is $\overline{\upsilon}$	(51.3 m/s, 0)
SOLID	
Density ρ_s	0.1 kg/m^3
E_s	2.50×10^6 N/m^2/s
ν_s	0.35

materials are used. The time discretization is carried out by a generalized HHT-α scheme with the following parameters: $\rho_\infty = 1/2$; $\beta = 4/9$; $\gamma = 5/3$; and $\alpha = 2/3$. At each iteration, the linear system is solved by the direct solver for asymmetric matrices.

The interface matching computation is carried out by DFMT-BGS solver, that can easily converge since the channel is open and the flow is not mainly driven by incompressibility. The Aitken relaxation technique is used with an initial value of 0.5, that rapidly increases to 1. No more than four iterations are required to reach the required tolerance on the interface displacement residual that is set to 1×10^{-7}. The results for the pressure and velocity field at several instants are given in Figure 8.2. The deformed shapes of the appendix also given in Figure 8.2, reveal the oscillations dominated by the first mode.

The displacement at the free-end of the appendix is plotted in Figure 8.3 for both Saint-Venant–Kirschoff and Neo- Hookean solid materials. As it can be seen the two results are very close, since the appendix deformations remain small, despite its large displacement and rotations.

The long term response (see Figure 8.4) indicates an almost harmonic response dominated by the first eigenfrequency of the structure. A comparison with the results from the literature in term of the maximum amplitude of motion is also presented in Figure 8.4. Despite the well-known sensitivity of the computed result with respect to the initial condition (Hübner, Walhorn and Dinkler 2004), the answers obtained are very close to the previous results from the literature based upon a FE discretization for both fluid and solid parts and obtained either by a monolithic (Dettmer and Perić 2007) or a partitioned approach (Matthies and Steindorf 2003; Wall, Mok and Ramm 1999).

Contrary to the lid-driven cavity with a flexible bottom, the small number of iterations required to solve the fluid– structure interaction problem of the oscillating appendix suggests that an explicit coupling can also be used for solving this problem. The results from explicit DFMT algorithm presented in Figure 8.5 for the free-end displacement compare to a reference solution obtained with an implicit computation.

Using better predictors is supposed to reduce the errors made in term of residual and energy. In Figure 8.6, the energy error time-history is represented for the zero, first and second-order predictor. All the results confirm the expected trends, with a decrease in errors when the predictor order increases.

The final study is then considered with respect to the size of time steps. In Figure 8.7, the maximum residual error on the time interval $t \in [0, 15\ s]$ is presented as a function of the time step size. The error is observed to decrease with a decreasing time step size. However, when the time steps become too small, the added mass effect triggers the divergence of the computation. Thus, only the less sensitive schemes with a zero order predictor are able to solve the coupled problem with the smallest time step.

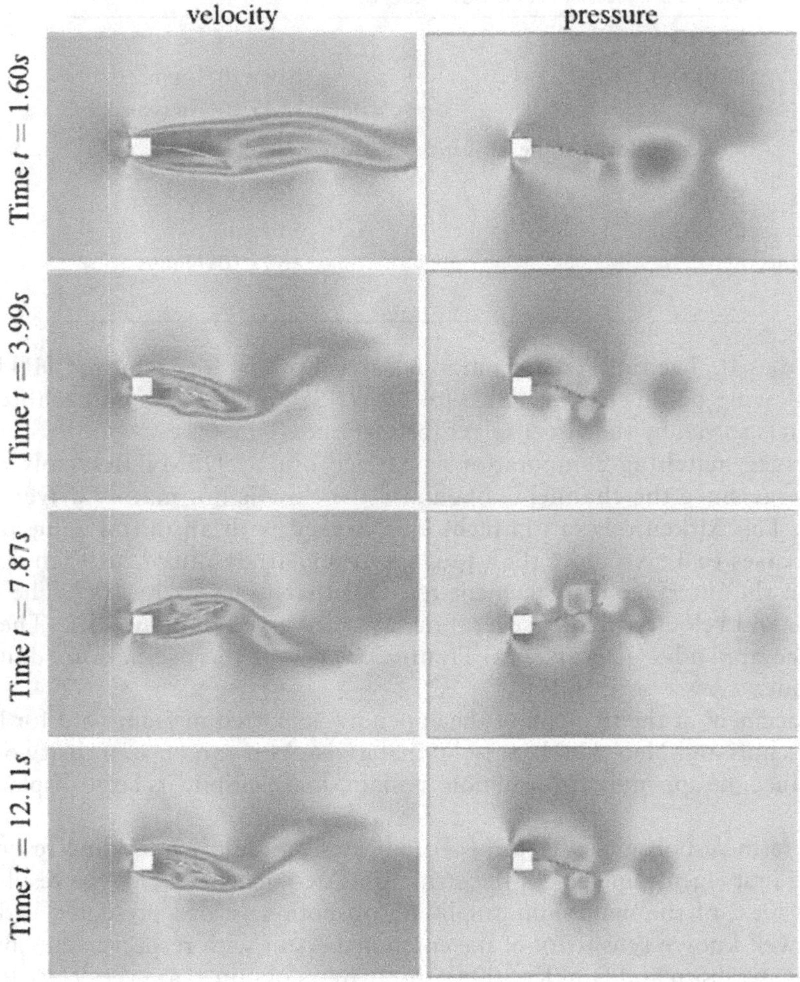

Figure 8.2 Problems oscillating appendix in flow: velocity and pressure field snapshots.

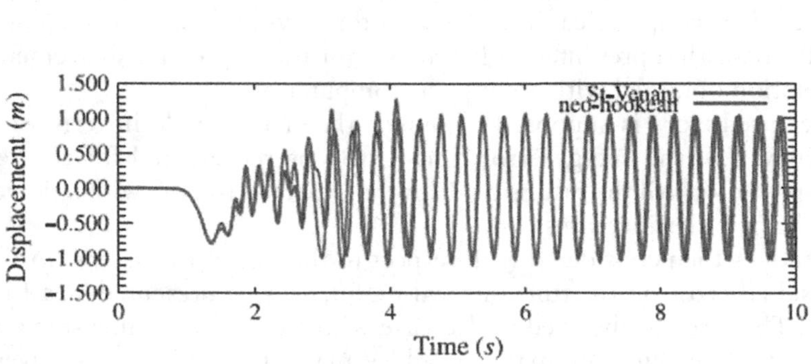

Figure 8.3 Implicit coupling: displacement of the appendix of the appendix extremity for two nonlinear materials (Saint-Venant and Neo-Hookean).

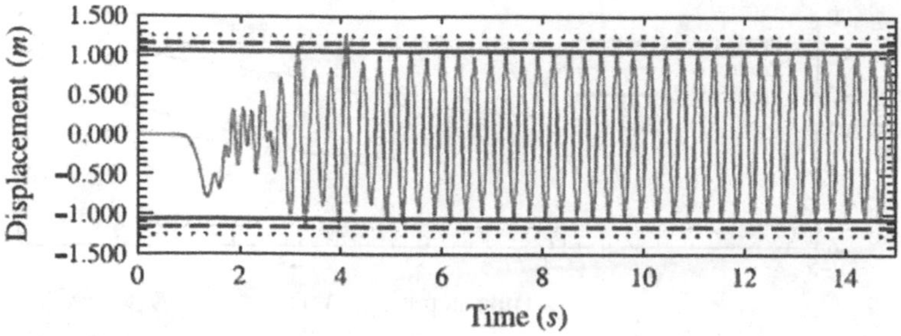

Figure 8.4 Implicit coupling: long term response with implicit coupling. Maximum amplitude comparison with (Dettmer and Perić 2007) (dotted-line), (Matthies and Steindorf 2003) (solid-line), (Wall, Mok and Ramm 1999) (dashed-line).

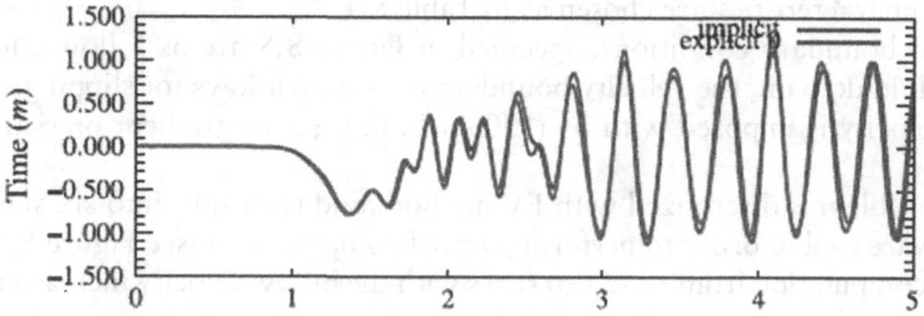

Figure 8.5 A comparison between explicit and implicit coupling algorithm-based computations for the displacement of the appendix.

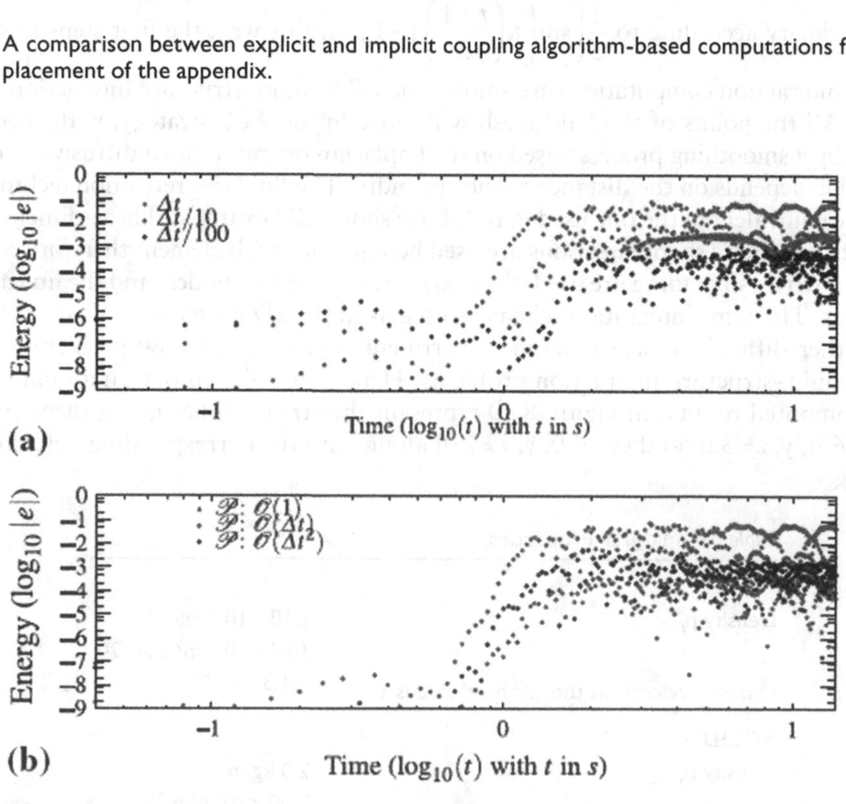

Figure 8.6 Energy error at the interface for different explicit coupling schemes (a) energy error for different time step size and (b) energy error for different predictors.

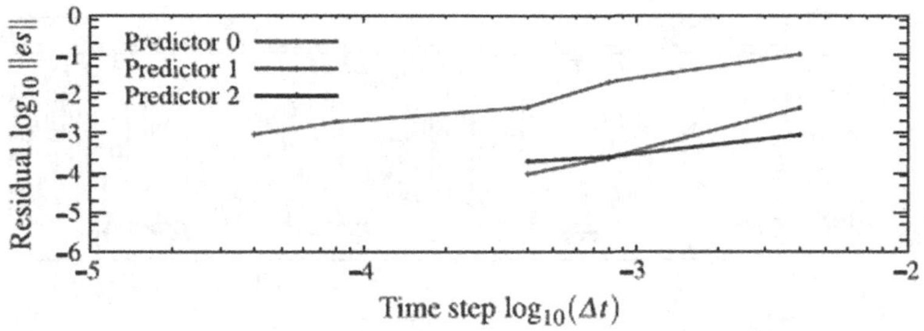

Figure 8.7 Maximum residual error for explicit coupling schemes with different time step sizes and predictors.

8.3.2 Example 1b — Flexible Appendix in a Flow—3d

The material characteristics are chosen as in Table 8.4.

The chosen boundary conditions, specified in Figure 8.8 are as follows: for the lateral walls of the fluid domain, the velocity boundary condition allows for slipping; at the inflow, a constant velocity is imposed with $\overline{v} = (100\ \text{cm/s}, 0, 0)$; zero gradient pressure is specified at the outflow.

The fluid problem is discretized with FV method, and then split into six sub-domains by METIS software tool in order to perform parallel computations (see Figure 8.9).

Fluid only computation from $t = -2$ s to $t = 0$ s with the inflow velocity increasing in a smooth way the velocity according to $\frac{1}{2}\left(\sin\left(\pi\left(\frac{t+1}{2}\right)\right)+1\right)$. In this way, the first steps of the fluid–structure interaction computations are smoothened. The fluid–structure interaction will start at $t = 0$ s. All the points of the fluid mesh will move in the ALE strategy, with their motion governed by a smoothing process based on the Laplacian operator and a diffusivity coefficient whose value depends on the distance to the appendix. The fluid discretization techniques and solvers are equivalent to the one used in the above shown 2D example. Three-dimensional elements with quadratic shape functions are used herein, with each element therefore containing 27 nodes. Two mesh grading are used: the coarse mesh with 663 nodes, and the fine mesh with 2475 nodes. The same integration scheme is used as in the 2D problem.

It is rather difficult to select the most pertinent result for 3D flow problems, and even more for fluid–structure interaction problems. Hence, in order to obtain a qualitative picture of computed results, in Figure 8.10 represent the stream-tubes going through the two lines (x = 6.0, y, z = 3.0) and (x = 6.0, y, z = 7.0) along with the corresponding deformed shape of the flag.

Table 8.4 Material Properties

FLUID	
Density ρ_f	1.18×10^{-3} kg/m^3
ν_f	1.54×10^{-1} m^2/s air 20°C
Imposed velocity at the left-hand side is $\overline{v}$	(51.3 m/s, 0)
SOLID	
Density ρ_s	2.0 kg/m^3
E_s	2.00×10^{6} N/m^2/s
ν_s	0.35

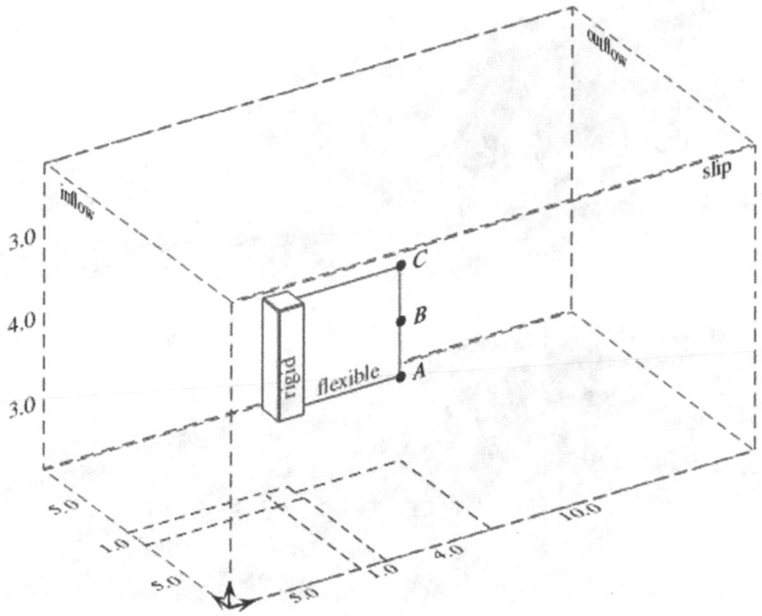

Figure 8.8 Boundary conditions.

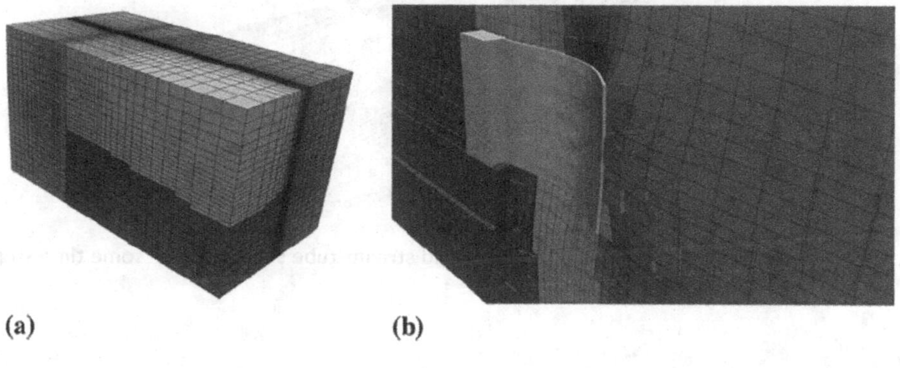

Figure 8.9 Flag in the wind: Decomposed fluid domain and zoom on the structure (a) fluid domain decomposed and (b) zoom on the deformed mesh.

The displacement of the points at the free-end represented in Figure 8.11 show that the motion of the flag corresponds to the first flexural mode. It is hard to predict the exact solution for such a complex three-dimensions flow in with a relatively high Reynolds number. These results are more in agreement with the ones provided for the 2D version of the same problem (e.g. Dettmer and Perić 2007; Matthies and Steindorf 2003; Wall, Mok and Ramm 1999) with a flexible appendix.

8.3.3 Example 2 — Three-Dimensional Sloshing Wave Impacting a Flexible Structure

This example is a simplified representation of a dam-breaking event that brings about the sloshing wave impact on a flexible structure standing in its way, as presented in Figure 8.12. At initial time t = 0 s, a three-dimensional water column starts falling down under the gravity

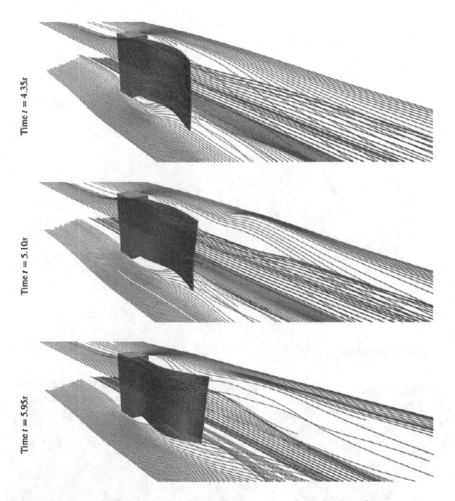

Figure 8.10 Flag in the wind: motion of the structure and stream-tube snapshots for some time steps.

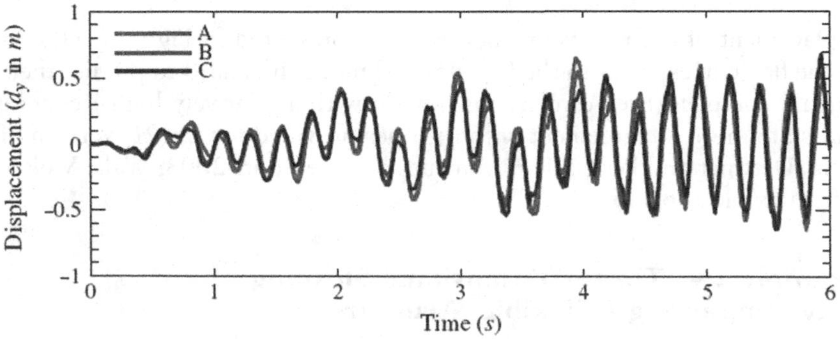

Figure 8.11 Displacement of structure extremity—Oscillating 3D Appendix: extremity displacements for points A =(10.0, 5.5, 3.0), B=(10.0, 5.5, 5.0) and C=(10.0, 5.5, 7.0).

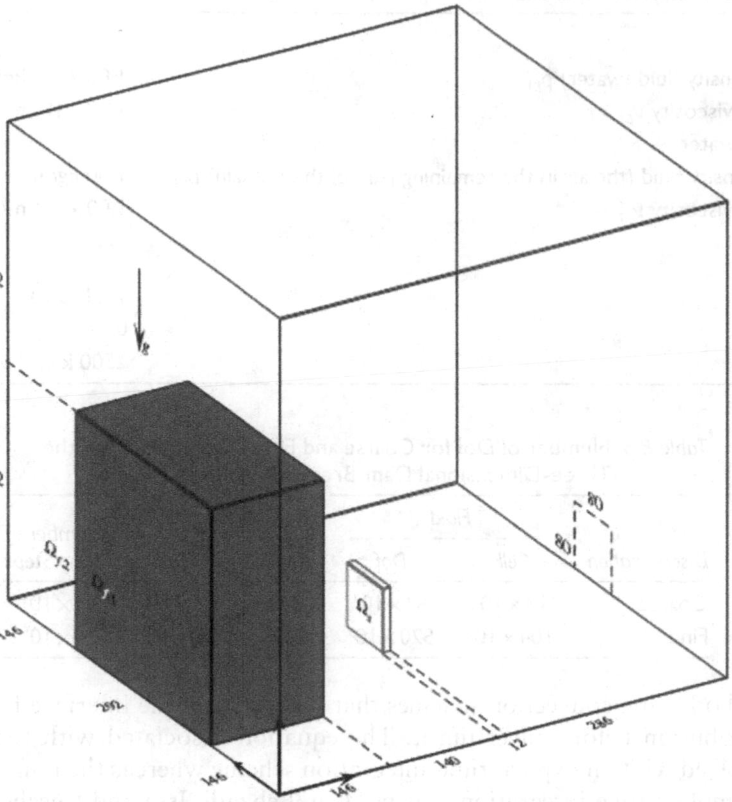

Figure 8.12 Three-dimensional wave impacting an obstacle: geometry and boundary conditions.

loading. Spreading further at a later time, it hits the obstacle which is a slender plate-like solid body made of elastic material that can undergo large deformations. The dimension of the problem and the imposed boundary conditions are given in Figure 8.12.

In order to prevent that the water will bounce-back and again hit the structure after breaking on the walls, only the left and bottom planes of the fluid domain are defined as non-slipping walls, while the others are defined with atmospheric boundary condition for the pressure. Material properties are given in Table 8.5.

The mesh motion problem is solved using a Laplacian smoothing material where the diffusion coefficient is a quadratic inverse function of the distance to the interface between structure and fluid.

The fluid domain is discretized with finite volume cells always covering the complete domain either by one phase or by the other. The computations are performed for two different meshes with the chosen discretization and the number of cells given in Table 8.6.

An explicit–implicit algorithm is used to compute the two-phase flow evolution from the corresponding Navier–Stokes equations. In the V.O.F. method used herein (e.g. Ghidaglia, Kumbaro and Le Coq 2001), an indicator function (volume fraction, level set or phase-field) is used to represent the phases; leaving only the remaining issue on how to convert the interface without diffusing, dispersing or wrinkling it. This is particularly troublesome when the volume fraction is chosen as an indicator function because the convection scheme has to guarantee that the volume fraction stays bounded, with the values that remain within its physical bounds of 0 and 1. The idea proposed in Behzadi, Issa and Rusche (2004) is used:

Table 8.5 Material Properties

FLUID	
High-density fluid (water) $\rho_{f,1}$	1.00×10^3 kg/m³
Kinetic viscosity $\nu_{f,1}$	1.00×10^5 m/s air 20°C
FLUID-water	
Low-density fluid (the air in the remaining part of the domain) $\rho_{f,2}$	1.00 kg/m³
Kinetic viscosity $\nu_{f,2}$	1.00×10^6 m/s air 20°C
SOLID	
E_s	1 × 106 Pa
ν_s	0
ρ_s	2500 kg/m³

Table 8.6 Number of Dof for Coarse and Fine Discretization of the Three-Dimensional Dam-Breaking Problem

	Fluid		*Solid*		
Discretization	*Cells*	*Dof*	*Nodes*	*Dof*	*Number of Time Steps*
Coarse	13×10^3	63×10^3	363	1.1×10^3	1×10^5
Fine	104×10^3	520×10^3	2205	6.6×10^3	1×10^5

a V.O.F. method using convection schemes that reconstructs the interface from the volume fraction distribution before adverting it. The equation associated with the characteristic function is solved with an explicit time-integration scheme whereas the remaining terms are solved with implicit time-integration schemes (e.g. Behzadi, Issa and Rusche 2004; Ubbink and Issa 1999). The fluid is handled by a second-order space discretization with a VanLeer limiter used for the advection terms, and the implicit Euler time-integration scheme. The small time steps are required for the explicit solution of the phase function indicator equation, as well as the half-implicit nature of the coupling between the momentum predictor and the pressure corrector. At this scale of modeling, it is not required to consider surface tension between the two fluids. The structure model is constructed by using three-dimensional elements with quadratic shape functions, where each element has 27 nodes. The time-integration is carried out by a generalized-α scheme as in the previous examples.

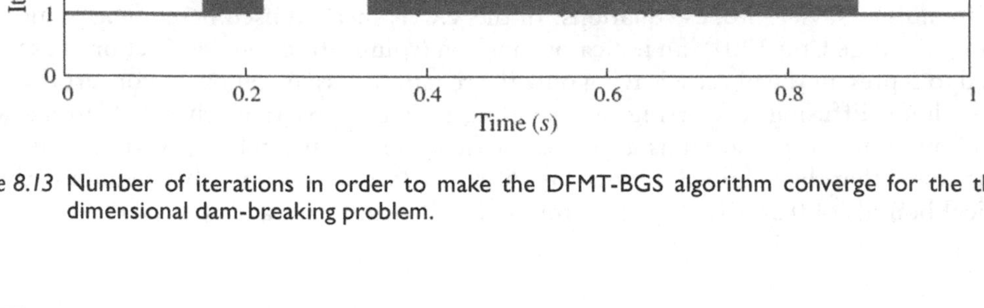

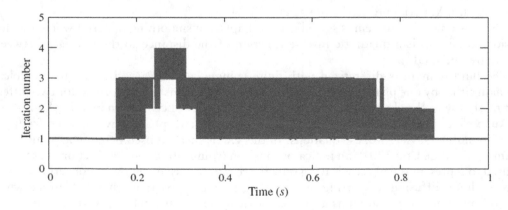

Figure 8.13 Number of iterations in order to make the DFMT-BGS algorithm converge for the three-dimensional dam-breaking problem.

The computation of the coupled problem is carried out by an iterative scheme. The results of fluid and structure computations are matched for a time step of 1×10^{-4} for the coarse and 2×10^{-5} for the fine discretization. The coupling scheme is DFMT-BGS with Aitken's relaxation with the initial parameter $\omega=0.25$ and the predictor of order 1. The absolute tolerance for coupled computation is equal to:

$$\left\| r_N^k \right\| \leq 1\times10^{-6} \tag{8.32}$$

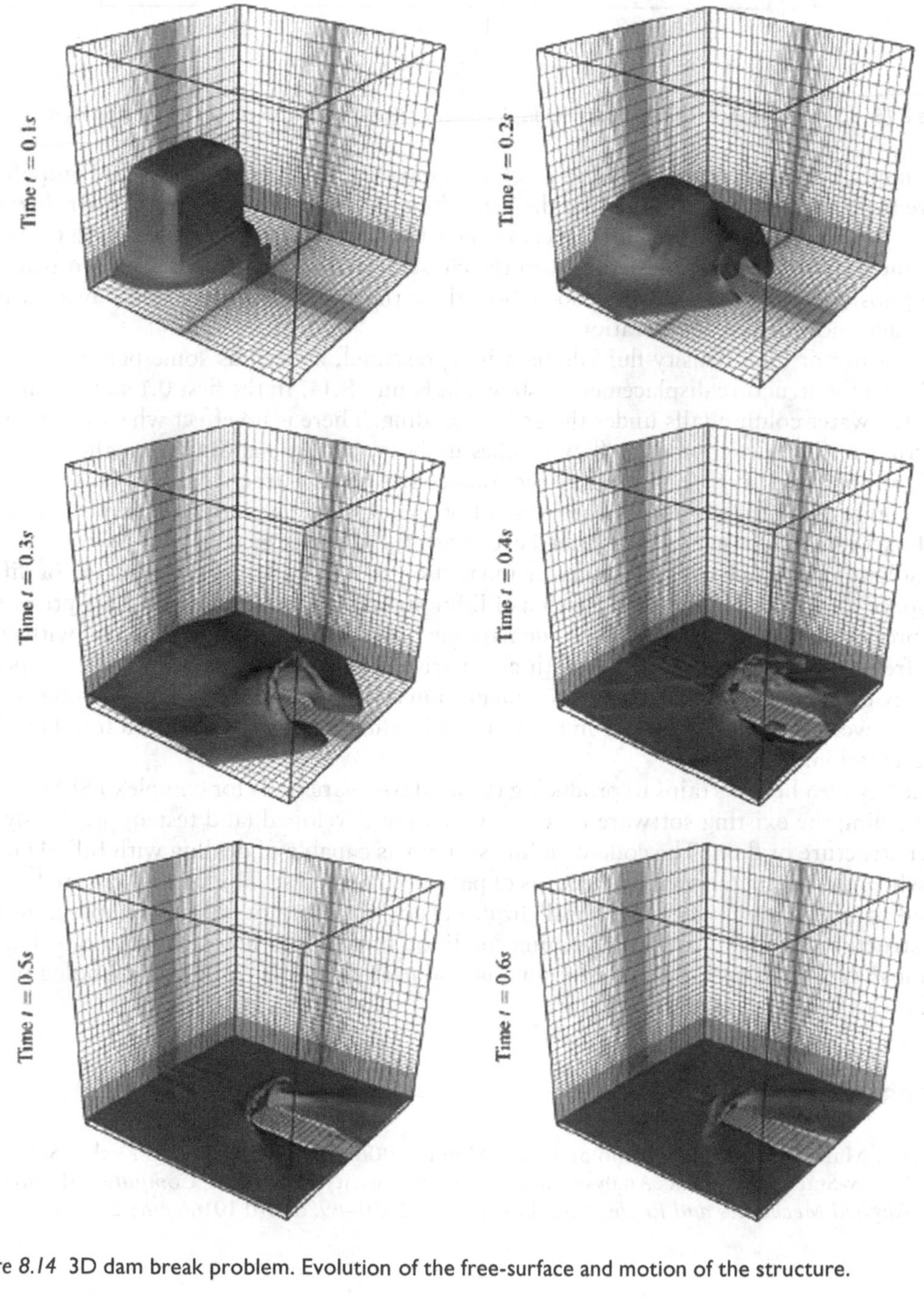

Figure 8.14 3D dam break problem. Evolution of the free-surface and motion of the structure.

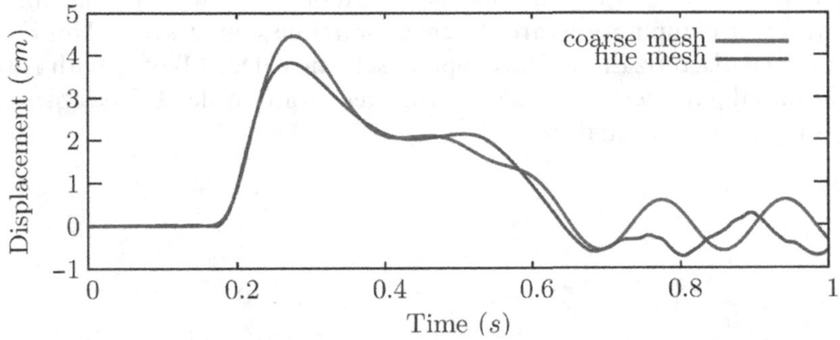

Figure 8.15 3D dam break example: obstacle displacement measured at the center of the top face.

The number of iterations required to reach the convergence criteria is given in Figure 8.13. There is no coupling iteration before the water hits the structure (the effect of air flow can almost be deemed negligible with respect to the structure). During the water–structure contact, the number of iteration depends on the chosen the discretization density. In reaching the opposite wall, the water does not rebound on the wall but simply flows away, which again does not require any iteration.

The water or high-density fluid domain is represented, as well as some part of the fluid mesh and the structure displacement is shown in Figure 8.14. In the first 0.1 s of the simulation, the water column falls under the gravity loading. There is no effect whatsoever on the structure until the high-density flow reaches its bottom. The maximum amplitude of the motion is obtained at t = 0.25 s before the solid comes back to free-vibration phase.

In Figure 8.15, the motion of the free-end of the obstacle is plotted. The motion of the solid part remains rather well described even with the coarsest grid.

A solution approach to fluid–structure interaction problems that allow coupling of different space discretization methods, such as FE for structures and FV for fluids is presented. The proposed strategy is applicable to demanding problems of this kind that deal with complex free-surface flows interacting with geometrically nonlinear structures. The proposed strategy can also employ different time-integration schemes, and interface matching conditions between fluid and structure in the spirit of implicit analysis and chosen interface field representation.

The key idea here pertains to producing the final software tools for complex FSI problems by coupling the existing software products that were developed (and tested) previously for either structure or fluid. The code-coupling strategy is capable of dealing with full 3D models, where the computational efficiency is of paramount interest. The latter is ensured by the nested parallelization, where the parallelization is carried out not only for coupling of fluid and structure through interface matching condition but also for flow computations that are the most expensive task. Both code-coupling and nested parallelization are handled by the CTL, with the latter as its new feature.

REFERENCES

Barcelos, Manuel, Henri Bavestrello and Kurt Maute. 2006. "A Schur–Newton–Krylov Solver for Steady-State Aeroelastic Analysis and Design Sensitivity Analysis". *Computer Methods in Applied Mechanics and Engineering* 195 (17–18): 2050–69. doi:10.1016/j.cma.2004.09.013.

Bathe, Klaus-Jürgen and Hou Zhang. 2009. "A Mesh Adaptivity Procedure for CFD and Fluid-Structure Interactions". *Computers & Structures* 87 (11–12): 604–17. doi:10.1016/j.compstruc.2009.01.017.

Bazilevs, Y., V. M. Calo, T. J. R. Hughes and Y. Zhang. 2008. "Isogeometric Fluid-Structure Interaction: Theory, Algorithms, and Computations". *Computational Mechanics* 43 (1): 3–37. doi:10.1007/s00466-008-0315-x.

Bazilevs, Y., V. M. Calo, Y. Zhang and T. J. R. Hughes. 2006. "Isogeometric Fluid–structure Interaction Analysis with Applications to Arterial Blood Flow". *Computational Mechanics* 38 (4–5): 310–22. doi:10.1007/s00466-006-0084-3.

Bazilevs, Y., J. R. Gohean, T. J. R. Hughes, R. D. Moser and Y. Zhang. 2009a. "Patient-Specific Isogeometric Fluid–structure Interaction Analysis of Thoracic Aortic Blood Flow Due to Implantation of the Jarvik 2000 Left Ventricular Assist Device". *Computer Methods in Applied Mechanics and Engineering* 198 (45–46): 3534–50. doi:10.1016/j.cma.2009.04.015.

Bazilevs, Yuri, M. C. Hsu, D. J. Benson, S. Sankaran and A. L. Marsedn. 2009b. "Computational Fluid–structure Interaction: Methods and Application to a Total Cavopulmonary Connection". *Computational Mechanics* 45 (1): 77–89. doi:10.1007/s00466-009-0419-y.

Bazilevs, Y., M. -C. Hsu, Y. Zhang, W. Wang, X. Liang, T. Kvamsdal, R. Brekken and J. G. Isaksen. 2009c. "A Fully-Coupled Fluid-Structure Interaction Simulation of Cerebral Aneurysms". *Computational Mechanics* 46 (1): 3–16. doi:10.1007/s004.

Behzadi, A., R. I. Issa and H. Rusche. 2004. "Modelling of Dispersed Bubble and Droplet Flow at High Phase Fractions". *Chemical Engineering Science* 59 (4): 759–70. doi:10.1016/j.ces.2003.11.018.

Belytschko, T. 1983. "An Overview of Semidiscretization and Time Integration Procedures. Computational Methods for Transient Analysis". Edited by Hughes T. J. R., Belytschko T. *Journal of Applied Mechanics* (North-Holland, Amsterdam), 1–65.

Belytschko, T., Kam W. Liu and B. Moran. 2000. *Nonlinear Finite Elements for Continua and Structures*. New York: Wiley.

Belytschko,T.,H.J.YenandR.Mullen.1979."MixedMethodsforTimeIntegration".*ComputerMethods in Applied Mechanics and Engineering* 17–18: 259–75. doi:10.1016/0045-7825(79)90022-7.

Causin, P., J. F. Gerbeau and F. Nobile. 2005. "Added-Mass Effect in the Design of Partitioned Algorithms for Fluid–structure Problems". *Computer Methods in Applied Mechanics and Engineering* 194 (42–44): 4506–27. doi:10.1016/j.cma.2004.12.005.

Chung, Jintai and Gregory M. Hulbert. 1994. "A Family of Single-Step Houbolt Time Integration Algorithms for Structural Dynamics". *Computer Methods in Applied Mechanics and Engineering* 118 (1–2): 1–11. doi:10.1016/0045-7825(94)90103-1.

Degroote, Joris, Klaus-Jürgen Bathe and Jan Vieren. 2009. "Performance of a New Partitioned Procedure Versus a Monolithic Procedure in Fluid–structure Interaction". *Computers & Structures* 87 (11–12): 793–801. doi:10.1016/j.compstruc.2008.11.01.

Demirdžić, I. and M. Perić. 1988. "Space Conservation Law in Finite Volume Calculations of Fluid Flow". *International Journal for Numerical Methods in Fluids* 8 (9): 1037–50. doi:10.1002/fld.1650080906.

Deparis, Simone, Marco Discacciati, Gilles Fourest and Alfio Quarteroni. 2006. "Fluid–structure Algorithms Based on Steklov–Poincaré Operators". *Computer Methods in Applied Mechanics and Engineering* 195 (41–43): 5797–812. doi:10.1016/j.cma.

Dettmer, W. G. and D. Perić. 2007. "A Fully Implicit Computational Strategy for Strongly Coupled Fluid–Solid Interaction". *Archives of Computational Methods in Engineering* 14 (3): 205–47. doi:10.1007/s11831-007-9006-6.

Farhat, Charbel, Philippe Geuzaine and Céline Grandmont. 2001. "The Discrete Geometric Conservation Law and the Nonlinear Stability of ALE Schemes for the Solution of Flow Problems on Moving Grids". *Journal of Computational Physics* 174 (2): 669–94. doi:10.1006/jcph.2001.6932.

Felippa, C. A. and K. C. Park. 2004. "Synthesis Tools for Structural Dynamics and Partitioned Analysis of Coupled Systems". Edited by Ibrahimbegovic A. and Brank B. *NATO Advanced Research Workshop* (IOS Press, The Netherlands), 50–111.

Ferziger, J. H. and M. Peric. 2002. *Computational Methods for Fluid Dynamics*. Berlin: Springler-Verlag.

Förster, C. H., W. A. Wall and E. Ramm. 2006. "On the Geometric Conservation Law in Transient Flow Calculations on Deforming Domains". *International Journal for Numerical Methods in Fluids* 50 (12): 1369–79. doi:10.1002/fld.1093.

Ghidaglia, Jean-Michel, Anela Kumbaro and Gérard Le Coq. 2001. "On the Numerical Solution to Two Fluid Models via a Cell Centered Finite Volume Method". *European Journal of Mechanics - B/Fluids* 20 (6): 841–67. doi:10.1016/s0997-7546(01)01150.

Ghidaglia, Jean-Michel and Frédéric Pascal. 2005. "The Normal Flux Method at the Boundary for Multidimensional Finite Volume Approximations in CFD". *European Journal of Mechanics - B/Fluids* 24 (1): 1–17. doi:10.1016/j.euromechflu.2004.05.003.

Glowinski, R. 2003. *Numerical Methods for Fluids (Part III). Handbook of Numerical Analysis*. Edited by Lions J. L., Ciarlet P. G. Vol. 9. North-Holland: Elsevier.

Hilber, Hans M., Thomas J. R. Hughes and Robert L. Taylor. 1977. "Improved Numerical Dissipation for Time Integration Algorithms in Structural Dynamics". *Earthquake Engineering & Structural Dynamics* 5 (3): 283–92. doi:10.1002/eqe.4290050306.

Hübner, Björn, Elmar Walhorn and Dieter Dinkler. 2004. "A Monolithic Approach to Fluid–structure Interaction Using Space–Time Finite Elements". *Computer Methods in Applied Mechanics and Engineering* 193 (23–26): 2087–2104. doi:10.1016/j.cma.2004.

Hughes, Thomas J. R., Wing Kam Liu and Thomas K. Zimmermann. 1981. "Lagrangian-Eulerian Finite Element Formulation for Incompressible Viscous Flows". *Computer Methods in Applied Mechanics and Engineering* 29 (3): 329–49. doi:10.1016/0045–7825.

Hughes, T. J. R. and W. K. Liu. 1978. "Implicit-Explicit Finite Elements in Transient Analysis: Stability Theory". *Journal of Applied Mechanics* 45 (2): 371. doi:10.1115/1.3424304.

Hughes, Thomas J. R., Karl S. Pister and Robert L. Taylor. 1979. "Implicit-Explicit Finite Elements in Nonlinear Transient Analysis". *Computer Methods in Applied Mechanics and Engineering* 17–18: 159–82. doi:10.1016/0045-7825(79)90086-0.

Ibrahimbegovic, Adnan. 2009. *Nonlinear Solid Mechanics: Theoretical Formulations and Finite Element Solution Methods*. New York: Springer.

Ibrahimbegovic, A. and B. Brank. 2005. *Engineering Structures Under Extreme Conditions: Multi-Physics and Multi-Scale Computer Models in Non-Linear Analysis and Optimal Design*. Amsterdam: IOS Press.

Kassiotis C., A. Ibrahimbegovic, H. Matthies. 2010. "Partitioned solution to fluid-structure interaction problem in application to free-surface flow". *European Journal of Mechanics, Part B: Fluids* 29: 510–21.

Kassiotis C., A. Ibrahimbegovic, R. Niekamp, H. Matthies. 2011a. "Partitioned solution to nonlinear fluid-structure interaction problems. Part I: implicit coupling algorithms and stability proof". *Computational Mechanics* 47: 305–23.

Kassiotis C., A. Ibrahimbegovic, R. Niekamp, H. Matthies. 2011b. "Partitioned solution to nonlinear fluid-structure interaction problems. Part II: CTL based software implementation with nested parallelization". *Computational Mechanics* 47: 335–57.

LeVeque, Randall J. 1996. "High-Resolution Conservative Algorithms for Advection in Incompressible Flow". *SIAM Journal on Numerical Analysis* 33 (2): 627–65. doi:10.1137/0733033.

Matthies, Hermann G., Rainer Niekamp and Jan Steindorf. 2006. "Algorithms for Strong Coupling Procedures". *Computer Methods in Applied Mechanics and Engineering* 195 (17–18): 2028–49. doi:10.1016/j.cma.2004.11.032.

Matthies, Hermann G. and Jan Steindorf. 2003. "Partitioned Strong Coupling Algorithms for Fluid–structure Interaction". *Computers & Structures* 81 (8–11): 805–12. doi:10.1016/s0045-7949(02)00409-1.

Oden, J. Tinsley, Ted Belytschko, Ivo Babuska and T. J. R. Hughes. 2003. "Research Directions in Computational Mechanics". *Computer Methods in Applied Mechanics and Engineering* 192 (7–8): 913–22. doi:10.1016/s0045-7825(02)00616-3.

Roshko, A. 1952. *Of the development of turbulent wakes from vortex streets.* Ph.D. thesis, Pasadena, CA: California Institute of Technology.

Ross, Michael R., Michael A. Sprague, Carlos A. Felippa and K. C. Park. 2009. "Treatment of Acoustic Fluid–structure Interaction by Localized Lagrange Multipliers and Comparison to Alternative Interface-Coupling Methods". *Computer Methods in Applied Mechanics and Engineering* 198 (9–12): 986–1005. doi:10.1016/j.cma.2008.11.006.

Ubbink, O. and R. I. Issa. 1999. "A Method for Capturing Sharp Fluid Interfaces on Arbitrary Meshes". *Journal of Computational Physics* 153 (1): 26–50. doi:10.1006/jcph.1999.6276.

Wall, W. A., D. P. Mok and E. Ramm. 1999. "Partitioned Analysis Approach of the Transient Coupled Response of Viscous Fluids and Flexible Structures". In: *Solids, Structures and Coupled Problems in Engineering, Proceedings of the European Conference on Computational*Mechanics ECCM'99, Munich, August/September 1999.

Wall, W. A. and E. Ramm. 1988. *Fluid-Structure Interaction Based Upon a Stabilized (ALE) Finite Element Method.* Sonderforschungsbereich 404, Institut für Baustatik und Baudynamik, German.

Zienkiewics, O. C. and R. L. Taylor. 2005. *The Finite Element Method.* 6th. New York: Elsevier/ Butterworth-Heinemann.

Index

Printed in the United States
by Baker & Taylor Publisher Services

Printed in the United States
by Baker & Taylor Publisher Services